Relevant Logics

Relevant logics are non-classical logics that require the antecedent and consequent of implications to be relevantly related. They are paraconsistent logics, that is, they are able to robustly handle contradictory information. The field of non-classical logics is rapidly expanding, particularly with the addition of modalities and quantifiers. This is the first book to systematically develop a basic frame theory of relevant logics that includes both modal and quantified extensions. It includes sections comparing features of relevant logics with other, more common logics used in philosophy; examples and exercises to make the material more accessible; and an extensive bibliography. It also includes philosophical discussion of many aspects of relevant logics and highlights several directions for future research, both philosophical and formal.

Shawn Standefer is an assistant professor of philosophy at North Carolina State University. He is a co-author of *Logical Methods* (2023) and co-editor of *New Directions in Relevant Logic* (2025).

Relevant Logics

Implication, Modality, Quantification

SHAWN STANDEFER
North Carolina State University

CAMBRIDGE
UNIVERSITY PRESS

CAMBRIDGE
UNIVERSITY PRESS

Shaftesbury Road, Cambridge CB2 8EA, United Kingdom

One Liberty Plaza, 20th Floor, New York, NY 10006, USA

477 Williamstown Road, Port Melbourne, VIC 3207, Australia

314–321, 3rd Floor, Plot 3, Splendor Forum, Jasola District Centre, New Delhi – 110025, India

Cambridge University Press is part of Cambridge University Press & Assessment, a department of the University of Cambridge.

We share the University's mission to contribute to society through the pursuit of education, learning and research at the highest international levels of excellence.

www.cambridge.org
Information on this title: www.cambridge.org/9781009422857
DOI: 10.1017/9781009422871

When citing this work, please include a reference to the DOI 10.1017/9781009422871

First published 2026

A catalogue record for this publication is available from the British Library

Library of Congress Cataloging-in-Publication Data
Names: Standefer, Shawn author
Title: Relevant logics : implication, modality, quantification / Shawn Standefer, North Carolina State University.
Description: Cambridge, United Kingdom ; New York, NY, USA : Cambridge University Press, 2026. | Includes bibliographical references and index.
Identifiers: LCCN 2025048222 (print) | LCCN 2025048223 (ebook) | ISBN 9781009422857 hardback | ISBN 9781009422901 paperback | ISBN 9781009422871 epub
Subjects: LCSH: Logic | Relevance (Philosophy) Classification: LCC BC71 .S73 2026 (print) | LCC BC71 (ebook)
LC record available at https://lccn.loc.gov/2025048222
LC ebook record available at https://lccn.loc.gov/2025048223

ISBN 978-1-009-42285-7 Hardback
ISBN 978-1-009-42290-1 Paperback

Contents

Preface

This is a book on the formal theory of relevant logics. Some of this material is well known, and some is not. Some of it is old and some new. The material is enough to bring the reader up to speed on the current state of the field, or a portion of it at least, and it provides a non-comprehensive picture of the contemporary research in the area.

In many ways, this book is a product of my time in Australia. I started working seriously on relevant logics during my postdoc at the University of Melbourne, and I started working on this material in the final year of my postdoc, writing notes as a way to systematize things I had learned there. Writing those notes helped me get through the first year and a half of COVID. At some point, it seemed feasible to turn the notes I was writing into a book. Many of the ideas for this work, and my understanding of relevant logics, developed during my time at the University of Melbourne working with Greg Restall. Greg and I talked about logic, including relevant logic, a lot. It would be hard to overstate how much I got out of those conversations. I greatly benefited from the regular interactions with the attendees of the Melbourne Logic Seminar. I also benefited from regular attendance of the Pukeko Logic meetings in New Zealand and the Australasian Association for Logic conferences. Apart from Greg, I'd like to thank four regular attendees in particular: Lloyd Humberstone, Rohan French, Ross Brady, and Ed Mares. Lloyd has given me feedback on more or less everything I have written since 2016, and he has done a great deal in shaping how I approach logic. Along with Lloyd, Rohan has given me lots of feedback on my work, and our regular logic chats have been immensely helpful for me. Ross and I have had many discussions, both in person and over email, that clarified my understanding. Lastly, Ed has been extremely encouraging and enthusiastic in our discussions of logic, and I think his influence on this work should be clear.

This book is also a product of my time at the University of Pittsburgh. I started learning about relevant logics during grad school, where I had the good fortune to learn logic from many great logicians on a wide range of topics, including material connected to this book. I am very grateful to them, and I would like to thank three in particular. As with many things from my time in graduate school, I am indebted to Anil Gupta. During my first year at Pitt, I did a directed reading with Anil on Dunn and Hardegree's *Algebraic Methods in Philosophical Logic* that, in retrospect, influenced me greatly. Two of my teachers, Nuel Belnap and Mike Dunn, have, unfortunately, passed away in recent years. I would like to thank Mike for the encouragement and discussions. Mike's explanation of the completeness proof for the ternary relational models was the first time that material clicked for me. I would like to thank Nuel for many things, not least of which is patiently walking through many proofs with me on a whiteboard and stopping me to ask, "Could you explain that step?" I hope that Mike and Nuel would have liked this book, even though their approaches to the topic differed somewhat from mine.

I am grateful to the Department of Philosophy at National Taiwan University for providing a wonderful environment for logic research. This research was supported by the National Science and Technology Council of Taiwan grant 111-2410-H-002-006-MY3.

Many people have given me feedback on this material at some stage of its development. Apart from the people above, I have gotten feedback from the following people, to each of whom I am very grateful: Guillermo Badia, Jc Beall, Katalin Bimbó, Patrick Blackburn, Ethan Brauer, Thomas Ferguson, Patrick Girard, Ole Hjortland, Tomasz Kowalski, Shay Logan, Tore Fjetland Øgaard, Hitoshi Omori, Hiroakira Ono, Mina Pedersen, Graham Priest, Ellie Ripley, Katsuhiko Sano, Igor Sedlár, An-Pang Tai, Tanyen Tsai, Joe Ulatowski, Ren-June Wang, Heinrich Wansing, Zach Weber, and Yale Weiss. Andrew Tedder and Nai-Chung Hou deserve special mention for in-depth comments on the full manuscript.

Finally, I would like to thank my partner, Suzie, for her encouragement and patience throughout this project. Without her support, this book would have been neither started nor finished.

1 The Heart of Logic

[W]e take the heart of logic to lie in the notion "if... then -"

Anderson and Belnap (1975), §1

1.1 Introduction

Relevant logics are logics with demanding conditionals, or implication connectives, that require a substantive connection between their antecedents and consequents.[1] There are different views of what this substantive connection amounts to.[2] For example, Anderson and Belnap (1975) think that it means that for a valid conditional $A \rightarrow B$, the antecedent must really be used in deriving the consequent. Routley and Routley (1972) think that it is a matter of the antecedent of a conditional being absolutely sufficient for the consequent.[3] Brady (2006b) takes it to be a matter of meaning containment, that the meaning of the antecedent contains the meaning of the consequent. While there is disagreement on the nature of the implication, there is universal agreement that any implication worth the name "relevant" has *Belnap's variable-sharing property*:[4] If $A \rightarrow B$ is derivable, then A and B share a propositional variable. In other words, in a valid implication, the antecedent and consequent share some non-logical content. The variable-sharing property imposes a minimal amount of *formal relevance* on implicational logical truths. That amount of formal relevance is a necessary condition on being a relevant logic. We will return to the variable-sharing property throughout this book.

To provide a general characterization of the class of relevant logics, we will add two conditions to Belnap's variable-sharing property. One is that the logics obey *modus ponens*, that is, if A and $A \rightarrow B$ are both logical truths, then so is B. The other is that, when conjunction ($\wedge$) is in the language, the logics obey adjunction, that is, if

1 While this book will provide an introduction to relevant logics, there are several excellent alternatives for the reader who would like some other options. Such a reader can consult Read (1988), Dunn and Restall (2002), Bimbó (2007b), Mares (2020), or Logan (2024).

2 For more discussion of these views, see Standefer (2024).

3 A note on naming conventions is in order. Both Richard and Val Routley adopted different names later in life. Richard Routley changed his name to Richard Sylvan. Val Routley changed her name to Val Plumwood. We will follow the convention of the area of citing both Sylvan's and Plumwood's work under the published names, although we will refer to them using the final names they adopted.

4 This was the variable-sharing property originally demonstrated by Belnap (1960).

A and B are logical truths, then so is $A \wedge B$. Those three give us the class of relevant logics.[5] This definition casts a wide net, and the focus of this book will be narrower. Our focus in this book is primarily on the major relevant logics that have been studied in the tradition of Anderson, Belnap, Meyer, Dunn, Sylvan, and Plumwood.[6] Those logics are, roughly, those between the logic B and the logic R, although there will be occasion to highlight some relevant logics that fall outside this range. Several logics in that range have been studied, and many have presentations in terms of axioms, sequent systems, frames, combinators, and matrices.[7] To borrow from Anderson and Belnap, being able to view the logics from many different perspectives makes them seem more natural or substantial.

The logics between B and R will be the primary focus of this book. While the grouping may seem slightly ad hoc, when taken as a whole a kind of unity emerges. For comparison purposes it will be useful to talk about logics outside that range, both relevant logics outside that range, such as BB, and non-relevant logics, such as classical logic, TV.[8]

The basic logical language that we will work with is built from the logical vocabulary $\{\rightarrow, \circ, \wedge, \vee, \neg, \mathsf{t}\}$. Some of these connectives, in particular implication ($\rightarrow$), negation ($\neg$), conjunction ($\wedge$), and disjunction ($\vee$), are likely to be somewhat familiar to the reader. The connective fusion ($\circ$) is sometimes called an intensional conjunction, and its conjunction-like behavior will be elaborated in later chapters.[9] In short, moving to a relevant logic splits the logical roles occupied by conjunction in classical logic, and those roles are distributed between fusion and conjunction. The Ackermann constant (t), also known as "little tee," represents logical truth, and it can be understood, informally, as the conjunction of all logical truths.[10] In classical logic, any tautology will work equally well to represent logical truth, but in relevant logics, no finite combination of them will work. In later chapters, we will extend this vocabulary with additional connectives, such as modal operators and quantifiers. There will be a set At of atoms, members of which will be denoted by $p, q, r, \ldots$, possibly with subscripts. The basic language $\mathfrak{L}$ will be built from the vocabulary in the usual way.

$$A := p \mid \mathsf{t} \mid B \rightarrow C \mid B \circ C \mid B \wedge C \mid B \vee C \mid \neg B.$$

5 See Standefer (2025b) for further discussion and defense of this definition.

6 To some extent, this tradition is artificial, as the focus of Anderson and Belnap, and (to some degree) Dunn, was on logics different from those of Meyer, Sylvan, and Plumwood. Orlov (2024) studied a relevant logic that fits in this tradition, as discussed by Došen (1992), although his work was unknown to the logicians listed in the previous sentence.

7 We won't cover combinators or sequent systems in detail in this book. The connections among these areas has been called "the Key to the Universe" by Meyer, as for example, in his joint work with Routley (Meyer & Routley, 1972, 423). For more on combinatory logics, the interested reader should see Bimbó (2009; 2011).

8 The label for classical logic, TV, is due to Anderson and Belnap. We are adopting it here, as the labels C and CL are going to be used for relevant logic purposes, and there are already too many Ks being considered.

9 Read (1988) extensively discusses and defends fusion as being fundamental to the relevant logic program.

10 With t in the language, the variable-sharing property needs a qualification, namely that neither A nor B contains t.

The primary defined connective that will be used throughout the book is the biconditional, $A \leftrightarrow B$, defined as $(A \rightarrow B) \wedge (B \rightarrow A)$. The binding convention will be that the singulary connectives bind tighter than binary connectives, with the defined biconditional binding less tightly than the other binary connectives.

We can illustrate some features of relevant logics using $\rightarrow$, $\neg$, $\wedge$, and $\vee$. One of the guiding aims of relevant logics is avoidance of the paradoxes of implication.

- $A \rightarrow (B \rightarrow A)$.
- $A \rightarrow (B \rightarrow B)$.

These formula schemes, valid in classical and intuitionistic logic, are anathema to the relevant logician. There seems not to be the right sort of connection between the parts of the implications. Take the first, for example. B need not have anything to do with A, and A need not *imply* that any old B implies A. Given A, B need not make any contribution to A holding. A formula merely being true is not enough to secure being implied by another formula. Let us turn to the second. While $B \rightarrow B$ is a logical truth, A seems to be merely along for the ride, so to speak. A may make no contribution to the validity of the consequent. Without any sort of substantive connection, $B \rightarrow B$ won't be implied by just any given formula. A formula being a logical truth is not enough to secure being implied by another formula.

The second paradox is, on its own, a violation of variable sharing, as $p \rightarrow (q \rightarrow q)$ is an instance of it.[11] The first paradox leads to a violation of variable sharing, provided that the implication obeys *modus ponens* and the first paradox is a theorem. Let $p \rightarrow (p \rightarrow p)$ be A, and let q be B. Since $A \rightarrow (B \rightarrow A)$ is a theorem, as is A, by *modus ponens* $B \rightarrow A$ is as well. That formula, however, violates variable sharing, as it is $q \rightarrow (p \rightarrow (p \rightarrow p))$. Thus, we can see that the paradoxes are incompatible with the guiding intuition of relevant logic, that there is a substantive connection between the antecedent and consequent of valid implications, as codified in Belnap's variable-sharing property.

Relevant logics reject the paradoxes and things from which they follow. As one might expect, rejecting the paradoxes has consequences for the other logical vocabulary. Given some plausible axioms and rules governing implication, adjustments must be made in the logical behavior of other connectives as well. For example, $A \rightarrow (\neg A \rightarrow B)$ must be rejected, along with $\neg(A \rightarrow A) \rightarrow B$, $(A \wedge \neg A) \rightarrow B$, and $A \rightarrow (B \vee \neg B)$. There are many more examples of classical tautologies that are rejected by relevant logics, either because they violate variable sharing or because, in the presence of other plausible principles, they lead to such violations. Some illustrative examples are presented in Section 2.4.

The paradoxes were highlighted by the logician C. I. Lewis as indicating that the classical conditional is not a good candidate for an entailment connective.[12] Lewis

[11] It is worth observing here that logics are required to be closed under uniform substitution. Uniform substitutions will be defined in Chapter 2, by definition 2.2.

[12] Mares (2024) provides a detailed history of entailment connectives in logic. For more on the paradoxes in connection to relevant logics, see also Anderson and Belnap (1975) and Kielkopf (1977).

wanted a connective in the object language that expresses entailment. His developments in modal logic were partially in the service of finding such a connective. He proposed some strict implications, which are necessary implications, which in contemporary notation would be written as $\Box(A \rightarrow B)$. The logic of strict implications varies, depending on the logic of the necessity operator ($\Box$) and the implication in its scope. Lewis's preferred strict implications avoided the first paradox, but strict implications tend to validate the second paradox or some of its relatives, such as $A \rightarrow (B \vee \neg B)$.

Lewis learned to live with some of the implicational paradoxes, but not everyone agreed with him that it was acceptable. Anderson and Belnap, building on the work of Ackermann (1956; 1958), defended logics that avoided the implicational paradoxes. The preferred logic of Anderson and Belnap was their logic E, which was meant to be the logic of entailment, combining relevance and necessity considerations. Along with E, they developed the logic R, the logic of relevance, as well as some other logics, such as the relevant logic T. Their student, Meyer, introduced NR, which added a necessity operator to R, with the goal of showing that the implication of E was the strict implication of R. We will return to this idea in Chapter 6. The development of models by Sylvan and Meyer opened the door to many more relevant logics, including the logic B, which is the basic logic adopted in this book.

It is worth pausing here to see the difficulty in avoiding the second paradox. The difficulty can be illustrated using *Kripke models*, which have been used with such success in the study of modal logics.[13] A Kripke model is a triple $\langle W, S, V \rangle$, where $W \neq \emptyset$, a non-empty set of worlds, $S \subseteq W \times W$, and V assigns atoms sets of worlds. We can define verification at a world in a model as follows, using standard classical conditions.

- $w \Vdash p$ iff $w \in V(p)$.
- $w \Vdash \Box B$ iff for all $u \in W$, if Swu, then $u \Vdash B$.
- $w \Vdash B \rightarrow C$ iff either $w \nVdash B$ or $w \Vdash C$.

The derived verification condition for the strict implication $\Box(B \rightarrow C)$ is

- $w \Vdash \Box(B \rightarrow C)$ iff for all $u \in W$, if Swu, then either $u \nVdash B$ or $u \Vdash C$.

We will say that a formula is valid if it is verified at all worlds in all models.

This simplified set up illustrates that there are two desiderata that are in tension.

- There are valid logical truths, such as $A \rightarrow A$.
- The second paradox of implication is invalid.

The relevant logician thinks that there are logical truths, such as $A \rightarrow A$. The relevant logician denies that every formula implies every logical truth, so, for example, $B \rightarrow (A \rightarrow A)$ is, generally, invalid. In the Kripke models, the validity of $A \rightarrow A$ requires that $A \rightarrow A$ be verified at every world in every model. The invalidity of $B \rightarrow (A \rightarrow A)$ would require a model with a world w where B holds at w but $A \rightarrow A$ does not, contradicting the validity of the latter. It is clear that using a strict implication,

[13] See Blackburn et al. (2002) for more on Kripke models for TV-based modal logics.

$\Box(B \rightarrow \Box(A \rightarrow A))$ will not change things in any important way, except that the world where B holds may not be required to be the world where $A \rightarrow A$ fails. The failure of $A \rightarrow A$ is still impossible.[14]

Models for relevant logics that permitted completeness results took some time to be developed. The issue that needed to be overcome is that the verification condition for the strict implication requires evaluating the antecedent and consequent at the *same* world. The insight for providing models for relevant logics is that, in order for both paradoxes to fail, one needs to be able, in some cases, to evaluate the antecedent and consequent of an implication at *different* worlds. A natural way to do this is using a ternary relation, which is the route followed by Sylvan and Meyer. In the ternary relational models, also known as Routley–Meyer models, one uses a ternary relation to interpret implications:[15]

- $w \Vdash B \rightarrow C$ iff for all x, y, if $Rwxy$ and $x \Vdash B$, then $y \Vdash C$.

As we will see, these models will ensure that there are logical truths, such as $A \rightarrow A$, while also ensuring that there are worlds at which logical truths can fail. Thus, there is a way to avoid the paradoxes of implication.[16] We will get into the details of the models in Chapter 3. Different logics can be obtained by adopting different conditions on the ternary relation, in a manner familiar from the study of Kripke models for modal logics.

Since the early work of Anderson and Belnap on relevant logics, there has been an interest in connections between necessity and relevance. Anderson and Belnap combined relevance and necessity in their logic E, where both were built into the implication, while Meyer considered also a primitive modality, separate from the (non-modal) implication. In the last couple of decades, a large amount of work has been done investigating the addition of primitive modal operators to relevant logics. If a relevant logic is to be one's all-purpose logic, then it will be important to understand how modal concepts work in a relevant logical setting. As one indication of how adopting a relevant modal logic could be useful, Blumson (2021) has shown that Ayer's verificationist criterion of meaning can be rescued from Lewis's triviality results by using a relevant modal logic, indeed even one of the stronger ones considered in this book.

Modalities in relevant logics can be modeled with the addition of a binary modal accessibility relation to the frames, in ways familiar from standard Kripke models. That is not the whole story, however, as there are interactions between the parts of the frames and the accessibility relation. The non-classical base logics open up new and interesting possibilities for modal notions, some of which we will examine in Chapter 8.

[14] One can introduce impossible worlds into the models as a way to invalidate the second paradox, but some of its relatives, such as $A \rightarrow (B \vee \neg B)$, remain valid. See Hughes and Cresswell (1996, ch. 11) or Priest (2008, ch. 4) for details and references.

[15] See Bimbó et al. (2018) and Ferenz (2018) for more on the early developments of the ternary relational models.

[16] An alternative approach, adopted by Urquhart (1972a) and Fine (1974), is to use an *operation* on worlds, so that evaluating the conditional at w requires checking the truth of the antecedent at u and the consequent at $w \cdot u$, a potentially distinct world. We will briefly see their approaches to frames in Chapter 3.

Quantification has been part of the relevant logical story for a long time, both the usual first-order quantifiers and propositional quantifiers. There have, however, been surprising difficulties modeling quantifiers. The difficulties arise on the straightforward approach of adding a domain to a frame and using the Tarskian verification condition, that a universal $\forall x A$ holds if A holds, for every value of the bound variable x. Fine (1989) showed that the standard form of quantified R is incomplete with respect to these models. Two alternative modellings for quantification have been developed in response, one by Fine and one by Mares and Goldblatt. The latter approach is adopted in Chapter 11. Many applications of relevant logics to areas of interest in philosophy, logic, and mathematics require quantification and identity, so we present the basics of these two topics.

1.2 Illustrative Examples

Most of the book will be devoted to exploring some of the best known relevant logics, roughly those studied in the broad tradition of Anderson and Belnap. We can, however, introduce the basics of relevant logics using some algebraic tools. We will return to the algebraic tools in more detail in Chapter 5.

The algebras can be viewed as many-valued logics. Each algebra has a set of values, with an ordering in the diagram, and when one adds a special subset of designated values, one obtains a *matrix*.[17] For a given matrix, a valuation v assigns the atoms a value, and then the tables are used to evaluate complex formulas. A formula A is valid on an matrix iff A takes a designated value, for every valuation.

The first matrix is the matrix $RM3$, presented in Table 1.1. It is a three-valued algebra, and the values are ordered $\mathbf{0} \leq \frac{1}{2} \leq \mathbf{1}$, with $\frac{1}{2}$ and $\mathbf{1}$ designated. The values of $v(A \wedge B)$ and $v(A \vee B)$ are, respectively, the minimum and maximum of $v(A)$ and $v(B)$. The negation and implication are given in Table 1.1. We can evaluate nested implications. In particular, we can show that the paradoxes of implication fail in this matrix. Take $p \rightarrow (q \rightarrow p)$. Let $v(p) = \frac{1}{2}$ and $v(q) = \mathbf{1}$. Then $v(q \rightarrow p) = \mathbf{0}$ and so $v(p \rightarrow (q \rightarrow p)) = \mathbf{0}$. To show that $p \rightarrow (q \rightarrow q)$ is invalid, let $v(p) = \mathbf{1}$ and $v(q) = \frac{1}{2}$. Then $v(q \rightarrow q) = \frac{1}{2}$, so $v(p \rightarrow (q \rightarrow q)) = \mathbf{0}$. Rejecting the paradoxes is possible

Table 1.1. Matrix *RM*3

$\rightarrow$	$\mathbf{0}$	$\frac{1}{2}$	$\mathbf{1}$	$\neg$
$\mathbf{0}$	$\mathbf{1}$	$\mathbf{1}$	$\mathbf{1}$	$\mathbf{1}$
$\frac{1}{2}$	$\mathbf{0}$	$\frac{1}{2}$	$\mathbf{1}$	$\frac{1}{2}$
$\mathbf{1}$	$\mathbf{0}$	$\mathbf{0}$	$\mathbf{1}$	$\mathbf{0}$

[17] The ordering is a partial ordering, meaning that it is reflexive, transitive, and anti-symmetric. The ordering is, in fact, a distributive lattice ordering. These terms will be defined in Chapter 5, where an explanation of the Hasse diagrams for the orderings can also be found.

Table 1.2. Belnap–Dunn matrix M_4

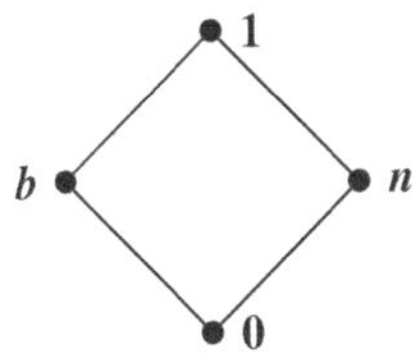

$\wedge$	**0**	b	n	**1**
0	**0**	**0**	**0**	**0**
b	**0**	b	**0**	b
n	**0**	**0**	n	n
1	**0**	b	n	**1**

$\vee$	**0**	b	n	**1**
0	**0**	b	n	**1**
b	b	b	**1**	**1**
n	n	**1**	n	**1**
1	**1**	**1**	**1**	**1**

$\neg$	
0	**1**
b	b
n	n
1	**0**

Table 1.3. Crystal lattice CL

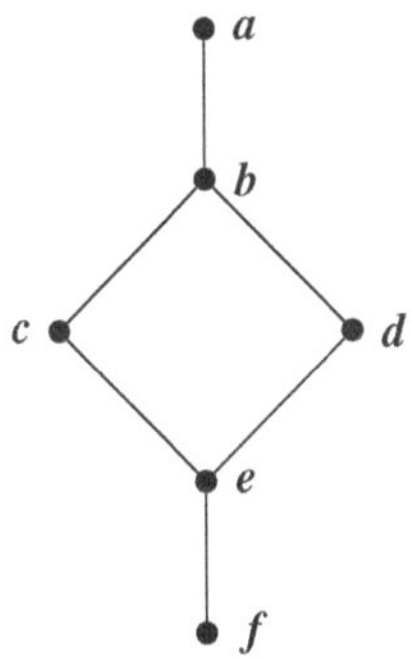

$\rightarrow$	f	e	c	d	b	a	$\neg$
f	a	a	a	a	a	a	a
e	f	e	c	d	b	a	b
c	f	f	c	f	c	a	c
d	f	f	f	d	d	a	d
b	f	f	f	f	e	a	e
a	f	f	f	f	f	a	f

in this three-valued matrix. As long as negation is not in the language, then the valid implications on this matrix will obey variable sharing. Unfortunately, once negation enters the picture, variable sharing will fail. For example, $\neg(p \rightarrow p) \rightarrow (q \rightarrow q)$ will be valid. Because of this, the logic of $RM3$ is not a relevant logic. To get a logic that obeys variable sharing with negation in the language, we will move to a four-valued matrix.

The logic of first-degree entailment, FDE, can be obtained from a four-valued matrix, given in Table 1.2. The four values are ordered $\mathbf{0} \leq \boldsymbol{b} \leq \mathbf{1}$ and $\mathbf{0} \leq \boldsymbol{n} \leq \mathbf{1}$, with $\boldsymbol{b}$ and $\mathbf{1}$ being designated. FDE focuses on the vocabulary $\{\wedge, \vee, \neg\}$, with implications holding just between formulas that lack implications. The logic FDE deals with formulas of the form $A \rightarrow B$, where neither A nor B contains an implication. While we have not supplied a table for implication, we will say that $v(A \rightarrow B)$ is designated if $v(A) \leq v(B)$ and undesignated otherwise. As we will see in Chapter 5, FDE obeys the variable-sharing property.

While this four-valued matrix gives a logic that obeys variable sharing in a restricted fragment, there are difficulties extending it with an implication table that permits nested implications. For that, we will use a six-valued matrix.

The six-valued matrix in Table 1.3 is known as the crystal lattice, CL. Its values are ordered $\boldsymbol{f} \leq \boldsymbol{e} \leq \boldsymbol{c} \leq \boldsymbol{b} \leq \boldsymbol{a}$ and $\boldsymbol{f} \leq \boldsymbol{e} \leq \boldsymbol{d} \leq \boldsymbol{b} \leq \boldsymbol{a}$, with everything but $\boldsymbol{f}$ designated. $v(A \wedge B)$ is evaluated as the minimum value of $v(A)$ and $v(B)$, with the caveat that $\boldsymbol{c} \wedge \boldsymbol{d} = \boldsymbol{e}$. $v(A \vee B)$ is evaluated as the maximum value of $v(A)$ and $v(B)$, with the caveat that $\boldsymbol{c} \vee \boldsymbol{d} = \boldsymbol{b}$. On the crystal lattice, if $A \rightarrow B$ is valid, then A and B share an atom. Thus, the logic of the crystal lattice obeys variable sharing with no restrictions. While the logic obeys variable sharing, some formulas are valid that many relevant logicians would want to reject, such as $A \vee (A \rightarrow B)$. This logic has a sublogic that is generally regarded as the strongest standard relevant logic, namely Anderson and Belnap's logic R. The logic R cannot be characterized by any matrix with only finitely many values.[18] In Chapter 2, we will introduce R axiomatically, and then in subsequent chapters, we will provide frames for R as well as a class of matrices for it, which includes some matrices with infinitely many values.

1.3 Why Relevant Logics?

Having seen some illustrative examples of relevant logics, the reader might wonder why one should be interested in them. There are philosophical reasons and there are formal reasons. We will begin with the philosophical reasons.

Different relevant logics have been motivated by different philosophical considerations. We will mention three, although there are others: the use criterion, meaning containment, and truthmaking.[19] As mentioned earlier, Anderson and Belnap thought that for an implication to be valid, one had to *really use* the antecedent in obtaining the consequent. Merely deriving the consequent, such as $p \vee \neg p$, under an assumption, such as q, is not enough to conclude that q implies $p \vee \neg p$. The antecedent needs to contribute in a substantive way to the derivation of the consequent. This is the use criterion of Anderson and Belnap, and it is taken to motivate some of the stronger relevant logics, particularly R and E.

The next motivation is meaning containment, championed by Brady. On this view, valid implications reflect meaning containments, namely that the meaning of the consequent is contained in that of the antecedent. Brady (1988; 1989a; 2006b) formalizes this idea via his content semantics. This idea of meaning containment is connected to an alternative form of variable sharing, depth relevance, which we will discuss in Section 5.5. Meaning containment is associated with the logic DJ.

The third motivating idea is truthmaking. With truthmakers, a claim is made true in virtue of a particular situation or state of affairs obtaining. On one view of truthmaking, the so-called exact view, a truthmaker must be wholly relevant to the claim made true. This leads to a natural relation of relevance between claims via shared truthmakers. Jago (2020; 2023) argues that R, in particular, has natural truthmaker models with a rich mereological structure on truthmakers.[20]

The most sustained, recent philosophical defense of a relevant logic is probably Mares's (2004) defense of R and interpretation of frames for R. We will return to this in Chapter 3, but in brief, Mares interprets the frames in informational terms. The points of the frames represent situations, and agents use the information in different situations to reason about what other sorts of situations are available. This is the theory

[18] R cannot be characterized by any class of finite-valued matrices either.

[19] See Standefer (2024) for more details and discussion.

[20] Majer et al. (2023) show how to generalize the formal structure of truthmaker models to cover more logics.

of situated inference, connecting relevant logics and the situation theory of Barwise and Perry (1983).

There are multiple philosophical views compatible with relevant logics. The philosophical foundations of relevant logics are undergoing a reevaluation, at the time of writing, and so there are likely to be alternative philosophical views supporting relevant logics in the future.

On the formal side, there are many reasons to be interested in relevant logics. As we will see, they have deep and interesting frame theory. They have important connections to algebras and matrices, and they have connections to combinatory logic.

In many non-classical logics, one needs to be careful with how various axioms are stated, since classically equivalent forms may be distinct in a non-classical setting. Relevant logics, with their rejection of the paradoxes of implication, draw many distinctions among classically equivalent formulas, having non-equivalent logical truths. This allows one to tease apart the contributions of the different ways of formulating principles and axioms.

Finally, relevant logics provide an illuminating contrast with the main non-classical logic that philosophers are familiar with, namely intuitionistic logic. While relevant logics and intuitionistic logic share some features, such as having a frame semantics, generally not being finite-valued, and having an implication not definable from disjunction and negation, they differ in important ways. Intuitionistic logic has a different approach to negation than the standard negation of relevant logics.[21] In intuitionistic logic, contradictions are *explosive*, meaning that they imply everything. This is not the case with relevant logics.[22] While the implication of intuitionistic logic is not the same as that of classical logic, they are fairly close in many ways. The implication of a relevant logic is much different to work with and makes salient the roles of different axioms and principles. Intuitionistic logic is typically thought of as a constructive logic. Relevant logics are not generally constructive, although some of them share some formal features of constructivity. While one can get to classical logic with a small addition to intuitionistic logic, once one moves to R or below, greater distance emerges with classical logic. One can get to classical logic from R by adding $A \to (B \to A)$, among other things, and for logics weaker than R, additional axioms need to be added to reach classical logic. A comparison between relevant logic and classical logic will be presented in Section 2.4.

In short, relevant logics are philosophically interesting, have rich algebraic and frame semantics, and provide an important contrast to the perhaps better known intuitionistic logic. These seem like good reasons for studying relevant logics. We hope that they provide enough motivation to read through the book to get a sense for relevant logics.

21 Relevant logics can adopt an intuitionistic style of negation, as implying a falsity constant, as studied by Robles and Méndez (2018). It is worth noting that in R, negation can already be understood as implying falsity, for more on which see discussion by Onishi (2016).

22 There is some nuance here, which we will return to in Section 9.8.

1.4 Overview of the Contents

This book splits into three parts. Part I, consisting of Chapters 2 to 5, develops the foundational material. These chapters provide an overview of the basic relevant language, axioms, frames, and some formal techniques and metatheory. Part II, consisting of Chapters 6 to 10, covers the modal extensions of relevant logics. These chapters build on the first part of the book. Part III, which is largely independent of the second part, consists of the remainder, Chapters 11 to 13. It covers quantified and first-order extensions.

The goal of this book is to provide the reader with a systematic introduction to relevant logics and their frames, providing up-to-date coverage of major developments in relevant logics. While the hope is that it can serve as an introduction to the area, the material is developed beyond an introductory level. To help readers absorb and explore the material, exercises have been provided for most chapters.

A few comments about the scope of the book are in order. The goal of the book is to give an up-to-date picture of relevant logics, in the broad tradition of Anderson and Belnap, focusing on frames.[23] Our focus is on ternary relational frames, although alternative frames will be discussed briefly. Logan (2024) provides an excellent introduction to one of the alternative frame theories. As far as proof systems go, we will primarily use axiom systems. There has been a lot of work done on proof systems, but adequately covering the different proof systems would take us beyond the scope of this book. We will briefly return to proof systems in Chapter 13.

While the focus is on formal results, the topics are those that should be of interest to philosophers and philosophical logicians. The topics are also ones that differ somewhat from the topics of interest in the earlier work on relevant logics, as discussed by Anderson and Belnap (1975) and Routley et al. (1982). Those volumes have some topics that will get comparatively little discussion here, such as independence results, conservative extension, and decidability.[24] These are all important topics, but they do not exhaust the interest of the area. They come in for some discussion, but they are not as central.

Finally, a comment on the name of the field. Some works use "relevant logics" and some use "relevance logics." The difference in name has, in the past, tended to signal a difference in approach to the field, with there being a roughly geographic split between the use of the two names. As far as we can tell, that viewpoint is largely a thing of the past, and the difference in name now signals nothing of substance. The ideological differences have largely fallen away as the field has developed and new logicians have joined in. We have opted for the name "relevant logics," since a large portion of the work in the area in the last 20 years has used that.[25]

23 There are other traditions of relevant logic, notably the core logic tradition of Tennant (2017).

24 Decidability, in particular, will barely be discussed. For more on decidability, mainly failures thereof, see the work of Urquhart (1984; 2016a). For some positive results, see Giambrone (1985) and Restall (2000, ch. 14).

25 We are also persuaded by Meyer's observation that "relevant" fits better with other nomenclature in logic, e.g., "modal logic" rather than "modality logic."

Part I

Foundations

2 Axioms

2.1 Introduction

As we said in Chapter 1, relevant logics are logics with demanding implication connectives. The sense of demandingness will be elaborated later, but a good way to get a feel for it is to work with some relevant logics to see what is provable and how.

In this chapter, we will look at provability in relevant logics. This will be done via Hilbert axiom systems, which are, for better or worse, the most common way to approach relevant logics. There are alternative kinds of proof systems for relevant logics, and in Chapter 13, we will briefly look at some. We will be primarily interested, throughout the book, in logics as sets of formulas.[1] There has been some work on more general logical frameworks, including consequence relations for logics, and we will comment on some of that work when appropriate.

We will begin, in Section 2.2, by defining the basic relevant logic, B. This is the weakest *standard* relevant logic, although it is not the *weakest* relevant logic. Many of the basic concepts used in the axiomatic study of relevant logics will be defined, in Section 2.3, in terms of B. We will then provide some common axioms and rules for extending B, and the basic concepts can be generalized to these extensions in natural ways. Following this, in Section 2.4, we will provide a detailed comparison between the relevant logic R, which is the strongest standard relevant logic, and classical logic, TV. Then, in Section 2.5, we will define several concepts related to logical theories, which are crucial in the study of relevant logics. Finally, in an appendix, we will present four sublogics of B that are worth highlighting: FDE, BB, F, and BM.

2.2 Basic logic

In this section, we will provide a list of the axioms and rules for the basic relevant logic, B. Unlike some axiom systems for classical logic, the system for B has many rules, as well as many axioms. In the stronger logics, to be presented in Section 2.3, some of these rules get incorporated as axioms. In the presentation of the rules, $A_1, \ldots, A_n \Rightarrow B$, the A_is are the premises of the rule, and B is the conclusion.

[1] This is the FMLA framework discussed by Humberstone (2011, 103ff.).

(A1) $A \rightarrow A$.
(A2) $(A \wedge B) \rightarrow A$, $(A \wedge B) \rightarrow B$.
(A3) $((A \rightarrow B) \wedge (A \rightarrow C)) \rightarrow (A \rightarrow (B \wedge C))$.
(A4) $A \rightarrow (A \vee B)$, $A \rightarrow (B \vee A)$.
(A5) $((B \rightarrow A) \wedge (C \rightarrow A)) \rightarrow ((B \vee C) \rightarrow A)$.
(A6) $(A \wedge (B \vee C)) \rightarrow ((A \wedge B) \vee (A \wedge C))$.
(A7) $\neg\neg A \rightarrow A$.
(A8) t.

(R1) $A, A \rightarrow B \Rightarrow B$ (Modus ponens).
(R2) $A, B \Rightarrow A \wedge B$ (Adjunction).
(R3) $A \rightarrow \neg B \Rightarrow B \rightarrow \neg A$ (Contraposition).
(R4) $A \rightarrow B \Rightarrow (B \rightarrow C) \rightarrow (A \rightarrow C)$ (Suffixing).
(R5) $A \rightarrow B \Rightarrow (C \rightarrow A) \rightarrow (C \rightarrow B)$ (Prefixing).
(R6) $A \rightarrow (B \rightarrow C) \Rightarrow (A \circ B) \rightarrow C$ (Importation).
(R7) $(A \circ B) \rightarrow C \Rightarrow A \rightarrow (B \rightarrow C)$ (Exportation).
(R8) $A \Rightarrow \mathsf{t} \rightarrow A$.

The logic B is defined using these axioms and rules. We can put the definition a bit more carefully.

DEFINITION 2.1 (Theorems and logic) *A is a* THEOREM *of the logic* B, *in symbols,* $\vdash_{\mathsf{B}} A$, *iff there is a sequence of formulas* $C_1, \ldots, C_n$, *for* $n \geq 1$, *such that for each* $i \leq n$, *either*

- C_i *is an axiom,*
- *there is* $j < i$ *such that* C_j *is the premise of a one-premise rule and* C_i *is the conclusion, or*
- *there are* $j, k < i$ *such that* C_j *and* C_k *are the premises of a two-premise rule and* C_i *is the conclusion, and*
- $C_n = A$.

Such a sequence is said to be a B*-proof of A.*

As a set of formulas, $\mathsf{B} = \{A \in \mathfrak{L} : \vdash_{\mathsf{B}} A\}$.

There is a more general notion of provability that permits *non-logical assumptions* to feature in the proof. Such proofs permit one to reason from assumptions that are not axioms. While important, this would introduce additional complications, so we will not use that notion much in this book.[2]

Before we proceed, it is worth commenting on the axioms and rules a bit. Axioms (A2)–(A6), together with rule (R2), ensure that conjunction and disjunction behave in more or less familiar ways. In the jargon, these connectives are the distributive lattice connectives. When considering just the interactions between conjunction and disjunction, they act like their classical counterparts. Axiom (A7) and rule (R3) ensure that

2 Among the complications is the distinction between a rule of inference and a rule of proof. For more on this, see Humberstone (2010). The interested reader should see Øgaard (2021d, 2022), for example, for details on consequence relations in relevant logics.

double negation introduction and elimination hold, and together with the conjunction and disjunction axioms, they ensure that all of the De Morgan equivalences with conjunction and disjunction hold, as we will see below.

Rules (R4)–(R5) are weak forms of transitivity for implication. With (A1) as the only purely implicational axiom, the implication of B is rather weak, and the extensions considered in Section 2.3 provide some standard axioms to strengthen the implication. The rules (R6)–(R7) are the only principles governing fusion. In intuitionistic or classical logic, conjunction is connected to implication via the import/export principles. In relevant logics, that job is taken over by fusion.[3]

The remaining principles to comment on, (A8) and (R8), provide the logical behavior of the Ackermann constant. It represents logical theoremhood. It is the weakest theorem, implying all others, which is given by (R8). It is distinctive in this regard, as there generally won't be any other single formula of the language without t that implies *all* the theorems. The Ackermann constant is technically useful, particularly in algebraic studies of relevant logics and in sequent proof systems for relevant logics. While we won't deal with the latter in any detail in this book, the former will be studied in later chapters.

With that commentary on the axioms and rules out of the way, we can turn to proving some things about the logics. We will begin with the definition of closure under a rule, as it will be useful to show that certain rules are available beyond the list described earlier. To present this carefully, we first need to define substitutions.

DEFINITION 2.2 (Substitutions) *A function $\sigma : \mathfrak{L} \mapsto \mathfrak{L}$ is a* SUBSTITUTION *iff it obeys the following conditions.*

- $\sigma(p) \in \mathfrak{L}$, *for* $p \in \mathsf{At}$.
- $\sigma(\mathsf{t}) = \mathsf{t}$.
- $\sigma(\neg B) = \neg\sigma(B)$.
- $\sigma(B \# C) = \sigma(B)\#\sigma(C)$, *for* $\# \in \{\rightarrow, \circ, \wedge, \vee\}$.

In Chapter 5, we will consider generalizations of substitutions.

DEFINITION 2.3 (Closure under a rule) *A logic is* CLOSED UNDER A RULE $A_1, \ldots, A_n \Rightarrow B$ *iff for all substitutions* σ, *whenever all of* $\sigma(A_1), \ldots, \sigma(A_n)$ *are theorems of the logic, then* $\sigma(B)$ *is as well.*

We will now list some rules under which B is closed.

THEOREM 2.4 B *is closed under the following rules.*

(i) $A \rightarrow B, B \rightarrow C \Rightarrow A \rightarrow C$.
(ii) $\neg A \rightarrow \neg B \Rightarrow B \rightarrow A$.
(iii) $A \rightarrow B \Rightarrow \neg B \rightarrow \neg A$.
(iv) $A \rightarrow B, C \rightarrow D \Rightarrow (B \rightarrow C) \rightarrow (A \rightarrow D)$.
(v) $A \rightarrow (B \rightarrow C), D \rightarrow B \Rightarrow A \rightarrow (D \rightarrow C)$.

[3] The connection between fusion and implication illustrates a general phenomenon of families of connectives. This is studied by gaggle theory. See Bimbó and Dunn (2008) for more.

(vi) $A \rightarrow (B \rightarrow C), C \rightarrow D \Rightarrow A \rightarrow (B \rightarrow D)$.
(vii) $C \rightarrow (A \vee B), B \rightarrow D \Rightarrow C \rightarrow (A \vee D)$.
(viii) $(A \wedge B) \rightarrow C, D \rightarrow B \Rightarrow (A \wedge D) \rightarrow C$.
(ix) $A \rightarrow (B \vee C), (C \wedge D) \rightarrow E \Rightarrow (A \wedge D) \rightarrow (B \vee E)$.
(x) $A \rightarrow B, \neg B \Rightarrow \neg A$.

Proof (i) Suppose $A \rightarrow B$ and $B \rightarrow C$ are theorems. By (R4) and the first assumption, $(B \rightarrow C) \rightarrow (A \rightarrow C)$ is a theorem. By the second assumption and (R1), $A \rightarrow C$ is a theorem.

(ii) Suppose $\neg A \rightarrow \neg B$ is a theorem. Then by (R3), $B \rightarrow \neg\neg A$. By the preceding and (A7), $B \rightarrow A$ is a theorem.

(iii) Suppose $A \rightarrow B$ is a theorem. By (A1), $\neg B \rightarrow \neg B$ is a theorem, so by (R3), so is $B \rightarrow \neg\neg B$. Using (i) and the assumption, $A \rightarrow \neg\neg B$ is a theorem. By (R3), $\neg B \rightarrow \neg A$ is a theorem as well.

The remainder are left as an exercise for the reader. □

There are a few more rules the logic is closed under that are straightforward to prove.

THEOREM 2.5 *The logic* B *is closed under the following rules.*

(i) $A \rightarrow B, A \rightarrow C \Rightarrow A \rightarrow (B \wedge C)$.
(ii) $A \rightarrow C, B \rightarrow C \Rightarrow (A \vee B) \rightarrow C$.
(iii) $A \rightarrow B, C \rightarrow D \Rightarrow (A \wedge C) \rightarrow (B \wedge D)$.

Now that we have some axioms and rules, we can list some theorems of the logic.

THEOREM 2.6 *The following are theorems of* B.

(i) $A \rightarrow \neg\neg A$.
(ii) $\neg(A \vee B) \rightarrow (\neg A \wedge \neg B)$.
(iii) $\neg(A \wedge B) \rightarrow (\neg A \vee \neg B)$.
(iv) $(\neg A \vee \neg B) \rightarrow \neg(A \wedge B)$.
(v) $(\neg A \wedge \neg B) \rightarrow \neg(A \vee B)$.
(vi) $A \rightarrow (B \rightarrow (A \circ B))$.

Proof The proofs are left to the reader. □

Next, we will prove that conjunction and disjunction behave in some of the expected ways.

THEOREM 2.7 (Associativity and commutativity) *The following are theorems of* B.

(i) $(A \wedge (B \wedge C)) \rightarrow ((A \wedge B) \wedge C)$.
(ii) $((A \wedge B) \wedge C) \rightarrow (A \wedge (B \wedge C))$.
(iii) $(A \vee (B \vee C)) \rightarrow ((A \vee B) \vee C)$.
(iv) $((A \vee B) \vee C) \rightarrow (A \vee (B \vee C))$.
(v) $(A \wedge B) \rightarrow (B \wedge A)$.
(vi) $(A \vee B) \rightarrow (B \vee A)$.

Proof (i) From (A2) we have $(A \wedge (B \wedge C)) \to A, (A \wedge (B \wedge C)) \to (B \wedge C), (B \wedge C) \to B$, and $(B \wedge C) \to C$. Using theorem 2.4 (i) and theorem 2.5 (i), we obtain $(A \wedge (B \wedge C)) \to ((A \wedge B) \wedge C)$.

(ii) This case is substantially similar to the preceding case.

(iii) This case is substantially similar to (i), except using (A4) and theorem 2.5 (ii).

(iv) This case is pretty similar to the preceding case.

(v) From (A2), we have $(A \wedge B) \to B$ and $(A \wedge B) \to A$, and by theorem 2.5 we have $(A \wedge B) \to (B \wedge A)$.

(vi) This is similar to the preceding case, using (A4) and theorem 2.5 (ii). □

To enable us to use the preceding theorem fully, we will need to prove one more thing. For that, we will need to introduce some notation.[4]

DEFINITION 2.8 (Formula context) *A* FORMULA CONTEXT (C, p) *is a pair of a formula* C *with a distinguished atom* p. *We will sometimes write contexts as* $C(p)$.

Given a formula context $C(p)$, $C(D)$ *is the result of (simultaneously) replacing all occurrences of* p *in* C *with the formula* D.

As an example, if $C(p)$ is $p \to (\neg p \vee q)$, then $C(r \wedge s)$ is $(r \wedge s) \to (\neg(r \wedge s) \vee q)$.

Formula contexts permit us to define substitution in formulas precisely, and this in turn lets us prove that one can substitute provable equivalents.

THEOREM 2.9 (Substitution of provable equivalents) *Let* $C(p)$ *be a formula context. If* $\vdash_{\mathsf{B}} A \leftrightarrow B$, *then* $\vdash_{\mathsf{B}} C(A) \leftrightarrow C(B)$.

Proof Suppose $\vdash_{\mathsf{B}} A \leftrightarrow B$. Therefore, $\vdash_{\mathsf{B}} A \to B$ and $\vdash_{\mathsf{B}} B \to A$. The proof is by induction on the construction of the formula context $C(p)$. We will assume that p occurs in the context in all cases. If it does not, then $C(A)$ is $C(B)$, and we are done in virtue of (A1) and (R2).

If $C(p)$ is an atom or t, then $C(A)$ is A and $C(B)$ is B, so $\vdash_{\mathsf{B}} C(A) \leftrightarrow C(B)$.

Suppose that $C(p)$ is $D(p) \wedge E(p)$. By the inductive hypothesis, $\vdash_{\mathsf{B}} D(A) \leftrightarrow D(B)$ and $\vdash_{\mathsf{B}} E(A) \leftrightarrow E(B)$. From some simple logical moves, it follows that $\vdash_{\mathsf{B}} C(A) \leftrightarrow C(B)$.

The case for disjunction is similar to the preceding.

Suppose that $C(p)$ is $\neg D(p)$. By the inductive hypothesis, $\vdash_{\mathsf{B}} D(A) \leftrightarrow D(B)$. Using (R3), $\vdash_{\mathsf{B}} \neg D(A) \leftrightarrow \neg D(B)$, as desired.

Suppose that $C(p)$ is $D(p) \to E(p)$. By the inductive hypothesis, $\vdash_{\mathsf{B}} D(A) \leftrightarrow D(B)$ and $\vdash_{\mathsf{B}} E(A) \leftrightarrow E(B)$. Using theorem 2.4 with $\vdash_{\mathsf{B}} (D(A) \to E(A)) \to (D(A) \to E(A))$, we obtain $\vdash_{\mathsf{B}} (D(A) \to E(A)) \to (D(B) \to E(B))$. Similarly, if we instead use $\vdash_{\mathsf{B}} (D(B) \to E(B)) \to (D(B) \to E(B))$, we obtain $\vdash_{\mathsf{B}} (D(B) \to E(B)) \to (D(A) \to E(A))$. Together they suffice to show $\vdash_{\mathsf{B}} C(A) \leftrightarrow C(B)$.

Suppose that $C(p)$ is $D(p) \circ E(p)$. By the inductive hypothesis, $\vdash_{\mathsf{B}} D(A) \leftrightarrow D(B)$ and $\vdash_{\mathsf{B}} E(A) \leftrightarrow E(B)$. By theorem 2.6 (vi), $\vdash_{\mathsf{B}} D(A) \to (E(A) \to (D(A) \circ E(A)))$ and $\vdash_{\mathsf{B}} D(B) \to (E(B) \to (D(B) \circ E(B)))$. Using theorem 2.4, we can show

[4] While the idea is fairly standard, we are defining it following Williamson (2006).

$\vdash_{\mathsf{B}} D(B) \to (E(B) \to (D(A) \circ E(A)))$ and $\vdash_{\mathsf{B}} D(A) \to (E(A) \to (D(B) \circ E(B)))$. Using (R7), we obtain both $\vdash_{\mathsf{B}} (D(B) \circ E(B)) \to (D(A) \circ E(A)))$ and its converse, $\vdash_{\mathsf{B}} (D(A) \circ E(A)) \to (D(B) \circ E(B)))$, which suffices for what we want. □

Theorem 2.9 will permit us to abuse notation and ignore bracketing on conjunction and disjunction. We will use $A \wedge B \wedge C$ to mean $A \wedge (B \wedge C)$, and similarly for disjunction, with all omitted parentheses inserted as associating to the right.

As we have said, B is a weak logic. Let us turn to some common extensions of it.

2.3 Extensions

The logic B is the weakest standard relevant logic. Although it is quite weak, there are some weaker logics that will be covered in the appendix to this chapter. B is of interest because B is the weakest logic that can be adequately modeled using ternary relational frames. There are many relevant logics, and these are obtained by adding new axioms or rules to the definition of B. In this section, we will present a few of the common ones. Much as with TV-based modal logics, there are infinitely many relevant logics one could study, but a few have been of sufficient interest to have gotten standard names. We will use L to stand for any logic that can be obtained by the addition of any combination of the axioms and rules for extending B.

The strongest standard relevant logic is Anderson and Belnap's logic R. While we will consider a few stronger logics, and even incomparable ones, R marks out a plausible upper bound to the primary family of relevant logics. Almost all of the best studied relevant logics fall between B and R. The logic R is obtained by adding to B the following axioms.

- $(A \to \neg B) \to (B \to \neg A)$.
- $(A \to B) \to ((B \to C) \to (A \to C))$.
- $A \to ((A \to B) \to B)$.
- $(A \to (A \to B)) \to (A \to B)$.

No new rules are needed for R. In fact, many of the rules of B become redundant, as they are traded for axioms.[5] The reason that R is singled out as the strongest relevant logic is because it is a strong, natural logic that has many different presentations and that enjoys Belnap's VARIABLE-SHARING PROPERTY.[6]

DEFINITION 2.10 (Variable-sharing property) *A logic* L *has the* VARIABLE-SHARING PROPERTY *iff whenever $A \to B$ is a theorem of* L *and* t *occurs in neither A nor B, then there is an atom, or propositional variable, that occurs in both A and B.*

[5] R can be axiomatized using only rules (R1) and (R2).

[6] There are other logics that properly extend R that enjoy the variable-sharing property. One example is Urquhart's semilattice logic, as shown by Weiss (2019). Dziobiak (1983) shows that there are $2^{\aleph_0}$ extensions of R, contained in the logic R-Mingle, aka RM, that satisfy the property. Swirydowicz (1999) shows that there are two maximal extensions of R that satisfy the property, one being the logic of the crystal lattice displayed in Table 1.3 and one being the logic of the matrix M_8, displayed in Table 5.8.

As noted in the introduction, variable sharing captures a sense of FORMAL RELEVANCE: If an implication is a theorem of the logic in which t does not occur, then the antecedent and consequent are relevant to each other in having at least one atom in common. Indeed, on our view, variable sharing is the key feature of a relevant logic.[7] We will say more about the variable-sharing property in Chapter 5, where we will see some ways to establish that a logic has the variable-sharing property as well as looking at stronger versions of the variable-sharing property. Variable sharing will play an important role in evaluating extensions of relevant logics with new connectives in later chapters, particularly Chapter 10.

There are two common ways to extend B while staying within the basic vocabulary. One way is to add additional *implicational* axioms, axioms that primarily or exclusively display the implication. The second way is adding *negation* axioms. Prominent examples of the first group are the following. For both groups of axioms, names common in the literature are included in parentheses.

(I1) $(A \wedge (A \rightarrow B)) \rightarrow B$ (Pseudo-modus ponens, WI).
(I2) $((A \rightarrow B) \wedge (B \rightarrow C)) \rightarrow (A \rightarrow C)$ (Conjunctive syllogism).
(I3) $(A \rightarrow B) \rightarrow ((C \rightarrow A) \rightarrow (C \rightarrow B))$ (Prefixing, B).
(I4) $(A \rightarrow B) \rightarrow ((B \rightarrow C) \rightarrow (A \rightarrow C))$ (Suffixing, B′).
(I5) $A \rightarrow ((A \rightarrow B) \rightarrow B)$ (Assertion, CI).
(I6) $(A \rightarrow (B \rightarrow C)) \rightarrow (B \rightarrow (A \rightarrow C))$ (Permutation, C).
(I7) $(A \rightarrow (A \rightarrow B)) \rightarrow (A \rightarrow B)$ (Contraction, W).
(I8) $(\mathsf{t} \rightarrow A) \rightarrow A$ (E axiom).
(I9) $A \rightarrow (A \rightarrow A)$ (Mingle, M).

The axiom (I8) does not have a standard name, despite its distinguished role in the logic E. Prominent examples of the second group of axioms, the negation axioms, include the following.

(N1) $(A \rightarrow \neg B) \rightarrow (B \rightarrow \neg A)$ (Contraposition).
(N2) $(A \rightarrow \neg A) \rightarrow \neg A$ (Reductio, RAA).
(N3) $A \vee \neg A$ (Excluded middle, LEM).

Some of these axioms are derivable from others, and some are interderivable. We can codify these ideas in a definition.

DEFINITION 2.11 (Derivability from, interderivability) *Let A and B be formula schemes.*

B IS DERIVABLE FROM *A in* L *iff if B is derivable in* L *extended with A.*

A and B are INTERDERIVABLE *in* L *iff both if A is derivable from B in* L *and B is derivable from A in* L.

For example, (I3) and (I4) can be obtained from each other if (N1) is in the logic. If (I3) and (I4) are in the logic, then (I5) and (I6) are interderivable. We are, generally, concerned only with derivability and interderivability in the underlying logic L when

[7] See Standefer (2025b) for an elaboration of this view.

neither formula scheme is a theorem of L. As should be clear from the definition, derivability from and interderivability are preserved under extensions with new axioms and rules.

Axiom (I9), the mingle axiom, is notable because its addition to R, which yields the logic RM, results in a violation of variable sharing. While its antecedent and consequent share all the same atoms, adding it to R results in the derivability of $\neg(A \rightarrow A) \rightarrow (B \rightarrow B)$. The mingle axiom is a special case of the first paradox of implication, one where B is identical to A. Despite this, the addition of (I9) to a logic does not *invariably* result in a violation of variable sharing, as has been demonstrated for the addition of (I9) to T.[8]

While RM does not have the variable-sharing property, it has another feature that is similar: the weak variable-sharing property.

DEFINITION 2.12 (Weak variable-sharing property) *A logic* L *has the* WEAK VARIABLE-SHARING PROPERTY *iff whenever* $\vdash_{\mathsf{L}} A \rightarrow B$ *where* A *and* B *have no occurrences of* t, *then either (i)* A *and* B *share an atom or (ii) both* $\vdash_{\mathsf{L}} \neg A$ *and* $\vdash_{\mathsf{L}} B$.

Meyer showed that RM has the weak variable-sharing property. The weak variable-sharing property is properly weaker than the variable-sharing property, as RM satisfies the former but not the latter. Despite being weaker, not all logics enjoy the weak variable-sharing property. For example, both TV and intuitionistic logic both lack the weak variable-sharing property.

An important feature of the variable-sharing property is that it is preserved to sublogics.

THEOREM 2.13 (Downward preservation of variable sharing) *Suppose* $\mathsf{L} \subseteq \mathsf{M}$. *If* M *has the variable-sharing property, then so does* L.

Proof Suppose L does not have the variable-sharing property. Then for some formulas A and B without t that do not share an atom, $\vdash_{\mathsf{L}} A \rightarrow B$. As $\mathsf{L} \subseteq \mathsf{M}$, it follows that $\vdash_{\mathsf{M}} A \rightarrow B$, so M lacks the variable-sharing property as well. □

The fact that variable sharing is preserved downwards is used frequently in work on relevant logics. Variable sharing is not, however, preserved upwards, as RM contains R but lacks variable sharing. By comparison, the weak variable-sharing property need not be preserved downwards or upwards. To see how preservation downwards could fail, note that the second disjunct of the weak variable-sharing property contains two provability statements. Provability is not preserved to sublogics, almost by definition, so one can see how both disjuncts of the weak variable-sharing property could fail.[9] Indeed, this possibility is realized by some relatives of RM, as noted by Øgaard (2023).

There is an equivalent form of axiom (N2) that will be useful in proving Completeness in Chapter 4. We will prove the equivalence here.

[8] See Méndez et al. (2012). See also Robles et al. (2011; 2010) for further discussion.

[9] Øgaard (2023) provides a sufficient condition, in terms of matrices, for a logic to have the weak variable-sharing property.

LEMMA 2.14 (Equivalent of (N2)) *Let* L *extend* B. *Then* (N2) *and* $(A \to B) \to (\neg A \vee B)$ *are interderivable in* L.

Proof Suppose L contains (N2). By (A2) and (R4) $(\neg B \to \neg(A \wedge \neg B)) \to ((A \wedge \neg B) \to \neg(A \wedge \neg B))$. From (N2) and theorem 2.4 (i), $(\neg B \to \neg(A \wedge \neg B)) \to \neg(A \wedge \neg B)$. By (A2) and theorem 2.4 (iii), $\neg A \to \neg(A \wedge \neg B)$ is a theorem, so by (R5), $(\neg B \to \neg A) \to (\neg B \to \neg(A \wedge \neg B))$ is a theorem. From theorem 2.4 (i), $(\neg B \to \neg A) \to \neg(A \wedge \neg B)$ is a theorem. Since this is a theorem for all formulas, A and B, we can substitute $\neg D$ and $\neg C$ for A and B, respectively. Therefore, $(\neg\neg C \to \neg\neg D) \to \neg(\neg D \wedge \neg\neg C)$ is a theorem. Using theorem 2.9 repeatedly, $(C \to D) \to \neg C \vee D$ is a theorem, as desired.

For the converse direction, suppose L contains $(A \to B) \to (\neg A \vee B)$. Then, it contains the instance with $\neg A$ substituted for B, $(A \to \neg A) \to (\neg A \vee \neg A)$. As $(\neg A \vee \neg A) \to \neg A$ is a B-theorem, $(A \to \neg A) \to \neg A$ is a theorem. □

A comment on implication axiom (I8) is worthwhile. Instead of it, one could adopt either

- $(((A \to A) \wedge (B \to B)) \to C) \to C$

or

- $((A \to A) \to B) \to B$ and
- $((A \to A) \to A) \wedge ((B \to B) \to B) \to (((A \wedge B) \to (A \wedge B)) \to (A \wedge B))$.

The latter pair was used by Anderson and Belnap, before the widespread recognition of the usefulness of t. How do these axioms relate? We can define a necessity operator for each formula A, $\Box^A B$, as $(A \to A) \to B$. Dropping the superscript for a moment, the final axiom can be rewritten as $(\Box A \wedge \Box B) \to \Box(A \wedge B)$, which says that necessity aggregates over conjunction.[10] Note, though, that the necessities involved in that axiom are not all the same, more fully written out as $(\Box^A A \wedge \Box^B B) \to \Box^{A \wedge B}(A \wedge B)$. We can define an alternative necessity operator for each formula A, using the basic logical resources as $\Box_A B$ as $A \to B$. Then $\Box_t A$ is a formula that says, roughly, A is logically necessary, that is, A is implied by the logic. Given some other axioms, one can show that, for each formula A, $\Box^A A \leftrightarrow \Box_t A$ is a theorem of E.

There are many axioms one could add to B, and there are additional rules one could add as well, although we have not explored those here. The selection earlier contains many of the standard axioms used to define some of the more common relevant logics. Table 2.1 lists several of the common logics together with their axioms. Beside each logic's name, there is another logic listed plus some axioms. To get the former logic, add to the latter logic the additional axioms. For example, to obtain DW, one adds (N1) to the axioms of B, and to obtain DJ, one adds to the axioms of DW the axiom (I2), which is to say one adds (N1) and (I2) to the axioms of B.

The nomenclature for the common relevant logics is less than ideally transparent, so we will comment on some of the names. The logic E is the logic of entailment defended by Anderson and Belnap. The logic R is Anderson and Belnap's logic of

[10] See Humberstone (2013) for more on aggregation.

Table 2.1. Some common relevant logics

DW	B + (N1)	DJ	DW + (I2)
TW	DW + (I3)–(I4)	RW	TW + (I5)
T	TW + (N2) + (I7)	R	T + (I5)
C	TW + (I1)	E	T + (I8)
LX	L + (N3)	LM	L + (I9)

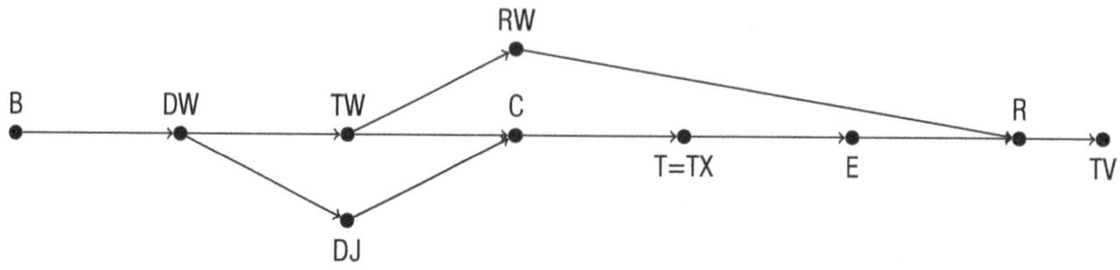

Figure 2.1 Relations of some major logics

relevance, and T is Anderson and Belnap's logic of ticket entailment. B is Sylvan and Meyer's basic logic. RW and TW are the logics R and T with axiom (I7) and its relatives, (I1) and (N2), removed.[11] The reason for this is that (I7) is sometimes called contraction or *W* in light of its connection to the combinator *W* of combinatory logic, which is tightly connected to the structural rule of contraction in sequent systems. The logics LX are so-called because (N3) is known as excluded middle, sometimes notated (X).[12] The logics LM are so-called because (I9) is known as the mingle axiom. Of the mingle logics, RM, the addition of (I9) to R, is the best known.[13]

The relations of some of the major logics are diagrammed in Figure 2.1. In the diagram, an arrow leading from one logic to another logic indicates inclusion. For example, B is included in DW, and TW is included in RW. The absence of a path between logics means they are incomparable. DJ and TW are incomparable strengthenings of DW. While many of these connections are immediate from the way the logics are defined, some of them are not. For example, C does not have (I2) as an axiom, but it does have it as a theorem. The extensions with (M) are left off, as they would complicate the diagram. The extensions with (X) are also left off, although from T upwards in strength, the addition of (X) adds nothing, that is, T is identical with TX.

In Section 2.5, we will define several notions in terms of B, such as B-theory, B-overlapping pair, and others. For each logic L, there is a corresponding notion obtained by changing "B" to "L" in each of the definitions. The lemmas and theorems proved there, and in the initial section, carry over to the context using L-based

[11] TW was known as P-W before the current name was standardized. The name P-W is sometimes used in connection with a particular formal problem in that logic. See Martin and Meyer (1982).
[12] See Slaney (1989).
[13] Meyer explored RM in Anderson and Belnap (1975, §29.3). See Avron (1987; 1990a,b; 1992) for a defense and exploration of RM.

concepts. The proofs carry over as well, using the same replacement as in the definitions.

2.4 Comparison

We will next make some comparisons between the relevant logics and classical logic, TV. For this comparison, we will treat TV as having a connective, $\rightarrow$, where $A \rightarrow B$ is equivalent to $\neg A \vee B$ and having a connective, $\circ$, where $A \circ B$ is equivalent to $A \wedge B$. This allows us to view the relevant logics, and in particular R, as sublogics of TV. The stronger logics, such as T and R, contain all the classical tautologies in the restricted vocabulary $\{\neg, \wedge, \vee\}$, which provides a sense in which those logics contain TV. It is important to note that, in contrast, R is a sublogic of TV in the full vocabulary, and in particular, we are interested in differences where $\rightarrow$ is concerned.

In the rest of this section, we will look at some classical tautologies that are rejected by R and weaker logics. At this point, however, we do not have models of any sort in the picture, so we cannot *show* that these are not theorems of the relevant logics. That will have to wait until later chapters.

We will begin with some classical tautologies in the $\{\rightarrow\}$-fragment of TV that are not valid in R.

(1) $A \rightarrow (B \rightarrow B)$.
(2) $A \rightarrow (B \rightarrow A)$.
(3) $((A \rightarrow B) \rightarrow A) \rightarrow A$.
(4) $A \rightarrow (A \rightarrow A)$.

Adding any of the first three to the $\{\rightarrow\}$-fragment of R will result in the corresponding fragment of TV.[14] In light of that, the first three will all generate violations of variable sharing when added to R. The first violates variable sharing on its own, while one can derive (1) from (2) using (I6). In addition, (2) generates violations of variable sharing on its own, just given (R1). In a roundabout way, one can get (1) from (3) as well, whence violations of variable sharing. As noted in the previous section, while one might think that the last one, axiom (I9), does not violate relevant scruples, it does lead to violations of variable sharing in the context of being added to the full logic R, in particular formula (15) below is a consequence witnessing this.[15]

Next, we will look at some classical tautologies from the $\{\rightarrow, \wedge, \circ\}$-fragment that are not valid in R.

(5) $((A \wedge B) \rightarrow C) \rightarrow (A \rightarrow (B \rightarrow C))$.
(6) $A \rightarrow (B \rightarrow (A \wedge B))$.
(7) $(A \rightarrow B) \rightarrow ((A \rightarrow C) \rightarrow (A \rightarrow (B \wedge C)))$.
(8) $(A \circ B) \rightarrow A$.

[14] Meyer (1990) showed the disastrous effects of adding Peirce's law to R.

[15] If negation is not in the language, then R plus I9 enjoys variable sharing. That logic is sometimes called RM0, or constructive Mingle. See Weiss (2022a) for more on RM0.

(9) $(A \circ B) \rightarrow (A \wedge B)$.
(10) $(A \rightarrow B) \rightarrow (A \rightarrow (A \wedge B))$.
(11) $(A \rightarrow B) \rightarrow ((A \wedge C) \rightarrow (B \wedge C))$.

The first five lead quickly to $A \rightarrow (B \rightarrow A)$ becoming derivable via appeal to the axiom $(A \wedge B) \rightarrow A$. The invalidity of (10) might be surprising, but it is invalid. It is worth noting that it has a valid relative, obtained by adding $A \rightarrow A$ as a conjunct to the antecedent, $((A \rightarrow B) \wedge (A \rightarrow A)) \rightarrow (A \rightarrow (A \wedge B))$. The final formula is known as (Factor), and it, along with its mate using disjunction rather than conjunction, is known to lead to violations of variable sharing.[16]

Next, we will look at the full language to see some classical tautologies that are not valid in R.

(12) $(B \wedge \neg B) \rightarrow A$.
(13) $A \rightarrow (B \vee \neg B)$.
(14) $(A \wedge (\neg A \vee B)) \rightarrow B$.
(15) $\neg(A \rightarrow A) \rightarrow (B \rightarrow B)$.
(16) $(\neg A \vee B) \rightarrow (A \rightarrow B)$.
(17) $((A \wedge B) \rightarrow C) \rightarrow ((A \wedge \neg C) \rightarrow \neg B)$.
(18) $(A \rightarrow B) \vee (B \rightarrow A)$.
(19) $\neg(A \rightarrow B) \rightarrow (A \wedge \neg B)$.
(20) $((A \rightarrow (B \vee C)) \wedge (B \rightarrow C)) \rightarrow (A \rightarrow C)$.

While the first two are clearly out, as they violate variable sharing, one can add them to R without winding up at TV.[17] While (14) is out, its rule form, known as (γ) or disjunctive syllogism,

$$(\gamma) \qquad A, \neg A \vee B \Rightarrow B,$$

is notable in relevant logics, and it will return in Chapter 9. There is a relative of (17) that is acceptable in relevant logics, namely the rule $(A \circ B) \rightarrow C \Rightarrow (A \circ \neg C) \rightarrow \neg B$, also known as rule antilogism.[18]

Some comment on the constant t is in order. For any classical tautology A, $\mathsf{t} \rightarrow A$ and $A \rightarrow \mathsf{t}$ are also tautologies in TV. The latter fails in the setting of relevant logics. Having both around, for all theorems A of a relevant logic, would lead straight away to violations of variable sharing.[19]

We have listed some notable classical tautologies that are not theorems of R. R, and the other stronger logics that contain (N2) and (N3), will contain all the classical tautologies in the vocabulary $\{\neg, \vee, \wedge\}$. In the weaker logics, where neither (N2) nor

[16] See Urbas and Sylvan (1989). It is worth noting that the rule form of (Factor), $A \rightarrow B \Rightarrow (A \wedge C) \rightarrow (B \wedge C)$, is a rule under which B is closed.
[17] The logic obtained by adding them to R is defended by Kerr (2019). This logic has some important connections to projective geometry, as discussed by Urquhart (1983; 2017; 2019; 2023).
[18] Read (1988, ch. 3.7) discusses this rule in more detail.
[19] Adding both to B would lead to a *monothetic* logic, in the terminology of Humberstone (2011, 221), where all theorems are equivalent. Relevant logics are, by contrast, *polythetic*, in the terminology of Standefer (2019), which is to say they have non-equivalent theorems.

(N3) is available, none of the tautologies in the vocabulary $\{\neg, \vee, \wedge\}$ will be derivable. As the logic gets weaker, further divergences from TV emerge. Of course, from the point of view of the relevant logician, these divergences may be viewed as features, rather than bugs.

2.5 Theories

Theories are sets of sentences obeying certain closure conditions. Theories are important for applications of a logic, and they are crucial in the development of the metatheory of relevant logics. In particular, they are used to define the canonical frames used in the proofs of completeness. Toward that end, in this section, we will develop several features of theories, and we will prove two lemmas that are important for the canonical model proof of Completeness in Chapter 4, namely the Full pair extension lemma (lemma 2.25) and the Squeezing lemma (lemma 2.39). For the reader that is not interested in the details of Completeness proofs, this section can reasonably be skipped.

At this point, we will focus on B-theories, which are sets of formulas closed under B and adjunction, rule (R2).[20] What it means to be closed under a logic is that whenever an implication is a theorem of the logic and the antecedent is in the theory, then so is the consequent.

DEFINITION 2.15 (B-theory) *A set of formulas X is a* B*-theory iff both of the following hold:*

(1) If $A \in X$ and $B \in X$ then $A \wedge B \in X$, and
(2) if $A \in X$ and $\vdash_{\mathsf{B}} A \rightarrow B$ then $B \in X$.

Theories are important in the general development of relevant logics. In particular, they are used in defining the canonical frame constructions for the completeness proof. Once other logics have been introduced, we can adjust the definition of B-theory to be appropriate to the other logics. The definition of an L-theory, for some logic L, is obtained by changing clause (2) from using the notion of B-theorem to that of L-theorems.

There is another important notion related to theories, that of a disjoint pair.

DEFINITION 2.16 (Disjoint pair) *Let X and Y be sets of formulas. Then (X, Y) is a* B-OVERLAPPING PAIR *iff there are $A_1, \ldots, A_n \in X$ and $B_1, \ldots, B_m \in Y$ such that*

$$\vdash_{\mathsf{B}} (A_1 \wedge \cdots \wedge A_n) \rightarrow (B_1 \vee \cdots \vee B_m).$$

(X, Y) is a B-DISJOINT PAIR *iff it is not a* B*-overlapping pair.*

Using disjoint pairs, we can provide an alternative definition of theory.

[20] This definition of theory is slightly different than another standard one, in the context of TV, that includes all theorems of the logic. See Mangraviti and Tedder (2023) for discussion of some of the differences.

DEFINITION 2.17 (B-inclusion theory) *A set of formulas X is a* B*-inclusion theory iff whenever $(X,\{A\})$ is a* B*-overlapping pair, then $A \in X$.*

The two definitions of theory coincide.

THEOREM 2.18 *Let X be a set of formulas. Then X is a* B*-theory iff X is a* B*-inclusion theory.*

Proof It is straightforward to verify that the empty set is both a B-theory and a B-inclusion theory. Let us assume, then, that X is a set of formulas.

Assume that X is a B-theory and suppose $(X,\{A\})$ is a B-overlapping pair. Then there are $B_1,\ldots,B_n \in X$ such that $(B_1 \wedge \cdots \wedge B_n) \rightarrow A$ is a theorem of B. Since X is closed under adjunction, which is condition (1) of being a B-theory, $B_1 \wedge \cdots \wedge B_n \in X$. Since X is closed under the logic B, which is condition (2) of being a B-theory, $A \in X$.

Assume that X is a B-inclusion theory. Suppose that $A \in X$ and $B \in X$. Then $(X,\{A \wedge B\})$ is a B-overlapping pair, since $\vdash_{\mathsf{B}} (A \wedge B) \rightarrow (A \wedge B)$, by axiom (A1), and the conjuncts of that conditional's antecedent are in X. Thus, $A \wedge B \in X$. Next, suppose $A \in X$ and $\vdash_{\mathsf{B}} A \rightarrow B$. It follows that $(X,\{B\})$ is a B-overlapping pair. Therefore, $B \in X$. □

This equivalence gives us an immediate corollary.

COROLLARY 2.19 *If X is a* B*-theory and $A \notin X$, then $(X,\{A\})$ is a* B*-disjoint pair.*

Let us say that the least B-theory over some set of formulas X is $\theta_{\mathsf{B}}(X)$. There may be other theories that properly contain $\theta_{\mathsf{B}}(X)$, but $\theta_{\mathsf{B}}(X)$ is the theory obtained by extending X with just enough formulas to be a theory.

DEFINITION 2.20 (Theory over axioms) *Let X be a set of formulas. Let $Y_0 = X$. Let Y_{n+1} be defined as follows.*

$$Y_{n+1} = \{A \wedge B \in \mathfrak{L} : \ A, B \in Y_n\} \cup \{B \in \mathfrak{L} : \ \exists A \in Y_n \vdash_{\mathsf{B}} A \rightarrow B\} \cup Y_n.$$

Set $\theta_{\mathsf{B}}(X) = \bigcup_{n<\omega} Y_n$.

We will now show that this definition works to define a B-theory and that such a theory is minimal in a certain sense.

LEMMA 2.21 *The following claims are true.*

(1) For a set of formulas X, $\theta_{\mathsf{B}}(X)$ is a B*-theory.*
(2) If Z is a B*-theory, then $\theta_{\mathsf{B}}(Z) = Z$.*
(3) Let U be a set of formulas. If $X \subseteq U$, then $\theta_{\mathsf{B}}(X) \subseteq \theta_{\mathsf{B}}(U)$.
(4) If Z is any B*-theory such that $X \subseteq Z$, then $\theta_{\mathsf{B}}(X) \subseteq Z$.*

Proof (1). Let X be a set of formulas. Suppose that $A \in \theta_{\mathsf{B}}(X)$ and $B \in \theta_{\mathsf{B}}(X)$. Let k be the first stage such that $A \in Y_k$ and $B \in Y_k$. From definition 2.20, $A \wedge B \in Y_{k+1}$, so $A \wedge B \in \theta_{\mathsf{B}}(X)$. Suppose that $A \in \theta_{\mathsf{B}}(X)$ and $\vdash_{\mathsf{B}} A \rightarrow B$. Let k be some stage such that $A \in Y_k$. Then $B \in Y_{k+1}$, so $B \in \theta_{\mathsf{B}}(X)$, as desired.

(2). Next, suppose Z is a B-theory. Obviously $Z \subseteq \theta_{\mathsf{B}}(Z)$. Suppose $A \in \theta_{\mathsf{B}}(Z)$. We will show by induction on the construction of $\theta_{\mathsf{B}}(Z)$ that $A \in Z$. For the base case, $A \in Z$. So we are done.

Suppose that n is the least stage such that $A \in Y_{n+1}$ and $A \notin Y_n$. There are two inductive cases. For the first, suppose that A is $B \wedge C$ and $B \in Y_n$ and $C \in Y_n$. By the inductive hypothesis, $B \in Z$ and $C \in Z$. As Z is a theory, $B \wedge C \in Z$. For the second inductive case, suppose $\vdash_{\mathsf{B}} B \rightarrow A$ and $B \in Y_n$. By the inductive hypothesis, $B \in Z$, so $A \in Z$, as Z is a theory. Therefore, we conclude that $\theta_{\mathsf{B}}(Z) \subseteq Z$.

(3). Next, suppose $X \subseteq U$ and that $A \in \theta_{\mathsf{B}}(X)$. We will show by induction on the construction of $\theta_{\mathsf{B}}(X)$ that $A \in \theta_{\mathsf{B}}(U)$. If $A \in X$, then $A \in U$ by assumption. The inductive cases are similar to those of (2), with the inductive hypothesis taking care of both cases.

Finally, suppose $X \subseteq Z$. By (3), $\theta_{\mathsf{B}}(X) \subseteq \theta_{\mathsf{B}}(Z)$. By (2), $\theta_{\mathsf{B}}(Z) = Z$. Thus, $\theta_{\mathsf{B}}(X) \subseteq Z$.[21] □

This lemma tells us that the operator $\theta_{\mathsf{B}}(\cdot)$ is a CLOSURE OPERATOR on sets of formulas, meaning it obeys the following conditions.

- $X \subseteq \theta_{\mathsf{B}}(X)$.
- If $X \subseteq Y$, then $\theta_{\mathsf{B}}(X) \subseteq \theta_{\mathsf{B}}(Y)$.
- $\theta_{\mathsf{B}}(X) = \theta_{\mathsf{B}}(\theta_{\mathsf{B}}(X))$.

There are several properties that theories could have. We will list some important ones.

DEFINITION 2.22 *A* B*-theory X is*

- *PRIME iff $A \vee B \in X$ only if $A \in X$ or $B \in X$;*
- B*-REGULAR iff* B $\subseteq X$*;*
- *CONSISTENT iff for no formula A is $A \in X$ and $\neg A \in X$;*
- *NORMAL iff X is prime,* B*-regular, and consistent.*

Prime theories play a particularly important role in the development of relevant logics. Not all theories are prime. For example consider $\theta_{\mathsf{B}}(\{p \vee q\})$. Nothing in the logic forces either p or q to be in $\theta_{\mathsf{B}}(\{p \vee q\})$. An important and fundamental result is that, given a B-disjoint pair (X, Y), we can always extend X to a prime theory $Z \supseteq X$ in a way that avoids Y, which is to say that (Z, Y) is a B-disjoint pair.

DEFINITION 2.23 (Pair extensions, full pairs) *A pair of sets of formulas (X, Y) is EXTENDED BY another pair of sets of formulas (U, V) iff $X \subseteq U$ and $Y \subseteq V$.*

A pair of sets of formulas (X, Y) is FULL iff $X \cup Y = \mathfrak{L}$.

We will use the notation $\bigwedge X = \{\bigwedge_n A_i : \text{ for some } n, 1 \leq n < \omega, A_1, \ldots, A_n \in X\}$ and $\bigvee X = \{\bigvee_n A_i : \text{ for some } n, 1 \leq n < \omega, A_1, \ldots, A_n \in X\}$.

[21] An alternative proof of this could proceed via the fact that $\theta_{\mathsf{B}}(X)$ is defined by an inductive definition, which guarantees the minimality described in (4). See Moschovakis (1974) for more on the general theory of inductive definitions.

LEMMA 2.24 (Single step extension lemma) *Let X and Y be any sets of formulas and C be a formula. If (X, Y) is a disjoint pair, then either $(X \cup \{C\}, Y)$ is a disjoint pair or $(X, Y \cup \{C\})$ is a disjoint pair.*

Proof Let (X, Y) be a disjoint pair. Suppose that $(X \cup \{C\}, Y)$ is not a disjoint pair. So there is some conjunction $A \in \bigwedge X$ and some disjunction $B \in \bigvee Y$ such that $\vdash_{\mathsf{B}} (C \wedge A) \rightarrow B$. Suppose that $(X, Y \cup \{C\})$ is not a disjoint pair. So there is some conjunction $D \in \bigwedge X$ and some disjunction $E \in \bigvee Y$ such that $\vdash_{\mathsf{B}} D \rightarrow (C \vee E)$. By theorem 2.4, this suffices for $\vdash_{\mathsf{B}} (A \wedge D) \rightarrow (B \vee E)$. But, $(A \wedge D) \in \bigwedge X$ and $(B \vee E) \in \bigvee Y$, so (X, Y) was not a disjoint pair, contradicting the assumption. Therefore, either $(X \cup \{C\}, Y)$ is a disjoint pair or $(X, Y \cup \{C\})$ is a disjoint pair. □

LEMMA 2.25 (Full pair extension lemma) *Suppose that (X, Y) is a* B*-disjoint pair. Then there is a full,* B*-disjoint pair (Z, V) such that (X, Y) is extended by (Z, V).*

Proof Suppose that (X, Y) is a B-disjoint pair. Let $C_1, \ldots, C_k, \ldots$ be an enumeration of the language. Define a sequence of pairs of sets of formulas as follows.

- $(X_0, Y_0) = (X, Y)$.
- $(X_{n+1}, Y_{n+1}) = \begin{cases} (X_n \cup \{C_{n+1}\}, Y_n) & \text{if } (X_n \cup \{C_{n+1}\}, Y_n) \text{ is a disjoint pair,} \\ (X_n, Y_n \cup \{C_{n+1}\}) & \text{otherwise.} \end{cases}$

Clearly for each $n \geq 0$, (X_n, Y_n) is extended by (X_{n+1}, Y_{n+1}). Set $Z = \bigcup_{n<\omega} X_n$ and $V = \bigcup_{n<\omega} Y_n$.

We need to verify that (Z, V) is a full, disjoint pair. To show that (Z, V) is full, let C be an arbitrary formula. Then for some m, C is C_{m+1} in the enumeration of the language. Therefore, either $C \in X_{m+1}$ or $C \in Y_{m+1}$, so either $C \in Z$ or $C \in V$, as desired.

To show that (Z, V) is a disjoint pair, suppose it is an overlapping pair. Then there is some $A \in \bigwedge Z$ and some $B \in \bigvee V$ such that $\vdash_{\mathsf{B}} A \rightarrow B$. There must be some stage i such that all the conjuncts of A are in X_i and some j such that all the disjuncts of B are in Y_j. Let $m = \max(i, j)$. Then (X_m, Y_m) is an overlapping pair. By lemma 2.24, for each $n \geq 0$, either $(X_n \cup \{C_{n+1}\}, Y_n)$ is a disjoint pair or $(X_n, Y_n \cup \{C_{n+1}\})$ is a disjoint pair. Therefore, (X_m, Y_m) is a disjoint pair, contradicting the assumption. □

LEMMA 2.26 (Full disjoint pairs yield prime theories) *If (X, Y) is a full,* B*-disjoint pair, then X is a prime* B*-theory.*

Proof Suppose that (X, Y) is a full, B-disjoint pair. First we show that X is a B-theory. Suppose that $(X, \{A\})$ is B-overlapping. Then there is some conjunction C of formulas from X such that $\vdash_{\mathsf{B}} C \rightarrow A$. Since (X, Y) is full, $A \in X$ or $A \in Y$. Suppose that $A \in Y$. By assumption, (X, Y) is B-disjoint, which is a contradiction. Therefore, $A \in X$, so X is a theory.

To show that X is prime, suppose $A \vee B \in X$. Toward a contradiction, assume $A \notin X$ and $B \notin X$. Then $A \in Y$ and $B \in Y$, so $A \vee B \in \bigvee Y$. But as $\vdash_{\mathsf{B}} A \vee B \rightarrow A \vee B$, (X, Y) is not B-disjoint, contradicting the initial assumption. Therefore either $A \in X$ or $B \in X$, so X is prime. □

THEOREM 2.27 (Prime extension theorem) *Suppose that (X, Y) is a* B*-disjoint pair. Then there is a* B*-theory $Z \supseteq X$ such that Z is prime and (Z, Y) is a disjoint pair.*

Proof Suppose that (X, Y) is a B-disjoint pair. By lemma 2.25, there is a full, disjoint pair (Z, V) extending (X, Y). Therefore, $X \subseteq Z$. By lemma 2.26, Z is prime theory. Since $Y \subseteq V$, (Z, Y) is a disjoint pair. □

While not every theory is prime, we can often find prime extensions of theories that have some properties we are after. In particular, theorem 2.27 tells us that if we have a disjoint pair (X, Y), we can always find a prime extension of X that avoids Y.

Not every theory is consistent. Some theories may have blatantly inconsistent axioms, or a more subtle inconsistency may be revealed by the logic. Relevant logics are PARACONSISTENT, meaning that it is not the case that for some formula A, $(A \wedge \neg A) \rightarrow B$ is a theorem for all formulas B. This contrasts with TV, for example, which is EXPLOSIVE, which is to say that it is not paraconsistent. Say that a theory is TRIVIAL iff it contains every formula. In explosive logics, inconsistency results in triviality. A difference with classical logic is that B, and relevant logics more generally, can have inconsistent but non-trivial theories.

There are some useful operations on theories that we will now discuss.

LEMMA 2.28 *Let X and Y be* B*-theories. Then the following are* B*-theories.*

- $X \cap Y$.
- $X \bullet Y = \{C \in \mathfrak{L} : \exists A \in X \exists B \in Y (\vdash_{\mathsf{B}} A \rightarrow (B \rightarrow C))\}$.
- $-X = \{A \in \mathfrak{L} : \neg A \notin X\}$, *provided X is prime.*

Proof For the first, suppose that $(X \cap Y, \{A\})$ is an overlapping pair. Then there is some conjunction $C \in X \cap Y$ such that $\vdash_{\mathsf{B}} C \rightarrow A$. As $C \in X \cap Y$, $C \in X$ and $C \in Y$. As X and Y are theories, $A \in X$ and $A \in Y$, so $A \in X \cap Y$.

For the second, suppose that $A \in X \bullet Y$ and $B \in X \bullet Y$. Then for some $D \in X$ and $E \in Y$, $\vdash_{\mathsf{B}} D \rightarrow (E \rightarrow A)$ and for some $F \in X$ and $G \in Y$, $\vdash_{\mathsf{B}} F \rightarrow (G \rightarrow B)$. It follows that $\vdash_{\mathsf{B}} (D \wedge F) \rightarrow ((E \wedge G) \rightarrow (A \wedge B))$. Further, $D \wedge F \in X$, as X is a theory, and $E \wedge G \in Y$, as Y is a theory. Therefore, $A \wedge B \in X \bullet Y$. Next, suppose that $A \in X \bullet Y$ and $\vdash_{\mathsf{B}} A \rightarrow B$. By assumption, there are $C \in X$ and $D \in Y$ such that $\vdash_{\mathsf{B}} C \rightarrow (D \rightarrow A)$. By theorem 2.4 (vi), $\vdash_{\mathsf{B}} C \rightarrow (D \rightarrow B)$, so $B \in X \bullet Y$.

For the third, suppose that X is prime. Suppose that $(-X, \{A\})$ is an overlapping pair. Then there is $C \in \bigwedge -X$ such that $\vdash_{\mathsf{B}} C \rightarrow A$. Suppose that $A \notin -X$. Then $\neg A \in X$, by definition. By theorem 2.4 (iii), $\vdash_{\mathsf{B}} \neg A \rightarrow \neg C$. Since X is a theory, $\neg C \in X$. For some formulas $D_1, \ldots, D_n$, C is $D_1 \wedge \cdots \wedge D_n$. Using theorem 2.6 (iii) and theorem 2.4 (i) and repeatedly, we get that $\neg D_1 \vee \cdots \vee \neg D_n \in X$. Since X is prime, for some i, $\neg D_i \in X$. But then $D_i \notin -X$, so $C \notin \bigwedge -X$, contradicting an assumption. Therefore, $A \in -X$. □

It need not be the case that $X \cup Y$ is a theory whenever X and Y are. We could have $A \in X, B \notin X, B \in Y$, and $A \notin Y$, so both A and B are in $X \cup Y$ with $A \wedge B$ being in neither X nor Y.

We can define two other useful operations on theories, *theory application*, $X \star Y$, and *theory fusion*, $X \odot Y$.

DEFINITION 2.29 (Theory application, theory fusion) *Let X and Y be* B*-theories. Then the THEORY APPLICATION, $X \star Y$, and THEORY FUSION, $X \odot Y$, are defined as follows.*

- $X \star Y = \{B \in \mathfrak{L} : \exists A \in \mathfrak{L}(A \rightarrow B \in X \text{ and } A \in Y)\}$.
- $X \odot Y = \{A \circ B \in \mathfrak{L} : A \in X \text{ and } B \in Y\}$.

When X and Y are theories, $X \bullet Y, X \star Y$, and $\theta_{\mathsf{B}}(X \odot Y)$ coincide.

THEOREM 2.30 *Let X and Y be* B*-theories. Then the following are identical.*

(1) $X \bullet Y$.
(2) $X \star Y$.
(3) $\theta_{\mathsf{B}}(X \odot Y)$.

Proof (2) $\subseteq$ (1). Suppose that $B \in X \star Y$. So, for some A, $A \rightarrow B \in X$ and $A \in Y$. As $\vdash_{\mathsf{B}} (A \rightarrow B) \rightarrow (A \rightarrow B)$, $B \in X \bullet Y$.

(1) $\subseteq$ (2). Suppose that $C \in X \bullet Y$. So for some $A \in X$ and $B \in Y$, $\vdash_{\mathsf{B}} A \rightarrow (B \rightarrow C)$. As X is a theory, $B \rightarrow C \in X$. So, $C \in X \star Y$.

(3) $\subseteq$ (1). Suppose that $C \in X \odot Y$. So C is $A \circ B$, for some $A \in X$ and $B \in Y$. By theorem 2.6 (vi), $\vdash_{\mathsf{B}} A \rightarrow (B \rightarrow (A \circ B))$. So, $A \circ B \in X \bullet Y$. So $X \odot Y \subseteq X \bullet Y$. Then, by lemma 2.21 and the fact that $X \bullet Y$ is a theory, $\theta_{\mathsf{B}}(X \odot Y) \subseteq X \bullet Y$.

(1) $\subseteq$ (3). Suppose that $C \in X \bullet Y$. Then there are $A \in X$ and $B \in Y$ such that $\vdash_{\mathsf{B}} A \rightarrow (B \rightarrow C)$. By (R6), $\vdash_{\mathsf{B}} (A \circ B) \rightarrow C$. As $A \circ B \in \theta_{\mathsf{B}}(X \odot Y)$, $C \in \theta_{\mathsf{B}}(X \odot Y)$. □

The preceding theorem shows that three apparently distinct ways of combining theories actually coincide. In light of that fact, we will freely move between these different ways of combining theories.

Next, we will record some more useful facts about theories. In particular, we will demonstrate some ways of generating theories.

LEMMA 2.31 (From formulas to theories) *Let A be a formula and let $[A)_{\mathsf{B}} = \{B \in \mathfrak{L} : \vdash_{\mathsf{B}} A \rightarrow B\}$, the set of formulas implied by A according to the logic. Then $[A)_{\mathsf{B}}$ is a* B*-theory.*

Proof Suppose $B, C \in [A)_{\mathsf{B}}$. Then $\vdash_{\mathsf{B}} A \rightarrow B$ and $\vdash_{\mathsf{B}} A \rightarrow C$. So, $\vdash_{\mathsf{B}} A \rightarrow B \wedge C$. Therefore $B \wedge C \in [A)_{\mathsf{B}}$. Next, suppose $B \in [A)_{\mathsf{B}}$ and $\vdash_{\mathsf{B}} B \rightarrow C$. By assumption, $\vdash_{\mathsf{B}} A \rightarrow B$. Using theorem 2.4 (i), $\vdash_{\mathsf{B}} A \rightarrow C$, so $C \in [A)_{\mathsf{B}}$. □

LEMMA 2.32 (Logic is implied by the Ackermann constant) $[\mathsf{t})_{\mathsf{B}} = \mathsf{B}$.

Proof Suppose $A \in [\mathsf{t})_{\mathsf{B}}$. Then $\vdash_{\mathsf{B}} \mathsf{t} \rightarrow A$. By axiom (A8) and (R1), $\vdash_{\mathsf{B}} A$, so $A \in \mathsf{B}$.

Suppose $A \in \mathsf{B}$. Then $\vdash_{\mathsf{B}} A$. By (R8), $\vdash_{\mathsf{B}} \mathsf{t} \rightarrow A$, so $A \in [\mathsf{t})_{\mathsf{B}}$. □

COROLLARY 2.33 *A* B*-theory X is* B*-regular iff* $\mathsf{t} \in X$.

Note what was used in the proof of the preceding lemma. We appealed to (A8), (R1), (R8), and the definition of proof. For logics extending B, the lemma will hold, adjusting the notion of proof to be that of the new logic.

We can almost go the other way, from formulas (with B-theories) to B-theories.

LEMMA 2.34 (From formulas and theories to theories) *Let X be a* B*-theory and A be an arbitrary formula. Define $X^{-A} = \{B \in \mathfrak{L} : A \to B \in X\}$. Then X^{-A} is a* B*-theory.*

Proof Let X be a B-theory. Suppose $B \in X^{-A}$ and $C \in X^{-A}$. Then $A \to B \in X$ and $A \to C \in X$. Using (A3) and the fact that X, is a B-theory, we have $A \to B \wedge C \in X$, so $B \wedge C \in X^{-A}$, as desired. Next, suppose $\vdash_{\mathsf{B}} B \to C$ and $B \in X^{-A}$. Then $A \to B \in X$. Using (R5), $\vdash_{\mathsf{B}} (A \to B) \to (A \to C)$. Since X is a B-theory, $A \to C \in X$, so $C \in X^{-A}$, as desired. □

Say that a set of formulas Z is CLOSED UNDER DISJUNCTION iff $A \in Z$ and $B \in Z$ implies $A \vee B \in Z$.

LEMMA 2.35 *If X is a theory, then $\bigwedge X \subseteq X$. If Y is closed under disjunction, then $\bigvee Y \subseteq Y$.*

Proof Suppose X is a theory. Suppose $A \in \bigwedge X$. Then A is $B_1 \wedge \cdots \wedge B_n$ for some formulas B_i from X. By the first condition on being a theory, $B_1 \wedge \cdots \wedge B_n \in X$. Thus, $\bigwedge X \subseteq X$.

Suppose Y is closed under disjunction. Suppose $A \in \bigvee Y$. Then A is $B_1 \vee \cdots \vee B_n$, for some B_i from Y. By the definition of being closed under disjunction, $B_1 \vee \cdots \vee B_n \in Y$, so $\bigvee Y \subseteq Y$. □

LEMMA 2.36 *Let X be a* B*-theory. Suppose Y is a set of formulas closed under disjunction and $X \cap Y = \emptyset$. Then (X, Y) is a* B*-disjoint pair.*

Proof Let X be a B-theory. Suppose Y is a set of formulas closed under disjunction and $X \cap Y = \emptyset$. Suppose that (X, Y) is a B-overlapping pair. Then there is $A \in \bigwedge X$ and $B \in \bigvee Y$ such that $\vdash_{\mathsf{B}} A \to B$. By the preceding lemma, $\bigwedge X \subseteq X$ and $\bigvee Y \subseteq Y$. So, $X \cap Y \neq \emptyset$, contradicting the assumption. Therefore, (X, Y) is a B-disjoint pair. □

LEMMA 2.37 *Let X be a* B*-theory. Suppose Y is a set of formulas closed under disjunction and $X \cap Y = \emptyset$. Then there is a prime* B*-theory $Z \supseteq X$ such that $Z \cap Y = \emptyset$.*[22]

Proof Let X be a B-theory. Suppose Y is a set of formulas closed under disjunction and $X \cap Y = \emptyset$. By the preceding lemma, (X, Y) is a B-disjoint pair. Using lemma 2.25, there is a full, B-disjoint pair (Z, V) extending (X, Y). To show $Z \cap V = \emptyset$, suppose $A \in Z \cap V$. As $\vdash_{\mathsf{B}} A \to A$, (Z, V) is not a B-disjoint pair, contradicting the assumption. Therefore, $Z \cap V = \emptyset$, so $Z \cap Y = \emptyset$. By lemma 2.26, Z is prime. □

LEMMA 2.38 (Theory closures) *Let X be a* B*-theory. The following hold.*

[22] This lemma is from Seki (2009).

(i) *If* $A \to B \in X$ *and* $\vdash_{\mathsf{B}} B \to C$, *then* $A \to C \in X$.
(ii) *If* $A \to B \in X$ *and* $\vdash_{\mathsf{B}} C \to A$, *then* $C \to B \in X$.
(iii) *If* $C(A) \to B \in X$ *and* $\vdash_{\mathsf{B}} A \leftrightarrow D$, *then* $C(D) \to B \in X$.
(iv) *If* $A \to C(B) \in X$ *and* $\vdash_{\mathsf{B}} B \leftrightarrow D$, *then* $A \to C(D) \in X$.

Proof Let X be a B-theory.

(i) Suppose $A \to B \in X$ and $\vdash_{\mathsf{B}} B \to C$. By (R5), $\vdash_{\mathsf{B}} (A \to B) \to (A \to C)$. Then $A \to C \in X$.
(ii) Suppose $A \to B \in X$ and $\vdash_{\mathsf{B}} C \to A$. By (R4), $\vdash_{\mathsf{B}} (A \to B) \to (C \to B)$. Then $C \to B \in X$.
(iii) Suppose $C(A) \to B \in X$ and $\vdash_{\mathsf{B}} A \leftrightarrow D$. By theorem 2.9, $\vdash_{\mathsf{B}} C(A) \leftrightarrow C(D)$. By (ii), $\vdash_{\mathsf{B}} C(D) \to B$.
(iv) Suppose $A \to C(B) \in X$ and $\vdash_{\mathsf{B}} B \leftrightarrow D$. By theorem 2.9, $\vdash_{\mathsf{B}} C(B) \leftrightarrow C(D)$. By (i), $\vdash_{\mathsf{B}} A \to C(D)$. □

Squeezing lemmas show that if some theories are in certain relations, then a prime extension of one of them can be found in the desired relation with the other theories. In other words, if the theories are related a certain way, then a prime theory can be squeezed in as well.

LEMMA 2.39 (Squeezing lemma) *Suppose* X *and* Y *are* B-*theories and* Z *is a prime* B-*theory such that* $X \bullet Y \subseteq Z$. *Then the following hold.*

(1) *There is a prime* B-*theory* U *such that* $X \subseteq U$ *and* $U \bullet Y \subseteq Z$.
(2) *There is a prime* B-*theory* V *such that* $Y \subseteq V$ *and* $X \bullet V \subseteq Z$.
(3) *There are prime* B-*theories* U *and* V *such that* $X \subseteq U$, $Y \subseteq V$, *and* $U \bullet V \subseteq Z$.

Proof Suppose X and Y are B-theories and Z is a prime B-theory such that $X \bullet Y \subseteq Z$. For (1), define

$$W = \{A \in \mathfrak{L} : \exists B, C \in \mathfrak{L}(\vdash_{\mathsf{B}} A \to (B \to C) \text{ and } B \in Y \text{ and } C \notin Z)\}.$$

We will show that W is closed under disjunction. Suppose $F, G \in W$. Then there are formulas B, C, D, and E such that $F \to (B \to C)$ and $G \to (D \to E)$ are B-theorems, $B \in Y$, $D \in Y$, $C \notin Z$, and $E \notin Z$. Since Y is a theory, $B \wedge D \in Y$. Since Z is a prime theory, $C \vee E \notin Z$. From theorem 2.4 (v) and (vi), $F \to ((B \wedge D) \to (C \vee E))$ and $G \to ((B \wedge D) \to (C \vee E))$ are theorems, so by theorem 2.5 (ii), $(F \vee G) \to ((B \wedge D) \to (C \vee E))$ is a theorem. From the definition of W, $F \vee G \in W$.

Next, we show that $X \cap W = \emptyset$. Suppose not, so there is $D \in X \cap W$, so for some $B, C \in \mathfrak{L}$, $D \to (B \to C)$ is a theorem, $B \in Y$, and $C \notin Z$. As $X \bullet Y \subseteq Z$, it follows that $C \in Z$, which contradicts the assumption that $D \in W$.

Since X is a theory, W is closed under disjunction, and $X \cap W = \emptyset$, we can apply lemma 2.37 to obtain a prime theory U such that $X \subseteq U$ and $U \cap W = \emptyset$. Finally, we need to show that $U \bullet Y \subseteq Z$. Suppose not, so there is some $D \in U$ such that, for some $B, C \in \mathfrak{L}$, $D \to (B \to C)$ is a theorem, $B \in Y$ and $C \notin Z$. Then $D \in W$, which is a contradiction. Therefore, $U \bullet Y \subseteq Z$.

For (2), define

$$T = \{B \in \mathfrak{L} : \exists C \in \mathfrak{L}(B \to C \in X \text{ and } C \notin Z)\}.$$

We will show that T is closed under disjunction. Suppose $A, B \in T$. Then there are C, D such that $A \to C \in X$, $B \to D \in X$, $C \notin Z$, and $D \notin Z$. By lemma 2.38 (i) and axiom (A4), $A \to (C \vee D) \in X$ and $B \to (C \vee D) \in X$. From the fact that X is a theory together with axiom (A5), $(A \vee B) \to (C \vee D) \in X$. Since Z is prime, $C \vee D \notin Z$. It then follows that $A \vee B \in T$.

Next, we will show that $Y \cap T = \emptyset$. Suppose there is some $B \in Y \cap T$. So, for some C, $B \to C \in X$ and $C \notin Z$. As $B \in Y$ and $X \bullet Y \subseteq Z$, $C \in Z$, contradicting an assumption. So, $Y \cap T = \emptyset$. We can then use lemma 2.37 to obtain a prime theory V such that $Y \subseteq V$ and $V \cap T = \emptyset$.

To show that $X \bullet V \subseteq Z$, suppose there is some $B \in V$ such that for some C, $B \to C \in X$ but $C \notin Z$. Then $B \in T$, so $V \cap T \neq \emptyset$, which is a contradiction. Therefore, $X \bullet V \subseteq Z$.

Combined with the assumptions, (1) and (2) imply (3). □

LEMMA 2.40 *Suppose X, Y, and Z are* B*-theories such that $X \bullet Y \subseteq Z$. Suppose $B \notin Z$. Then there are prime* B*-theories V and Z' such that $Y \subseteq V$, $X \bullet V \subseteq Z'$, and $B \notin Z'$.*

Proof Suppose X, Y, and Z are B-theories, $X \bullet Y \subseteq Z$, and $B \notin Z$. As Z is a theory and $B \notin Z$, $(Z, \{B\})$ is a disjoint pair. From theorem 2.27, there is a prime theory Z' such that $Z \subseteq Z'$ such that $(Z', \{B\})$ is a disjoint pair, and so $B \notin Z'$. Since $Z \subseteq Z'$, it follows that $X \bullet Y \subseteq Z'$.

Next, as in the previous proof, define

$$T = \{B \in \mathfrak{L} : \exists C \in \mathfrak{L}(B \to C \in X \text{ and } C \notin Z')\}.$$

As before, T is closed under disjunction and $Y \cap T = \emptyset$. By lemma 2.37, there is a prime theory V such that $Y \subseteq V$ and $V \cap T = \emptyset$.

To show that $X \bullet V \subseteq Z'$, suppose there is $B \in V$ such that for some C, $B \to C \in X$ but $C \notin Z'$. Then $B \in T$, contradicting an assumption. Therefore, $X \bullet V \subseteq Z'$. □

COROLLARY 2.41 *Suppose X, Y, and Z are* B*-theories such that $X \bullet Y \subseteq Z$. Then there are prime* B*-theories U, V, and Z' such that $X \subseteq U$, $Y \subseteq V$, $Z \subseteq Z'$, and $U \bullet V \subseteq Z'$.*

Further, suppose that $B \notin Z$. It then follows that $B \notin Z'$.

That concludes the basic features of B-theories. While we have presented the definitions and proofs of this section in terms of B, they carry over to any relevant logic L extending B.

There are a few more types of closure on theories that are worth considering and have been considered in the literature. The first is closure not just under provable implications but under the implications of a particular theory.

DEFINITION 2.42 (T-L-theories) *Let T be an* L*-theory. A T-*L*-THEORY X is a set of formulas such that both*

- *if $A \in X$ and $B \in X$, then $A \wedge B \in X$, and*
- *if $A \in X$ and $A \rightarrow B \in T$, then $B \in X$.*

If L contains every instance of (I1), then any L-theory T will be a T-L-theory. Even so, given a particular L-theory T, it is not the case that every L-theory is a T-L-theory. T-L-theories are used in proving completeness with respect to reduced frames. Analogs of many of the lemmas proved earlier, in particular the Full pair extension lemma and the Squeezing lemma, can be proved for T-L-theories.

Apart from closure under implications of a particular theory, one may want to consider closing theories under additional rules or conditions. As one example, Slaney (1987a) considers theories that are closed under various conditions, such as a form of contraposition, if $A \rightarrow B \in X$ then $\neg B \rightarrow \neg A \in X$. We will see these theories in Section 4.4. As another example, Routley et al. (1982, ch. 4.8) and Seki (2009) consider logics with disjunctive rules, such as $C \vee A, C \vee (A \rightarrow B) \Rightarrow C \vee B$, $C \vee (A \rightarrow B) \Rightarrow C \vee (\neg B \rightarrow \neg A)$, and $C \vee A \Rightarrow C \vee \Box A$. For a logic L with disjunctive rules, L-theories, in general, are required to be closed under provable implications and adjunction, but not the disjunctive rules. While L-theories need not respect the disjunctive rules, L-*regular* L-theories must. L-regular L-theories are theories containing all the theorems of L as well as being closed under the disjunctive rules, that is, if the premises of the rule are contained in the theory, then the conclusion must be as well. As with T-L-theories, many of the important lemmas can be extended to theories in logics with disjunctive rules.

Further development of theories will wait until we get to the canonical frame construction, following the introduction of ternary relational frames. At this point, we have the means to show that a formula is in a theory, but we do not have many ways of showing that a formula is *not* in a theory. In Chapter 5, we will introduce algebraic models, which provide one particularly nice method for showing that a formula is not in a theory.

2.6 Appendix: FDE, BB, F, and BM

In this appendix, we will provide a brief overview of four relevant logics that are weaker than B: FDE, BB, F, and BM. For this appendix, we will set aside t and fusion.

In the introductory chapter, we used the logic of first-degree entailment (FDE) to illustrate some basic features of relevant logics. FDE has a prominent place in the development of relevant logics, and it has blossomed into its own research area, separate from relevant logics. The name of FDE comes from its connection to the logic E of entailment.[23] The degree of a formula is the depth of nesting of implications in it. The implication of E represents entailment, and FDE is the logic of entailments

[23] See Anderson and Belnap (1975, ch. 3) or Omori and Wansing (2017) for more on FDE.

between formulas that contain no entailments, in other words, first-degree formulas. This means that FDE is concerned with implications of the form $A \rightarrow B$ where A and B are built from $\{\wedge, \vee, \neg\}$. FDE was found to be the first-degree fragment of E, and in fact it is the first-degree fragment of R, B, and every standard relevant logic. It is also the only logic we will discuss with degree restrictions. Note that while $A \vee \neg A$ is a theorem of E, it is not included as a theorem of FDE, as it is not an implicational formula.

FDE has the following axioms.

(FDE1) $A \rightarrow A$.
(FDE2) $(A \wedge B) \rightarrow A, (A \wedge B) \rightarrow B$.
(FDE3) $A \rightarrow (A \vee B), B \rightarrow (A \vee B)$.
(FDE4) $\neg\neg A \rightarrow A, A \rightarrow \neg\neg A$.
(FDE5) $(A \wedge (B \vee C)) \rightarrow ((A \wedge B) \vee (A \wedge C))$.

It also has the following rules,

(RFDE1) $A \rightarrow B, B \rightarrow C \Rightarrow A \rightarrow C$.
(RFDE2) $A \rightarrow B, A \rightarrow C \Rightarrow A \rightarrow (B \wedge C)$.
(RFDE3) $A \rightarrow C, B \rightarrow C \Rightarrow (A \vee B) \rightarrow C$.
(RFDE4) $A \rightarrow B \Rightarrow \neg B \rightarrow \neg A$.

Using these, one can obtain much of the desirable logical behavior of conjunction, disjunction, and negation, including the De Morgan equivalences.

As mentioned, FDE is a sublogic of B. B, without t and fusion, is obtained from FDE by trading the rules (RFDE2) and (RFDE3) for their axiom forms, and adding rules, namely (R1), (R2), (R4), and (R5).

A logic, BB, intermediate between B and FDE, is sometimes studied.[24] BB is obtained from FDE by adding the rules (R4) and (R5) of B, but not adding the axiom forms of (RFDE2) and (RFDE3).

It is important that BB does not have fusion in its language. The reason, as shown by Lavers (1985), is that together with the other axioms and rules of BB, fusion with the rules (R6) and (R7) renders (A3) and (A5) provable. That would bump the logic up to be identical with B.

Lavers shows that one can define a family of relatives of BB. To obtain these other logics, BL, one adds axioms and rules to BB. There is one restriction, namely that the axioms and rules added can only display the logical vocabulary $\{\rightarrow, \neg\}$. The idea behind this definition is that the axioms and rules of BB specify the logical behavior of conjunction and disjunction. Any additional axioms or rules should only deal with the other logical connectives on pain of altering the basic logical behavior of conjunction and disjunction. For example, (N1) can be added to a BL logic to obtain another such logic, while (I1) cannot. Adding (N1) to BB results in the logic BDW. One can obtain a

24 BB was introduced by Lavers (1985). Tedder and Ferenz (2022) provide neighborhood frames for BB, among other logics. Logan (2025a) argues that the positive fragment of BB has a robust computational interpretation.

range of logics that are properly weaker than their non-BL-counterparts. Exceptions come, for example, with BRW and BR, which are, in fact, identical to RW and R, respectively. This is because of the definability of fusion in terms of negation and implication in those logics.

Although BB is weak, it is not the weakest relevant logic that has been considered. The logic F, studied by Goble (2003), Tedder (2021b), and Tedder and Ferenz (2022), drops the double negation elimination axiom (A7) and contraposition rule (R3) from BB, as well as the rules (R4) and (R5). Instead, it adds the following.

(F1) $\neg(A \wedge B) \rightarrow (\neg A \vee \neg B)$.
(F2) $(\neg A \wedge \neg B) \rightarrow \neg(A \vee B)$.
(F3) $A \rightarrow B \Rightarrow \neg B \rightarrow \neg A$.
(F4) $A \leftrightarrow B \Rightarrow (D \rightarrow B) \rightarrow (D \rightarrow A)$.
(F5) $A \leftrightarrow B \Rightarrow (B \rightarrow C) \rightarrow (A \rightarrow C)$.

We will not further study BB or F here, as the ternary relational frames used from Chapter 3 are too strong for them. Instead, one must use neighborhood frames, which will be described briefly in Section 7.12.

One more sublogic of B that is worth a mention is the logic BM. While not discussed a lot, it is a natural weakening of B, where one weakens negation slightly.[25] One drops the double negation axiom (A7) and contraposition rule (R3) of B and replaces them with the following.

(BM1) $\neg(A \wedge B) \rightarrow (\neg A \vee \neg B)$.
(BM2) $(\neg A \wedge \neg B) \rightarrow \neg(A \vee B)$.
(BM3) $A \rightarrow B \Rightarrow \neg B \rightarrow \neg A$.

BM lacks the double negation behavior of B, but it does retain the De Morgan laws for conjunction and disjunction. Unlike BB, it is not difficult to adapt the frames of later chapters to be adequate for BM. Nonetheless, we will not further study BM here.

2.7 Exercises

(1) Show that in RW, $(A \circ B) \leftrightarrow \neg(A \rightarrow \neg B)$ is a theorem.
(2) Prove the remaining parts of theorem 2.4.
(3) Provide proofs for the remaining parts of theorem 2.5.
(4) Provide proofs for the remaining parts of theorem 2.6.
(5) Show that the following interderivability facts hold.
  - In DW, (I3) and (I4) are interderivable.
  - In TW, (I5) and (I6) are interderivable.
  - In T, (I8) is interderivable with $(((A \rightarrow A) \wedge (B \rightarrow B)) \rightarrow C) \rightarrow C$.
  - In T, (I8) is interderivable with the conjunction of
    - $((A \rightarrow A) \rightarrow C) \rightarrow C$ and

[25] See Furhmann (1988; 1990), Brady (2003, 10ff.), and Norman and Sylvan (1989, ch. 19).

- $((D \to D) \to D) \wedge ((E \to E) \to E) \to (((D \wedge E) \to (D \wedge E)) \to (D \wedge E))$.

(6) Show that for any logic L extending B, the following are equivalent.
- L is closed under the rule $A \to \neg A \Rightarrow \neg A$.
- $A \vee \neg A$ is a theorem of L.

(7) Show that for any logic L extending B, if L contains (I8), then L is closed under the rule $A \Rightarrow (A \to B) \to B$.

(8) Show that for any logic L extending B, if L contains (I7), then L both
- contains (I1), and
- is closed under the rule $A \to (B \to C) \Rightarrow (A \wedge B) \to C$.

(9) Show that for any logic L extending TW and containing (I1), then L is closed under the rule $A \to (B \to C) \Rightarrow (A \wedge B) \to C$.

(10) Show that for any logic L extending B, if L contains (I1), then L is closed under the rule $A \to (A \to C), B \to (B \to C) \Rightarrow (A \vee B) \to ((A \vee B) \to C)$.

(11) Show that in E, if A is a theorem, then $(A \to A) \to A$, that is, $\Box A$, is a theorem.

(12) Show that in E, $\Box A \leftrightarrow (\mathsf{t} \to A)$ is derivable, for all A.

(13) Show that the following logics have the indicated theorems.
- RW – $(A \circ B) \to (B \circ A)$.
- TW – $((B \circ A) \circ C) \to (A \circ (B \circ C))$.
- TW – $((A \circ B) \circ C) \to (A \circ (B \circ C))$.
- C – $A \to (A \circ A)$.
- T – $(A \circ B) \to ((A \circ B) \circ B)$.
- E – $A \to (A \circ \mathsf{t})$.
- RM – $(A \circ A) \to A$.

3 Frame Theory

3.1 Introduction

Relevant logics, much like TV-based modal logics, were initially investigated primarily via proof theory, rather than via frames or models. Sylvan and Meyer developed a flexible frame theory for relevant logics, one which equally accommodated a range of logics outside the family of relevant logics.[1] Their insight was to use a *ternary relation* to interpret the implication connective, and by extension the fusion connective. While there were other frames presented for relevant logics, Sylvan and Meyer's ternary relational frame theory was adopted as the more or less standard frame theory for relevant logics.[2] The ternary relational frames, or Routley–Meyer frames as they are often called, will be the primary frame theory adopted for this book. In an appendix to this chapter, we will introduce one of the important variations of the ternary relational frames, one that uses an incompatibility relation rather than the Routley star to interpret negation.[3] In two other appendices, we will briefly present two of the alternative frame theories for relevant logics, one due to Fine and one due to Urquhart. There are some frame theories that we will not go into here, such as the collection frames of Restall and Standefer (2023a).[4]

We will begin this chapter by presenting the basic ternary relational frames for relevant logics in Section 3.2. We will start with the frames for B, as those have the fewest conditions, and then additional frame conditions will be introduced to obtain frames for the stronger logics. In Section 3.3, we will consider rules for logics determined by frames, and then in Section 3.4, we will discuss propositions in models. Propositions will be useful, particularly once we get to quantification, as the approach to modeling quantification adopted from Chapter 11 onwards uses propositions in a fundamental way. Section 3.5 presents several examples of ternary relational frames satisfying various conditions. Section 3.6 presents the reduced frames, which have been important in the development of relevant logics. Finally, there are three

1 The early work was published as Routley and Meyer (1972a; 1972b; 1973). See Bimbó et al. (2018) for an overview of the early development of the theory.

2 Other models were developed by Maksimova (1969), Urquhart (1972a; 1972b), and Fine (1974; 1988), some of which will be discussed in appendices to this chapter. See also Bimbó and Dunn (2018) for more on Maksimova's contributions to the area.

3 The Routley star originates with Routley and Routley (1972) and it is named after Richard Sylvan and Valerie Plumwood, albeit using their earlier names.

4 Onishi (2024) provides one development of collection frames.

appendices that present some alternative approaches to frame theory: compatibility frames for negation in Section 3.7, Urquhart's semilattice and operational frames in Section 3.8, and Fine's operational frames in Section 3.9.

3.2 Basic Frames

We will begin by presenting the basic ternary relational frames for the logic B. After we define the frames and make some comments on them, we will present some common frame conditions used to define frames for stronger logics.

DEFINITION 3.1 (Ternary relational frames) *A* TERNARY RELATIONAL FRAME, *or* ROUTLEY-MEYER FRAME, *for* B *is a quadruple* $\langle K, N, R, ^* \rangle$ *where* $K \neq \emptyset, N \subseteq K, R \subseteq K \times K \times K, ^*: K \mapsto K$, *and which obeys the following conditions, for all* $a, b, c, d, e, f \in K$, *where* $a \leq b =_{Df} \exists x \in N, Rxab$*:*

(i) $a \leq a$,
(ii) *if* $a \leq b$ *and* $b \leq c$, *then* $a \leq c$,
(iii) *if* $a \in N$ *and* $a \leq b$, *then* $b \in N$,
(iv) *if* $a \leq b$, *then* $b^* \leq a^*$,
(v) $a^{**} = a$, *and*
(vi) *if* $Rabc, d \leq a, e \leq b$, *and* $c \leq f$, *then* $Rdef$.

The defined ordering, $\leq$, is called the HEREDITY ORDERING. One could add it as a primitive frame component and change the definition to an additional frame condition. Nothing hinges on that choice, so we will proceed with the heredity ordering as defined.

A few explanatory comments about the parts of the frames are in order. The ternary relation, R, will be used to interpret implication and fusion. The Routley star, *, will be used to interpret negation. The set of regular, or normal, points, N, will be used in the definition of counterexample, a point to which we will return below.[5] The heredity ordering, $\leq$, will play a role in several definitions, including those of models and propositions. The conditions relating the heredity ordering and the other parts of the frame are used in proving the Heredity lemma, lemma 3.3 below.

By way of motivation, one can think of the set K as a collection of situations, which are partial (and potentially contradictory). The heredity ordering can then be viewed as the relation of one situation being contained in another. The situation composed of my office in Taipei settles things going on in that room, but it does not settle the facts concerning the weather in Tainan. Conversely, the more expansive situation composed of the whole of Taiwan does settle the issues concerning the weather in Tainan as well as the goings on in my office, but it doesn't settle anything regarding the traffic in Wellington. The set of regular points, N, contains the situations where logical truths hold, and for situations outside it, logical truths can fail. The ternary relation can be

[5] We would call these "normal points," except that there is a clash with the terminology of normal theories. They are similar to the normal worlds of some Kripke models for modal logics.

understood in different ways, but one interpretation of it is that $Rabc$ expresses that the situation a is an informational channel from the situation b to the situation c.[6] This is to say that the situation a is a channel that transmits information from situation b to situation c.

There are different philosophical interpretations of the ternary relational frames. We will mention three here. Mares (2004; 2009; 2021) interprets the frames in informational terms. The points of the frames are interpreted as situations that carry information. Mares offers a theory of *situated inference*, according to which we reason from the information in some given situations to the existence of a situation that carries other information. The ternary relation models informational connections among situations that are located in the same world. Mares's theory of situated inference works best for R, although it has been extended to cover frames for additional logics such as B and DJ.

Beall et al. (2012) provide several different interpretations of the ternary relation. They argue that, depending on how it is interpreted, it neatly captures one of three different conceptions of conditionality. These different interpretations involve viewing the ternary relation as a *binary* relation between a point and a pair of points. The first interpretation pairs the second and third points, $Ra\langle b,c\rangle$ and it yields the ABSENCE OF COUNTEREXAMPLES CONCEPTION of conditionality. The points b and c can function as potential counterexamples from the point of view of a. The second interpretation pairs the first two points, $R\langle a,b\rangle c$ to yield the OPERATOR CONCEPTION of conditionality. On this conception, c contains at least as much information as the fusion of the information contained in each of the points in the initial pair. The third interpretation groups the first and third points, $Ra\rangle b\langle c$, to obtain the CONDITIONAL LOGIC CONCEPTION of conditionality. In conditional logics, such as counterfactual logics, one can view the antecedent of the conditional as a kind of formula-relativized necessity. On this view, the antecedent is supposed to hold at b, and this fixes a modal accessibility relation between the points where the conditional is to hold and the points where the consequent does.

These three different conceptions of conditionality are unified by viewing the ternary relation as a relation of *relative* relative possibility, similarly to how the binary modal accessibility relation of Kripke models is viewed as a relation of relative possibility. A largely unexplored question regarding the three conceptions and their generalization is whether any privileges a certain logic or naturally supports certain frame conditions. As an example, the absence of counterexamples interpretation plausibly provides a reason against (FI5), the condition that if $Rabc$, then $Rbac$, as it swaps points out of the salient pair of counterexample points. On the other hand, the operator conception may provide some support for (FI5).

More recently, Golan (2023) has offered an interpretation of the ternary relation in terms of compatibility, connecting the ternary relation to the sorts of compatibility

[6] The idea of information channels originates in the work of Barwise (1993) in situation theory. For more on the information channel idea in the context of relevant logics, see Restall (2005), Mares et al. (2010), Punčochář (2020), and Tedder (2021b; 2023).

Table 3.1. Axioms and their frame conditions

Name	Axiom	Name	Condition
(I1)	$(A \wedge (A \to B)) \to B$	(FI1)	$Raaa$
(I2)	$((A \to B) \wedge (B \to C)) \to (A \to C)$	(FI2)	$Rabc \Rightarrow Ra(ab)c$
(I3)	$(A \to B) \to ((C \to A) \to (C \to B))$	(FI3)	$Rabcd \Rightarrow Ra(bc)d$
(I4)	$(A \to B) \to ((B \to C) \to (A \to C))$	(FI4)	$Rabcd \Rightarrow Rb(ac)d$
(I5)	$A \to ((A \to B) \to B)$	(FI5)	$Rabc \Rightarrow Rbac$
(I6)	$(A \to (B \to C)) \to (B \to (A \to C))$	(FI6)	$Rabcd \Rightarrow Racbd$
(I7)	$(A \to (A \to B)) \to (A \to B)$	(FI7)	$Rabc \Rightarrow Rabbc$
(I8)	$(\mathsf{t} \to A) \to A$	(FI8)	$\exists x \in N, Raxa$
(I9)	$A \to (A \to A)$	(FI9)	$Rabc \Rightarrow a \leq c$ or $b \leq c$
(N1)	$(A \to \neg B) \to (B \to \neg A)$	(FN1)	$Rabc \Rightarrow Rac^*b^*$
(N2)	$(A \to \neg A) \to \neg A$	(FN2)	Raa^*a
(N3)	$A \vee \neg A$	(FN3)	$\forall a \in N, a^* \leq a$

relations discussed in appendix 3.7. Hornischer and Berto (2025) have offered an interpretation of the ternary relation in terms of dynamical systems.

The class of ternary relational frames for B does not need any other conditions. One can obtain frames appropriate to other logics by restricting the class of frames to those that obey certain conditions. To state the conditions, we will introduce some standard notation.[7]

- $Rabcd =_{Df} \exists x \in K(Rabx$ and $Rxcd)$
- $Ra(bc)d =_{Df} \exists x \in K(Raxd$ and $Rbcx)$

The frame conditions will be named as (FY), where (Y) is the name of some axiom from Chapter 2. The conditions are presented in Table 3.1, and they should be understood as universally closed. In addition, "$\Rightarrow$" is used for the metalinguistic conditional to save space.

Ternary relational frames that obey all the frame conditions corresponding to the axioms for the logic L will be called L-frames. For example, a DW-frame is a ternary relational frame that obeys condition (FN1).

In Chapter 1, we presented two paradoxes of implication, and we can give frame conditions appropriate for them.

(PI1) $A \to (B \to A)$	(FPI1) If $Rabc$, then $a \leq c$,
(PI2) $A \to (B \to B)$	(FPI2) If $Rabc$, then $b \leq c$.

These two frame conditions are presented here to illustrate two things. First, the ternary relational frames are very powerful and general, able to capture even

[7] Although we will not exploit it here, these definitions allow one to define an n-place relation, $n \geq 2$, out of chaining together R related points, with a stipulation about $n = 2$, e.g., R^2ab means $a \leq b$. These R-chains are useful for investigating some of the stronger relevant logics. See Mares (2004, 210) or Meyer and Routley (1974, 183) for presentations of this idea. This thought leads, fairly naturally, to some of the collection frames of Restall and Standefer (2023a).

Table 3.2. Verification conditions

$a \Vdash p$	iff	$V(p,a) = 1$
$a \Vdash \mathsf{t}$	iff	$a \in N$
$a \Vdash \neg B$	iff	$a^* \nVdash B$
$a \Vdash B \wedge C$	iff	$a \Vdash B$ and $a \Vdash C$
$a \Vdash B \vee C$	iff	$a \Vdash B$ or $a \Vdash C$
$a \Vdash B \rightarrow C$	iff	$\forall b,c \in K$(if $Rabc$ and $b \Vdash B$, then $c \Vdash C$)
$a \Vdash B \circ C$	iff	$\exists b,c \in K$($Rbca$ and $b \Vdash B$ and $c \Vdash C$)

non-relevant logics such as TV. Second, axiom (I9) was described as a special case of weakening, and this connection is further illustrated by the corresponding frame condition: Condition (FI9) and conditions (FP1) and (FP2) are close in form.

We can add a valuation to an L-frame in order to obtain an L-model.

DEFINITION 3.2 (Valuation, models) *A valuation V for a frame F is a function from* At $\times K$ *to* $\{0,1\}$.

A TERNARY RELATIONAL MODEL *for* L *is a pair $\langle F, V\rangle$ of an* L*-frame F and a valuation V where if $a \leq b$ and $V(p,a) = 1$ then $V(p,b) = 1$.*

A ternary relational model $\langle F, V\rangle$ is said to be built on the frame F.

We can extend the valuation to a verification relation, $\Vdash$, on the whole language as listed in Table 3.2.

Verification for arbitrary formulas, not just atoms, is preserved up the heredity ordering. This fact is an important lemma in the frame theory.

LEMMA 3.3 (Heredity lemma) *Let M be an* L*-model. For all $a,b \in K$ and $A \in \mathfrak{L}$, if $a \Vdash A$ and $a \leq b$, then $b \Vdash A$.*

Proof Let M be an L-model. Suppose $a \Vdash A$ and $a \leq b$. The proof is by induction on the structure of A.

If A is an atom, then the result follows from the definition of a model.

If A is t, then the result follows from condition (iii) of definition 3.1 that if $a \leq b$ and $a \in N$ then $b \in N$.

Suppose A is $B \wedge C$. Then, from the verification condition, $a \Vdash B$ and $a \Vdash C$. By the inductive hypothesis, $b \Vdash B$ and $b \Vdash C$, so $b \Vdash B \wedge C$.

Suppose A is $B \vee C$. Then, from the verification condition, $a \Vdash B$ or $a \Vdash C$. By the inductive hypothesis, $b \Vdash B$ or $b \Vdash C$, so $b \Vdash B \vee C$.

Suppose A is $B \rightarrow C$. Suppose $Rbcd$ and $c \Vdash B$. From condition (vi) of definition 3.1 and the assumption that $a \leq b$, $Racd$. From the verification condition, it follows that $d \Vdash C$, so $b \Vdash B \rightarrow C$, as desired.

Suppose A is $B \circ C$. Then $a \Vdash B \circ C$. By the verification condition, there are $c,d \in K$ such that $c \Vdash B$, $d \Vdash C$, and $Rcda$. By condition (vi) of definition 3.1, $Rcdb$, so $b \Vdash B \circ C$. □

Now that we have models, we can define *counterexamples* and *validity*.

DEFINITION 3.4 (Counterexamples, holding, validity) *A formula B has a* COUNTEREXAMPLE *in an* L*-model* $\langle F, V\rangle$ *iff there is* $a \in N$ *such that* $a \not\Vdash B$.

A formula B HOLDS *in an* L*-model* $\langle F, V\rangle$ *iff B has no counterexamples in the model.*

A formula B is VALID *on an* L*-frame F iff B holds in all* L*-models built on F.*

A formula B is VALID *in a class $\mathcal{C}$ of* L*-frames iff for all* $F \in \mathcal{C}$, *B is valid on F.*

When B is valid in a class $\mathcal{C}$ of frames, we write $\models_{\mathcal{C}} B$.

Where $\mathcal{C}$ is the class of L*-frames and B is valid in $\mathcal{C}$, we write* $\models_{\mathsf{L}} B$.

Let $\mathsf{L}(\mathcal{C}) = \{A \in \mathfrak{L} : \models_{\mathcal{C}} A\}$.

It is worth emphasizing some differences with Kripke models for TV-based modal logics. A Kripke model $\langle W, S, V\rangle$ is a triple of a non-empty set of worlds W, a binary relation S on W, and a valuation V. For Kripke models, the concepts of counterexample and validity are defined with respect to all points of the model. The situation with validity is different for ternary relational models. The set of regular points, N, is used in the definitions of counterexample, holding, and validity. Valid formulas are required to hold with respect to the regular points but need not hold with respect to *all* points in a model.[8] Even valid formulas can fail at some points in some models, as they must to avoid validating implicational paradoxes where validities are implied by every formula.

There is a terminological warning we would like to make. Earlier we have defined valid formulas, and we will sometimes refer to them as *validities*. When logics defined in terms of frames are being discussed, we will usually called members of those logics "validities" to emphasize the originating definition for the logic. Sometimes we will call their members *theorems*. While "theorem" is often reserved for a member of a logic defined in terms of proofs, we are using it as a neutral term for a member of a logic, however the latter is defined.

In Section 3.5, we will present several examples of models satisfying different conditions. We will use a tabular notation for presenting the ternary relation and the Routley star. In this frame, $K = \{a,b,c,d\}$, $N = \{b,c\}$, the star and R are given in the Table 3.3. This example will be repeated in Section 3.5. The row header indicates the first spot in the ternary relation, the column header the second, and the inner spots in the table indicate value(s) for the third. So, the b-row, a-column entry, abc, which is underlined in the table, means that $Rbaa$, $Rbab$, and $Rbac$ hold. The $*$ column indicates values for the star operation, so $a^* = b$, $b^* = a$, $c^* = d$, and $d^* = c$.

Table 3.3. Frame example

R	a	b	c	d	$*$
a	$abcd$	$abcd$	bc	$abcd$	b
b	$\underline{abc}$	bc	c	$abcd$	a
c	c	c	c	$abcd$	d
d	$abcd$	$abcd$	$abcd$	$abcd$	c

[8] Robles and Méndez (2014d) study what happens when validity is defined with respect to all points. One notable difference is that a rule form of weakening, $A \Rightarrow B \to A$ is validated. Gabbay (1976) does something similar, albeit in conditional logics over TV.

The heredity relation can be computed from the table, and in this frame, the heredity ordering is $d \leq a \leq b \leq c$.

To show that a formula holds in a model, one must verify that it holds at all regular points. When the formula is an implication, there is another method of verification, which is very useful.

LEMMA 3.5 (Verification lemma) *Let M be an* L*-model. The following are equivalent.*

- $A \to B$ *holds in* M.
- *For all* $b \in K$, *if* $b \Vdash A$, *then* $b \Vdash B$.

Proof Suppose that $A \to B$ holds in M. Then for all $a \in N$, $a \Vdash A \to B$. Suppose $b \in K$ and $b \Vdash A$. There is some $x \in N$ such that $Rxbb$, by condition (i) of definition 3.1. Since $x \Vdash A \to B$, by assumption, $b \Vdash B$, by the verification condition for implication.

Suppose that for all $b \in K$, if $b \Vdash A$, then $b \Vdash B$. Suppose $a \in N$, $Rabc$ and $b \Vdash A$. By assumption, $b \Vdash B$. By assumption, $b \leq c$, so by lemma 3.3, $c \Vdash B$. Therefore, $a \Vdash A \to B$, as desired. □

The Verification lemma simplifies the proofs of Soundness, as we will see in Chapter 4. Note that the Verification lemma establishes that an implication is true in all *regular* points. It does *not* establish that that conditional is true at *all* points in the model. For the latter argument, one would need to assume that $a, b, c \in K$ are arbitrary points such that $Rabc$ and the antecedent holds at b and show that the consequent holds at c. This generally does not follow from an appeal to the Verification lemma. For example, the Verification lemma immediately establishes the validity of $p \to p$, but $p \to p$ cannot hold at all points in all B-models, as then $q \to (p \to p)$ would be valid, which is not the case. We will return to this point in Chapter 10.4.

3.3 Rules and Models

In this section, we will briefly highlight some features of rules that emerge in the context of the models and frame theory for relevant logics. The issues of rules in relevant logics are subtle and interesting. Since we are working in the FMLA framework, some of the subtleties will not emerge for us, such as the distinction between *rules of proof*, which can only be applied to theorems, and *rules of inference*. Nonetheless, it will be useful to highlight some features of rules here.

DEFINITION 3.6 (Closure under rules) *The logic* L($\mathcal{C}$) *is* CLOSED UNDER THE RULE $A_1, \ldots, A_n \Rightarrow B$ *iff for all subsittutions* σ, *if* $\models_{\mathcal{C}} \sigma(A_1), \ldots, \models_{\mathcal{C}} \sigma(A_n)$, *then* $\models_{\mathcal{C}} \sigma(B)$.

We can say that the rule $A_1, \ldots, A_n \Rightarrow B$ PRESERVES VALIDITY in $\mathcal{C}$ iff L($\mathcal{C}$) is closed under the rule. This is to say that whenever instances of the premises are valid, an instance of the conclusion is valid as well. These rules are sometimes said to be ADMISSIBLE.

LEMMA 3.7 (Rule contraposition) *Suppose* $\mathsf{L}(\mathcal{C})$ *is closed under the rule* $A \Rightarrow B$. *Let* M *be a model with a point* $a \in N$ *such that* $a \not\Vdash B$. *Then* $a \not\Vdash A$.

Proof Suppose $\mathsf{L}(\mathcal{C})$ is closed under the rule $A \Rightarrow B$. Let M be a model built on a frame $F \in \mathcal{C}$ with a point $a \in N$ such that $a \not\Vdash B$. Suppose $a \Vdash A$. From the definition of being closed under a rule, B must hold at all regular points, which is a contradiction. Therefore, $a \not\Vdash A$. □

While this lemma is called "Rule contraposition," it should not be mistaken as establishing the rule $A \rightarrow \neg B \Rightarrow B \rightarrow \neg A$. It is, rather, a fact about contraposing in the metalanguage.

Brady (1993; 1994) discusses differences among rules in great detail. One important difference is the distinction between a rule preserving validity, as in definition 3.6, and a rule preserving verification at a point.

DEFINITION 3.8 (Preservation of verification in $\mathcal{C}$) *A rule* $A_1, \ldots, A_n \Rightarrow B$ *PRESERVES VERIFICATION IN MODELS BUILT ON FRAMES FROM* $\mathcal{C}$ *iff for all substitutions* σ, *for any model* M *built on a frame in* $\mathcal{C}$, *for any* $a \in K$, *if* $a \Vdash \sigma(A_1), \ldots, a \Vdash \sigma(A_n)$, *then* $a \Vdash \sigma(B)$.

One feature of rules that preserve verification is that they correspond to valid implications.

THEOREM 3.9 *The following are equivalent.*

- *The rule* $A_1, \ldots, A_n \Rightarrow B$ *preserves verification in models built on frames from* $\mathcal{C}$.
- $(A_1 \wedge \cdots \wedge A_n) \rightarrow B \in \mathsf{L}(\mathcal{C})$.

Proof Both directions are more or less immediate from the Verification lemma. □

Some rules, such as (R2) (adjunction), preserve verification at a point. Other rules need not do so in B-frames, such as (R1) (modus ponens). In some classes of frames, however, (R1) does preserve verification, such as the class of R-frames. One rule that does not preserve holding at a point, but does preserve validity for some classes of frames, is the rule (γ), $A, \neg A \vee B \Rightarrow B$. We will come back to discuss (γ) more in Chapter 9. For now, we turn to propositions in models.

3.4 Propositions

Formulas have meanings, relative to models. We will call these meanings PROPOSITIONS. A proposition is a set of points, but it isn't an *arbitrary* set of points. Rather, a proposition needs to respect the heredity ordering.

DEFINITION 3.10 (Hereditary set, proposition, truth set) *Let* F *be an* L*-frame and* M *be a model built on* F.

A set of points $X \subseteq K$ *is HEREDITARY in* F *iff for all* $a, b \in K$, *whenever* $a \in X$ *and* $a \leq b$, *then* $b \in X$.

Let $\mathcal{H}_F = \{X \in \wp(K) : X$ *is hereditary in* $F\}$.

A PROPOSITION $X \subseteq K$ is a hereditary set.

The TRUTH SET $|A|_M$ of a formula A in a model M is defined as $|A|_M = \{a \in K : a \Vdash A\}$.

We will drop the subscript on $\mathcal{H}$ when the frame is clear from context. Condition (iii) of definition 3.1 says that the set N is a proposition. It is the truth set of t, and it is the is a strongest logical proposition in the sense that it is the truth set of some theorem and it is contained in the truth sets of all theorems.

It is a useful fact that all truth sets are propositions, as we will prove.

LEMMA 3.11 *Let M be an L-model. For all formulas A, $|A|_M \in \mathcal{H}$.*

Proof This is immediate from lemma 3.3. □

As one might expect, there are some models in which some propositions are not truth sets. This will be an important point in some of the canonical model constructions of later chapters.

One can define some operations on propositions that yield new propositions.

LEMMA 3.12 *Let F be an L-frame and let X and Y be propositions. Then the following are propositions.*

(i) $X \cap Y$.
(ii) $X \cup Y$.
(iii) $X \Rrightarrow Y = \{a \in K : \forall b, c \in K(\text{if } Rabc \text{ and } b \in X, \text{ then } c \in Y)\}$.
(iv) $X \otimes Y = \{a \in K : \exists b, c \in K(Rbca \text{ and } b \in X \text{ and } c \in Y)\}$.
(v) $-X = \{a \in K : a^* \notin X\}$.

Proof Let F be an L-frame and let X and Y be propositions.

(i) Suppose $a \in X \cap Y$ and $a \leq b$. Then $a \in X$ and $a \in Y$. Since X and Y are propositions, $b \in X$ and $b \in Y$, so $b \in X \cap Y$.
(ii) Suppose $a \in X \cup Y$ and $a \leq b$. Then $a \in X$ or $a \in Y$. Since X and Y are propositions, $b \in X$ or $b \in Y$, so $b \in X \cup Y$.
(iii) Suppose $a \in X \Rrightarrow Y$ and $a \leq b$. So, for all $c, d \in K$, if $Racd$ and $c \in X$, then $d \in Y$. Suppose that $Rbcd$ and $c \in X$. By condition (vi) of definition 3.1, $Racd$. Therefore, $d \in Y$, so $b \in X \Rrightarrow Y$.
(iv) Suppose $a \in X \otimes Y$ and $a \leq b$. So, there are $c, d \in K$ such that $Rcda$, $c \in X$, and $d \in Y$. By condition (vi) of definition 3.1, $Rcdb$. Therefore, $b \in X \otimes Y$.
(v) Suppose $a \in -X$ and $a \leq b$. So, $a^* \notin X$. By condition (iv) of definition 3.1, $b^* \leq a^*$, so $b^* \notin X$, which suffices for $b \in -X$, as desired. □

There is a close connection between the operations of the previous lemma and truth sets for formulas.

LEMMA 3.13 *Let M be an L-model. Let A and B be formulas. Then the following identities hold.*

- $|A \wedge B|_M = |A|_M \cap |B|_M$.

- $|A \vee B|_M = |A|_M \cup |B|_M$.
- $|A \rightarrow B|_M = |A|_M \Rightarrow |B|_M$.
- $|A \circ B|_M = |A|_M \otimes |B|_M$.
- $|\neg A|_M = -|A|_M$.

Proof For the first identity, we reason as follows.

$$\begin{aligned} |A \wedge B|_M &= \{a \in K : a \Vdash A \text{ and } a \Vdash B\} \\ &= \{a \in K : a \in |A|_M \text{ and } a \in |B|_M\} \\ &= |A|_M \cap |B|_M. \end{aligned}$$

The second is similar.

$$\begin{aligned} |A \vee B|_M &= \{a \in K : a \Vdash A \text{ or } a \Vdash B\} \\ &= \{a \in K : a \in |A|_M \text{ or } a \in |B|_M\} \\ &= |A|_M \cup |B|_M. \end{aligned}$$

For the third identity, we reason as follows.

$$\begin{aligned} |A \rightarrow B|_M &= \{a \in K : a \Vdash A \rightarrow B\} \\ &= \{a \in K : \forall b, c \in K(Rabc \text{ and } b \Vdash A \Rightarrow c \Vdash B)\} \\ &= \{a \in K : \forall b, c \in K(Rabc \text{ and } b \in |A|_M \Rightarrow c \in |B|_M)\} \\ &= |A|_M \Rightarrow |B|_M. \end{aligned}$$

The fourth identity is verified by the following.

$$\begin{aligned} |A \circ B|_M &= \{a \in K : a \Vdash A \circ B\} \\ &= \{a \in K : \exists b, c \in K(Rbca \text{ and } b \Vdash A \text{ and } c \Vdash B)\} \\ &= \{a \in K : \exists b, c \in K(Rbca \text{ and } b \in |A|_M \text{ and } c \in |B|_M)\} \\ &= |A|_M \otimes |B|_M. \end{aligned}$$

The final identity is verified as follows.

$$\begin{aligned} |\neg A|_M &= \{a \in K : a \Vdash \neg A\} \\ &= \{a \in K : a^* \nVdash A\} \\ &= \{a \in K : a^* \notin |A|_M\} \\ &= -|A|_M. \end{aligned}$$

□

Models give a way to evaluate formulas recursively at points. We could, instead, use propositions and interpret formulas via propositions and operations on them. The preceding lemma tells us that these two ways of proceeding will agree. The latter approach has important connections to algebraic techniques and we will see it again when quantifiers are introduced, in particular when the Mares–Goldblatt semantics for quantifiers is introduced in Chapter 11. It will be useful to say a bit more about sets of propositions.

Suppose $\mathcal{X} \subseteq \mathcal{H}_F$ is a set of propositions on the frame F such that $N \in \mathcal{X}$ and if $X, Y \in \mathcal{X}$, then $X \cap Y, X \cup Y, -X, X \Rightarrow Y, X \otimes Y \in \mathcal{X}$. If M is a model such that $|p|_M \in \mathcal{X}$ for all $p \in \mathsf{At}$, then for all $A \in \mathfrak{L}$, $|A|_M \in \mathcal{X}$. This idea is important enough to make precise in a definition.

DEFINITION 3.14 (Model-like sets of propositions, admissible valuations) *Let F be a frame and let $\mathcal{X} \subseteq \mathcal{H}_F$. The set $\mathcal{X}$ is MODEL-LIKE iff all of the following hold.*

- $N \in \mathcal{X}$.
- *If* $X, Y \in \mathcal{X}$, *then* $X \cap Y \in \mathcal{X}$ *and* $X \cup Y \in \mathcal{X}$.
- *If* $X, Y \in \mathcal{X}$, *then* $-X \in \mathcal{X}$, $X \Rrightarrow Y \in \mathcal{X}$, *and* $X \otimes Y \in \mathcal{X}$.

Suppose $\mathcal{X} \subseteq \mathcal{H}_F$ is model-like. Then a valuation V is ADMISSIBLE with respect to $\mathcal{X}$ iff $|p|_{\langle F,V\rangle} \in \mathcal{X}$ for all $p \in$ At.

A model M is admissible with respect to $\mathcal{X}$ iff M is $\langle F, V\rangle$ and V is admissible with respect to $\mathcal{X}$.

LEMMA 3.15 *Let $\mathcal{X} \subseteq \mathcal{H}_F$ be model-like and V be admissible with respect to $\mathcal{X}$. Then for all $A \in \mathfrak{L}$, $|A| \in \mathcal{X}$.*

Proof The proof is by induction on the structure of A. The atomic base case is handled by the definition of admissibility, and the Ackermann constant base case is handled by the first condition on being model-like. The inductive cases are handled by the latter two conditions on being model-like together with lemma 3.13. □

One can always view a given frame F as accompanied by the full set of propositions, that is, $\mathcal{H}_F$. All valuations on a frame are admissible with respect to $\mathcal{H}_F$. Usually one does not need *all* of the propositions in $\mathcal{H}_F$. In fact, when dealing with quantifiers, it is important to be able to have a restricted set of propositions available. Fine (1989) showed that, using the Tarskian verification conditions for the quantifiers, R extended with quantifier axioms and rules is incomplete with respect to the validity defined in a way that uses all propositions. Mares and Goldblatt (2006) show that using an alternative verification condition, one motivated by relevance considerations, one can obtain a completeness result, provided that one can vary the propositions available over a base frame. If one requires that the full set of propositions is always available, then incompleteness returns. We will say more about this in Chapter 11.

In later chapters, we will deal with ternary relational frames equipped with a model-like set of propositions, Prop, as well as a set of propositional functions, about which more will be said later. Rather than considering all valuations on a frame, we will only consider valuations admissible with respect to Prop. It will be useful to define this notion here.

DEFINITION 3.16 (Propositional frames, model) *Let $F = \langle K, N, R, {}^*\rangle$ be an* L*-frame and let* Prop *be a model-like set of propositions on F. Then the pair $\langle F, \mathsf{Prop}\rangle$ is a propositional* L*-frame.*

A model M on a propositional L*-frame $\langle F, \mathsf{Prop}\rangle$ is a pair of $\langle\langle F, \mathsf{Prop}\rangle, V\rangle$ where V is a valuation admissible with respect to* Prop.

The definitions of counterexample, holding, and validity carry over in a straightforward way.

Propositional L-frames are akin to the general frames used in the study of TV-based modal logics.[9] General frames also come with a set of propositions, where models are constrained to assign atoms propositions from the given set. In the case of TV-based modal logics, any set of points is a proposition, which is not the case here.

We will not develop propositional L-frames more in this chapter. We will return to them in later chapters, as they will be useful in the study of modal logics and essential for the study of quantified logics.

3.5 Examples

We have introduced many logics and many classes of frames. In this section, we will look at a few example frames and see how they give rise to counterexamples to various principles. For most of the principles we will provide counterexamples to, the main connective is an implication. This means that we can use the Verification lemma to simplify the counterexample, providing a single point b where the antecedent holds but the consequent does not. When we use the Verification lemma in this way, b need not be in N, which is required for a counterexample. This is because the reasoning, when spelled out in full, is that for such a point b, we know that there is a $\exists x \in N$ such that $Rxbb$, and any witness to this existential quantifier is the counterexample point required. We won't explicitly appeal to the Verification lemma in the arguments, but keep in mind that it is being appealed to. These arguments are very different from showing that for some point b, A holds at b while B does not and concluding that $A \rightarrow B$ fails at b. This argumentation requires that $Rbbb$, which may not hold in a given model, and, if $b \notin N$, does not provide a counterexample to $A \rightarrow B$. Keeping in mind the warning about the Verification lemma, let us proceed to the example frames and counterexamples.

For the example of a B-frame, let $K = \{a,b,c,d\}$, $N = \{b,c\}$, and define the star and R as in Table 3.4.

This B-frame yields a counterexample to $(p \rightarrow \neg q) \rightarrow (q \rightarrow \neg p)$. Take a model M on this frame with $|p|_M = \{c\}$ and $|q|_M = \{b,c\}$. As the reader can work out, $a \Vdash p \rightarrow \neg q$. To show that $a \nVdash q \rightarrow \neg p$, we need to find a q-point that is related to a and a point where $\neg p$ fails. As $Rabd$ and $b \Vdash q$, it will be enough if $d \nVdash \neg p$. Since $d^* = c$ and $c \Vdash p$, we have $d \nVdash \neg p$, so $a \nVdash q \rightarrow \neg p$.

Table 3.4. B-frame

R	a	b	c	d	$*$
a	$abcd$	$abcd$	bc	$abcd$	b
b	abc	bc	c	$abcd$	a
c	c	c	c	$abcd$	d
d	$abcd$	$abcd$	$abcd$	$abcd$	c

[9] See Blackburn et al. (2002, ch. 5.5) for general frames in (TV-based) modal logics. See also Goldblatt (2011) for a recent treatment of general frames, there called admissible semantics.

Table 3.5. DW-frame

R	a	b	c	d	$*$
a	abc	bc	c	$abcd$	b
b	bc	c	c	$abcd$	a
c	c	c	c	$abcd$	d
d	$abcd$	$abcd$	$abcd$	$abcd$	c

Table 3.6. DJ-frame

R	a	b	c	d	$*$
a	$abcd$	$abcd$	abc	$abcd$	b
b	abc	bc	c	$abcd$	a
c	c	c	c	$abcd$	d
d	$abcd$	$abcd$	$abcd$	$abcd$	c

Table 3.7. TW-frame

R	a	b	c	d	$*$
a	abc	bc	c	$abcd$	b
b	abc	bc	c	$abcd$	a
c	c	c	c	$abcd$	d
d	$abcd$	$abcd$	$abcd$	$abcd$	c

For the example of a DW-frame, let $K = \{a,b,c,d\}$, $N = \{a,b,c\}$, and define the star and R as in Table 3.5.

For a counterexample to $((p \to q) \wedge (q \to r)) \to (p \to r)$, take the model M on this DW-frame with $|p|_M = \{a,b,c\}$, $|q|_M = \{b,c\}$, and $|r|_M = \{c\}$. It is relatively straightforward to see that $b \Vdash (p \to q) \wedge (q \to r)$. To show that $b \nVdash p \to r$, observe that $Rbab$, $a \Vdash p$, but $b \nVdash r$.

For the example of a DJ-frame, let $K = \{a,b,c,d\}$, $N = \{b,c\}$, with the star and R defined in Table 3.6. For a counterexample to $p \to (q \to q)$ on this DJ-frame, take a model M with $|p|_M = \{a,b,c\}$ and $|q|_M = \{b,c\}$. Since we have $a \Vdash p$, we need to show that $a \nVdash (q \to q)$. Since $Raba$ with $b \Vdash q$ but $a \nVdash q$, $a \nVdash q \to q$.

For the example of a TW-frame, let $K = \{a,b,c,d\}$, $N = \{b,c\}$, and define the star and R as in Table 3.7.

For a counterexample to $(p \to \neg p) \to \neg p$, take a model M based on this TW-frame with $|p|_M = \{b,c\}$. In this model, we have $a \Vdash p \to \neg p$, but we also have $a \nVdash \neg p$, as $a^* = b$ but $b \Vdash p$.

For the example of a RW-frame, let $K = \{a,b,c,d\}$, $N = \{a,b,c\}$, and define the star and R as given in Table 3.8.

For a counterexample to $(p \to (p \to q)) \to (p \to q)$, take a model M on the RW-frame with $|p|_M = \{b,c\}$ and $|q|_M = \{c\}$. It turns out that $a \Vdash p \to (p \to q)$, as the reader can verify, but $a \nVdash p \to q$. To show the latter, we have $Rabb$, $b \Vdash p$, but $b \nVdash q$. Therefore, $a \nVdash p \to q$, as desired.

For the example of a T-frame, let $K = \{a,b,c,d\}$, $N = \{b,c\}$, and define the star and R as in Table 3.9.

Table 3.8. RW-frame

R	a	b	c	d	$*$
a	abc	bc	c	$abcd$	b
b	bc	c	c	$abcd$	a
c	c	c	c	$abcd$	d
d	$abcd$	$abcd$	$abcd$	$abcd$	c

Table 3.9. T-frame

R	a	b	c	d	$*$
a	$abcd$	$abcd$	$abcd$	$abcd$	b
b	abc	bc	c	$abcd$	a
c	c	c	c	$abcd$	d
d	$abcd$	$abcd$	$abcd$	$abcd$	c

Table 3.10. R-frame

R	a	b	c	d	$*$
a	abc	abc	c	$abcd$	b
b	abc	bc	c	$abcd$	a
c	c	c	c	$abcd$	d
d	$abcd$	$abcd$	$abcd$	$abcd$	c

Table 3.11. E-frame

R	a	b	c	d	$*$
a	$abcd$	$abcd$	$abcd$	$abcd$	a
b	abc	bc	c	$abcd$	b
c	c	c	c	$abcd$	d
d	$abcd$	$abcd$	$abcd$	$abcd$	c

We can get a counterexample to $p \to ((p \to q) \to q)$ on this T-frame with a model M that has $|p|_M = \{a,b,c\}$ and $|q|_M = \{c\}$. By assumption, $a \Vdash p$ and $a \nVdash q$. To show that $a \nVdash (p \to q) \to q$, we will verify that it is the case that $c \Vdash p \to q$. If we consider the points in $|p|_M$ that occupy the second slot of the ternary relation with c in the first slot, we can see that these only relate to c, namely $Rcac$, $Rcbc$, and $Rccc$. Since $c \Vdash q$, we have $c \Vdash p \to q$. This fact together with the fact that $Raca$, however, suffices for $a \nVdash (p \to q) \to q$.

For the example of an R-frame, let $K = \{a,b,c,d\}$, $N = \{b,c\}$, and define the star and R as in Table 3.10.

The R-frame can provide a counterexample to $p \to (q \to p)$ by taking a model M with $|p|_M = \{b,c\}$ and $|q|_M = \{a,b,c\}$. As $Rbaa$, $a \Vdash q$, and $a \nVdash p$, $b \nVdash q \to p$. But, since $b \Vdash p$, we have a counterexample.

For the example of an E-frame, let $K = \{a,b,c,d\}$, $N = \{b,c\}$, and define the star and R as in Table 3.11.

This frame exhibits a failure of (FI5), as $Racb$ but not $Rcab$. The E-frame provides a counterexample to $p \to ((p \to q) \to q)$ by taking a model M with $|p|_M = \{a,b,c\}$

Table 3.12. C-frame

R	a	b	c	d	$*$
a	ac	bc	c	$abcd$	b
b	$abcd$	$abcd$	$abcd$	$abcd$	a
c	c	c	c	$abcd$	d
d	$abcd$	$abcd$	$abcd$	$abcd$	c

Table 3.13. Another R-frame

R	a	b	c	d	$*$
a	abc	abc	c	$abcd$	a
b	abc	abc	c	$abcd$	b
c	c	c	c	$abcd$	d
d	$abcd$	$abcd$	$abcd$	$abcd$	c

Table 3.14. Read's E-frame

R	a	b	c	d	e	$*$
a	a	b	c	d	be	a
b	a	b	c	d	be	e
c	abc	b	c	acd	abe	d
d	ad	b	acd	d	be	c
e	abc	be	c	acd	abe	b

and $|q|_M = \{c\}$. In this model, $c \Vdash p \rightarrow q$. To see this, consider any $x \in |p|_M$. If $Rcxy$, then $y = c$. As $c \Vdash q$, it follows that $c \Vdash p \rightarrow q$. We have, however, $Racb$ and $b \nVdash q$, so $a \nVdash (p \rightarrow q) \rightarrow q$. Since $a \Vdash p$, we can conclude that this model is a counterexample to $p \rightarrow ((p \rightarrow q) \rightarrow q)$.

For the example of a C-frame, let $K = \{a,b,c,d\}$, $N = \{a,c\}$, and define the star and R as in Table 3.12.

The C-frame supplies a counterexample to $p \vee \neg p$. Take a model M with $|p|_M = \{b,c\}$. Then we have a point $a \in N$ such that $a \nVdash p$ and $a \nVdash \neg p$. For the latter, note that $a^* = b$, but $b \Vdash p$.

Finally, here is another example of an R-frame, let $K = \{a,b,c,d\}$, $N = \{b,c\}$, and define the star and R as in Table 3.13.

The examples of frames so far are idiosyncratic in several ways. For example, they are all reduced, in the sense that their sets of regular points have a unique minimal point, which will be discussed in Section 3.6. As an example of a non-reduced frame, a frame where the set N does not have a unique minimal point, we can consider this E-frame, from Read (1988, 94), presented in Table 3.14. In this example, $N = \{a,b\}$.

The frames considered in this section have all been finite, and, in fact, rather small, based on no more than five points. One can, of course, find larger finite frames for each of the logics described earlier, and there are infinite frames. An example of an infinite L-frame is the canonical L-frame, to be constructed in Chapter 4. Nonetheless, this collection of frames should be of use to the reader in getting a feel for the different logics. Let us turn now to more on reduced frames.

3.6 Reduced Frames

One of the features of ternary relational frames is that there can be more than one regular point. Frames in which there is only one regular point have some attractions. Such frames are called *reduced*. If there is only one regular point, then one can view it as representing something like the "real" or "actual" world.[10] Additionally, the implications that hold in reduced models say more. Let us say that A ENTAILS B in a model M iff for all points $a \in K$, if $a \Vdash A$, then $a \Vdash B$. If $A \rightarrow B$ holds at the regular point in a reduced model, then A entails B in that model. The conditional expresses entailment in the model. By contrast, if $A \rightarrow B$ holds at a regular point in a non-reduced model, we cannot conclude that A entails B in that model. The reason is that there may be another regular point in the model, one at which $A \rightarrow B$ fails, which implies that A does *not* entail B in the model. One of the aims of relevant logic, at least in the early days, was to find an implication connective that expressed entailment in the object language, so the coincidence of implications that hold in a reduced model and entailment (in that model) could be viewed as an indication that the aim is being realized.[11]

The definition of a reduced model that we will adopt is slightly different than described earlier.

DEFINITION 3.17 (Reduced frames, models) *An* L*-frame F is REDUCED iff there is a unique minimal point $a \in N$ such that $N = \{b \in K : a \leq b\}$.*

An L*-model M is REDUCED iff it is built on a reduced frame.*

The minimal normal point in a reduced frame will be denoted by 0.

Where C is a class of L*-frames, $r(C)$ is the class of reduced frames in C.*

Other definitions of reduced frames either require that N be a singleton or they replace the set N with a single point, 0. The former definition is, generally, in conflict with condition (iii) on B-frames, which requires the set N be closed upwards under the heredity ordering. Both of the alternative definitions require that the verification condition for t be changed to the following to maintain the Heredity lemma.

- $a \Vdash \mathsf{t}$ iff $0 \leq a$.

In reduced frames, defined in the alternative ways listed earlier, one defines the heredity ordering $a \preceq b$ as $R0ab$. Definition 3.17 permits this to be done as well.

LEMMA 3.18 *Suppose $\langle K, N, R, ^* \rangle$ is a reduced* L*-frame. Let $\leq$ be the usual heredity ordering. Define $a \preceq b$ as $R0ab$. Then, $a \leq b$ iff $a \preceq b$. Further, if $Rabc$ and $d \preceq a$, $e \preceq b$, and $c \preceq f$, then $Rdef$.*

Proof To establish this, suppose $a \leq b$. Then, for some $x \in N$, $Rxab$. As $0 \leq x$, by the final condition on frames, $R0ab$, so $a \preceq b$.

Suppose $a \preceq b$. then $R0ab$. Since $0 \in N$, $a \leq b$.

[10] See for example, Routley and Meyer (1973). We will return to this topic in Chapter 8.

[11] Meyer (1971a).

To establish the further clause, we make the assumptions, apply the equivalence between $\leq$ and $\preceq$ repeatedly, and appeal to the final condition on frames. □

Thus, in a reduced frame, 0 does the job of the regular points. One could suggestively write reduced frames as $\langle K, [0), R, {}^* \rangle$ to emphasize the role of 0 with respect to the regular points.

Let us now turn to the issue of entailment in a model. We will show that holding at 0 in a model on a reduced frame is equivalent to entailment in the model.

LEMMA 3.19 *Let $\langle K, [0), R, {}^* \rangle$ be a reduced* L*-frame. Then the following are equivalent.*

- $0 \Vdash A \to B$.
- *For all $a \in K$, if $a \Vdash A$, then $a \Vdash B$.*

Proof Suppose that $0 \Vdash A \to B$. Suppose further that $a \Vdash A$. By lemma 3.18, $R0aa$, so $a \Vdash B$.

Suppose that for all $a \in K$, if $a \Vdash A$, then $a \Vdash B$. Suppose $0 \nVdash A \to B$, so there are points $b, c \in K$ such that $R0bc$, $b \Vdash A$, but $c \nVdash B$. By lemma 3.18, $R0bc$ implies $c \Vdash A$. By the initial assumption, it then follows that $c \Vdash B$, which is a contradiction. □

Holding at 0 is then equivalent to entailment in the model. The point 0 also has a special role with respect to counterexamples.

LEMMA 3.20 *Let $\langle K, N, R, {}^* \rangle$ be a reduced* L*-frame. Suppose A has a counterexample in a model built on this frame. Then $0 \nVdash A$.*

Proof Suppose for some $a \in N$, $a \nVdash A$. By definition, $0 \leq a$, so from $a \nVdash A$, it follows that $0 \nVdash A$. □

Given the distinctive role that 0 has, one might wonder why not just set the reduction condition to be that $N = \{0\}$? Indeed, this is the definition used by most logicians that have worked with reduced frames. It appears to fit better with their motivating comments that 0 somehow represents the real or actual world, although this is an issue that we will return to in Section 8.2.3. The primary difference, as far as we can tell, is that defining reduced frames in terms of a singleton set of regular points results in some additional complications in defining the canonical frames for Completeness. For some logics, those complications are required anyway, but for some logics they are not.

Since $r(\mathcal{C})$ is a subclass of $\mathcal{C}$, one would expect the logic of reduced frames satisfying some conditions to be, generally, stronger than logic of the frames satisfying those conditions. After all, there may be some counterexamples that depend on having more than one regular point available. In general, this is true. For some logics L, however, it turns out that the logic of L-frames coincides with the logic of reduced L-frames, that is, $\models_{\mathsf{L}} A$ iff $\models_{r(\mathsf{L})} A$. This is not the case for all logics that we are considering. For some of those logics, for example, BX, the logic of reduced frames for BX properly extends BX. Disjunction plays a pivotal role in this fact.

To see the issue, it will be useful to prove a theorem about disjunctive rules.

THEOREM 3.21 (Reduced frames and disjunctive rules) *If* $\mathsf{L}(\mathcal{C})$ *is closed under the rule* $A \Rightarrow B$, *then* $\mathsf{L}(r(\mathcal{C}))$ *is closed under the rule* $C \vee A \Rightarrow C \vee B$.

Proof Suppose $\mathsf{L}(\mathcal{C})$ is closed under the rule $A \Rightarrow B$. Suppose $C \vee A$ is valid in $\mathsf{L}(r(\mathcal{C}))$ but there is a countermodel to $C \vee B$ in $\mathsf{L}(r(\mathcal{C}))$. Then, by lemma 3.20, there is a model M built on a frame $\langle K, [0), R, ^* \rangle$ in $\mathcal{C}$ such that $0 \not\Vdash C \vee B$. So, $0 \not\Vdash C$ and $0 \not\Vdash B$. By lemma 3.7, $0 \not\Vdash A$. So, $0 \not\Vdash C \vee A$, contradicting an assumption. Therefore, there is no countermodel to $C \vee B$ in $\mathsf{L}(r(\mathcal{C}))$. □

Reduced frames are importantly connected to disjunctive rules of the sort in the previous theorem.[12]

COROLLARY 3.22 *The following formula schemes are valid in* $\mathsf{L}(r(\mathsf{BX}))$, *that is, the logic of the reduced* BX*-frames.*

- $\neg(A \to B) \vee (\neg B \to \neg A)$.
- $\neg(A \to B) \vee ((C \to A) \to (C \to B))$.
- $\neg(A \to B) \vee (\neg A \vee B)$.

Disjunctive rules can hold with respect to the reduced frames of a class, but they need not hold with respect to the class of frames itself, as illustrated by the following lemma.

LEMMA 3.23 *The following formulas are not theorems of* BX.

- $\neg(p \to q) \vee (\neg q \to \neg p)$.
- $\neg(p \to q) \vee ((r \to p) \to (r \to q))$.
- $\neg(p \to q) \vee (\neg p \vee q)$.

Proof We will provide a countermodel to the first and third, leaving a countermodel for the second to the reader.

The frame for the countermodel for the first will be $\langle K, N, R, ^* \rangle$, where $K = \{a, b, c\}$, $N = \{a, b\}$, and R and * are in Table 3.15. For the valuation, set $V(p) = \{c\}$ and $V(q) = \emptyset$. Then, we have $b^* = c \Vdash p \to q$, so $b \not\Vdash \neg(p \to q)$. It is also the case that $c \not\Vdash q$, so $b \Vdash \neg q$, while $c \Vdash p$ so $b \not\Vdash \neg p$. As $Rbbb$, $b \not\Vdash \neg q \to \neg p$

Table 3.15. Frame for counterexample to $\neg(p \to q) \vee (\neg q \to \neg p)$

	a	b	c	$*$
a	a	b	cb	a
b		b		c
c				b

[12] See the introduction of *Relevant Logics and their Rivals, Volume 2* by Brady (2003) for more on this point.

Table 3.16. Frame for counterexample to $\neg(p \rightarrow q) \vee (\neg p \vee q)$

	1	0	*
1	1 0		1
0			0

The frame for the countermodel for the third will be $\langle K, N, R, ^* \rangle$, where $K = \{0,1\} = N$ and R and * are in Table 3.16.[13] For the valuation, set $V(p) = \{0\}$ and $V(q) = \emptyset$. Then $0 \Vdash p \rightarrow q$, so $0 \nVdash \neg(p \rightarrow q)$, while both $0 \nVdash \neg p$ and $0 \nVdash q$. □

THEOREM 3.24 BX $\subsetneq$ L(r(BX)).

The logic of reduced frames for BX is properly stronger than the logic BX. BX is an example of a logic that is not prime, that is, a logic with a valid disjunction neither of whose disjuncts are valid. Reduced frames have some important surprising consequences when considering logics that are not prime. We will return to reduced frames once we have the machinery of Completeness and canonical frames on the table.

3.7 Appendix: Compatibility Frames for Negation

The Routley star, *, is used in ternary relational frames to interpret negation.[14] It has its proponents but it also has its detractors.[15] There have been some questions about how to interpret the star, philosophically.[16] The star also builds in a lot of formal assumptions, for example, that $a^{**} = a$ and that * is a function. One way of understanding what * is saying in a model is to consider the verification condition for negation. In a model, $a \Vdash \neg A$ iff $a^* \nVdash A$. So, $\neg A$ holds at a point just in the case that A does not hold at the star of that point. In other words, personifying the points somewhat, a point a can assert $\neg A$ just when its star-mate does not assert A. A point says as much as it can without contradicting what its star-mate says. Considerations along these lines have led some logicians to consider other ways of modeling negation.[17]

There are two primary approaches to modeling negation in relevant logics. One generalizes the Routley star. Instead of a function, one uses a binary relation C, sometimes described as a COMPATIBILITY relation. The move to a binary relation is natural since a unary function is easily understood as a binary relation obeying some additional constraints. This is the so-called AUSTRALIAN PLAN approach to negation.[18] On this approach, negation is a kind of negative modal operator. The other approach eschews the use of the Routley star and compatibility relations and uses both truth and

[13] This frame is provided by Slaney (1987a, 406). It satisfies further conditions beyond (FN3), such as (FI2), (FI3), (FI4), and (FN1).
[14] The Routley star goes back to the early work of Sylvan and Plumwood (1972).
[15] Copeland (1979; 1983).
[16] Restall (1999).
[17] See Goldblatt (1974), Routley (1984), Došen (1986; 1999), and Dunn (1993).
[18] See Restall (1999), Berto (2015), and Berto and Restall (2019).

falsity conditions for all connectives. Negation then toggles back and forth between true and false. This is the so-called AMERICAN PLAN approach to negation.[19] In certain contexts, the two approaches yield equivalent results. There are, however, some important differences between them, both formal and philosophical. In the book so far, we have adopted an Australian plan approach to negation, and we will present the compatibility relation generalization of it here.

With that background in place, let us define compatibility frames.[20]

DEFINITION 3.25 (Compatibility frame) *A COMPATIBILITY L-FRAME* $\langle K, N, R, C\rangle$ *is a quadruple where* $K \neq \emptyset, N \subseteq K, R \subseteq K \times K \times K, C \subseteq K \times K$, *and which obeys the following conditions for all* $a, b, c, d, e, f \in K$:

(i) $a \leq a$,
(ii) *if* $a \leq b$ *and* $b \leq c$ *then* $a \leq c$,
(iii) *if* $a \in N$ *and* $a \leq b$, *then* $b \in N$,
(iv) *if* aCb, $c \leq a$, *and* $d \leq b$, *then* cCd, *and*
(v) *if* $Rabc, d \leq a, e \leq b$, *and* $c \leq f$, *then* $Rdef$.

The definitions of model, counterexample, holding, and validity all carry over in a straightforward way. The verification condition for negation changes to the following.

- $a \Vdash \neg B$ iff for all $b \in K$ such that aCb, $b \nVdash A$.

For a negation to hold at a point, the negated formula must fail at all compatible points. The Routley star is a special case of this, when a is compatible with exactly one point.

Compatibility frames trade the two conditions on the Routley star for one condition governing the interaction of the heredity ordering and the compatibility relation. This is enough to prove that the Heredity lemma carries over to the present setting.

LEMMA 3.26 *Let* M *be a model on a compatibility frame. If* $a \Vdash A$ *and* $a \leq b$, *then* $b \Vdash A$.

Proof The only new case is negation. Suppose $a \Vdash \neg A$ and $a \leq b$. Suppose that $b \nVdash \neg A$. Then there is some $d \in K$ such that bCd and $d \Vdash A$. From the frame condition on compatibility and the assumption that $a \leq b$, aCd. Thus, $a \nVdash \neg A$, contradicting an assumption. Therefore, $b \Vdash \neg A$. □

The conditions on compatibility, so far, permit the case where a is compatible with b but b is not compatible with a. This opens the possibility of adding another negation, $-$, to the language with the verification condition

- $a \Vdash -A$ iff for all $b \in K$ such that bCa, $b \nVdash A$.

[19] See Routley (1984), De and Omori (2018), and Omori and De (2022).

[20] We note that condition (iv) is stronger than is strictly required. The stronger condition is satisfied by what Došen (1986) calls "strictly condensed frames." On this point, see also Onishi (2015). I would like to thank Masanobu Toyooka for pointing out these references.

Table 3.17. Conditions on compatibility

Axiom	Frame condition
$\neg A \leftrightarrow -A$	if xCy, then yCx
$\neg(A \wedge B) \rightarrow (\neg A \vee \neg B)$	if $(xCy \wedge xCz)$, then $\exists u(y \leq u$ and $z \leq u$ and $xCu)$
$\neg\neg A \rightarrow A$	$\forall x \exists y(xCy$ and $\forall z($if yCz, then $z \leq x))$
$(\neg A \wedge A) \rightarrow B$	xCx

The two negations are forced to coincide if symmetry is required, that is, if aCb then bCa. One could do something similar with the ternary relation and adding another conditional, $\leftarrow$, a point to which we will return in Section 10.2.

We close this section by listing some negation axioms and their compatibility frame conditions in Table 3.17. For further discussion and generalizations of compatibility frames, see Tabakci (2022).

3.8 Appendix: Semilattice and Operational Frames

Uruqhart (1972a; 1972b) presented operational models for a positive fragment of relevant logics, namely the $\{\rightarrow, \wedge, \vee\}$-fragment of $\mathfrak{L}$, although he did consider the addition of fusion. In particular, Urquhart's semilattice models provided a way to model R^+, and with the addition of a structure of possible worlds a way to model E^+.[21] The $\{\rightarrow\}$- and $\{\rightarrow, \wedge\}$-fragments of R are complete with respect to the semilattice models, but Completeness is lost with the addition of disjunction. A complete axiom system for the semilattice logic was developed by Fine (1976) and Charlwood (1978; 1981). Before getting into more details, let us present the semilattice models.

DEFINITION 3.27 (Semilattice frame) *A SEMILATTICE FRAME is a triple $\langle P, \sqcup, 0\rangle$, with $0 \in P$ and an operation $\sqcup \colon P \times P \mapsto P$ that obeys the following conditions, for all $x, y, z \in P$.*

(S1) $0 \sqcup x = x$.
(S2) $x \sqcup y = y \sqcup x$.
(S3) $x \sqcup (y \sqcup z) = (x \sqcup y) \sqcup z$.
(S4) $x \sqcup x = x$.

These conditions make $\sqcup$ a join operation in a semilattice. Other operational frames can be obtained from dropping some of the conditions (S2)–(S4). We get models by adding a valuation to a frame.

DEFINITION 3.28 (Semilattice model) *A SEMILATTICE MODEL is a pair of a semilattice frame $\langle P, \sqcup, 0\rangle$ together with a valuation $V \colon \mathsf{At} \mapsto \wp(P)$.*

A verification relation $\Vdash$ is a binary relation between points and formulas defined inductively as follows.

[21] For the remainder of this section, "semilattice" should be understood as meaning join semilattice.

Table 3.18. Examples of semilattices

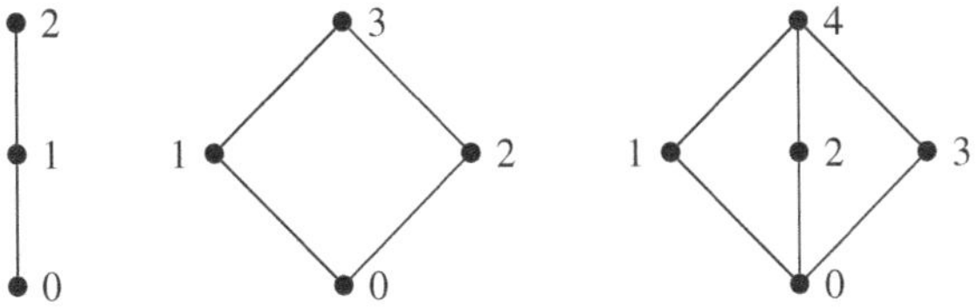

- $x \Vdash p$ *iff* $x \in V(p)$.
- $x \Vdash B \wedge C$ *iff* $x \Vdash B$ *and* $x \Vdash C$.
- $x \Vdash B \vee C$ *iff* $x \Vdash B$ *or* $x \Vdash C$.
- $x \Vdash B \to C$ *iff for all* $y \in P$, *if* $y \Vdash B$, *then* $x \sqcup y \Vdash C$.

DEFINITION 3.29 (Holds, validity)
A formula A HOLDS *in a semilattice model* $\langle P, \sqcup, 0, V\rangle$ *iff* $0 \Vdash A$.
A formula A is VALID *for semilattice frames iff A holds in all semilattice models.*
We will write $\models_{\mathsf{SL}} A$ *when A is valid for semilattice frames.*
The logic SL *is the set of all A such that* $\models_{\mathsf{SL}} A$.

There are two particularly pleasing features of semilattice frames: One is that it is rather easy to generate a semilattice frame, and the other is that they have nice diagrams, some examples of which can be found in Table 3.18. Let us look at two examples of semilattice models. For the first, we will use the leftmost frame from Table 3.18 to obtain a counterexample to $p \to (q \to p)$. Let $V(p) = \{1\}$ and $V(q) = \{2\}$. Since $1 \Vdash p$ and $0 \sqcup 1 = 1$, we need to verify that $1 \nVdash q \to p$. Note that $2 \Vdash q$, $1 \sqcup 2 = 2$, and $2 \nVdash p$. It then follows that $1 \nVdash q \to p$, which suffices for $0 \nVdash p \to (q \to p)$.

For the second, we will use the middle frame from Table 3.18 to obtain a counterexample to $p \to (q \to q)$. Let $V(p) = \{1\}$ and $V(q) = \{2\}$. To see $0 \nVdash p \to (q \to q)$, we need to check all points where p holds, in this case only 1. Since $0 \sqcup 1 = 1$, we need $1 \nVdash q \to q$. By assumption, $2 \Vdash q$ and $1 \sqcup 2 = 3$, but $3 \nVdash q$. Therefore, $1 \nVdash q \to q$, so $0 \nVdash p \to (q \to q)$.

The semilattice models obey a variant of the Verification lemma from Section 3.2.

LEMMA 3.30 (Verification lemma) *For all semilattice models* $\langle P, \sqcup, 0, V\rangle$ *and for all formulas* A, B, $0 \Vdash A \to B$ *iff for all* $a \in P$, $a \Vdash A$ *only if* $a \Vdash B$.

Proof For the left to right direction, suppose $0 \Vdash A \to B$ and $a \Vdash A$. By the verification condition for the conditional, $0 \sqcup a \Vdash B$, but as $0 \sqcup a = a$, $a \Vdash B$.

For the right to left direction, suppose for all $a \in P$, $a \Vdash A$ only if $a \Vdash B$ but $0 \nVdash A \to B$. Then there is a point $x \in P$ such that $x \Vdash A$ but $0 \sqcup x \nVdash B$, which implies $x \nVdash B$ by (S1). It follows from an assumption that $x \Vdash B$, but this is a contradiction. Therefore, $0 \Vdash A \to B$. □

It is worth noting a connection between semilattice frames and ternary relational frames for R. For every semilattice frame, one can define a corresponding ternary relational frame, without the * operation. The definition sets $K = P$, $N = \{0\}$, and $Rabc$ iff $a \sqcup b = c$. Similarly, if one has an R-frame where $Rabc$ and $Rabd$ implies $c = d$, then one can define a semilattice frame, using the preceding definition.

To return to an earlier point, there are formulas valid for semilattice frames that are not theorems of R.

LEMMA 3.31 *The following formula schemes are validities of* SL.

- $((A \to (B \vee C)) \wedge (B \to C)) \to (A \to C)$.
- $(A \to ((A \to A) \vee A)) \to (A \to A)$.

Proof For the first, suppose that there is a counterexample model $\langle P, \sqcup, 0, V\rangle$. So, for some $x \in P$, $x \Vdash (A \to (B \vee C)) \wedge (B \to C)$ but $x \nVdash A \to C$. So, there is $y \in P$ such that $y \Vdash A$ but $x \sqcup y \nVdash C$. From the assumption, $x \sqcup y \Vdash B \vee C$, so $x \sqcup y \Vdash B$ or $x \sqcup y \Vdash C$. From the former and the fact that $x \Vdash B \to C$, $(x \sqcup y) \sqcup y \Vdash C$. By (S3) and (S2), $(x \sqcup y) \sqcup y = x \sqcup y$, so we can conclude $x \sqcup y \Vdash C$, which is a contradiction.

For the second, suppose that there is a counterexample model $\langle P, \sqcup, 0, V\rangle$. So, for some $x \in P$, $x \Vdash A \to ((A \to A) \vee A)$ but $x \nVdash A \to A$. The latter implies that for some $y \in P$, $y \Vdash A$ but $x \sqcup y \nVdash A$. From the verification condition for the conditional, $x \sqcup y \Vdash (A \to A) \vee A$, so either $x \sqcup y \Vdash A \to A$ or $x \sqcup y \Vdash A$. In the first case, it follows that $(x \sqcup y) \sqcup y \Vdash A$, which by (S3) and (S2), implies $x \sqcup y \Vdash A$. We can then conclude $x \sqcup y \Vdash A$, which is a contradiction. □

LEMMA 3.32 *The following are not theorems of* R.

- $((p \to (q \vee r)) \wedge (q \to r)) \to (p \to r)$.
- $(p \to ((p \to p) \vee p)) \to (p \to p)$.

The proof is left to the reader. Algebraic counterexamples can be found using the matrix M_8, introduced in Section 5.4.

COROLLARY 3.33 R^+ *is a proper sublogic of* SL.

One can, following Urquhart, modify the semilattice frames to obtain frames that agree with the $\{\to\}$- and $\{\to, \wedge\}$-fragments of E and T as well. To obtain frames for E^+, one adds a set of worlds W to the frames along with a binary relation on worlds that is reflexive and transitive. Formulas are evaluated at pairs of points and worlds. To obtain operational frames for T^+, we add a transitive binary relation, $\preceq$, that obeys two conditions, for all $x, y, z \in P$:

- if $x \preceq y$, then $x \sqcup z \preceq y \sqcup z$, and
- $0 \preceq x$.

The verification condition for implication is then changed to the following.

- $x \Vdash B \to C$ iff for all $y \in P$, if $x \preceq y$ and $y \Vdash B$, then $x \sqcup y \Vdash C$.

The logics of the resulting operational frames, respectively, outstrip E^+ and T^+, with the earlier examples working in both cases. One can drop condition (S4) to obtain operational models on which RW^+ and TW^+ are sound. In both cases, with respect to the $\{\to\}$- and $\{\to, \wedge\}$-fragments, the logics of the operational models agree, respectively, with RW^+ and TW^+. It is unknown whether the addition of disjunction leads to a divergence in either case.

One route to operational models for contractionless logics is dropping condition (S4). There is another route, the disjoint models, developed by Giambrone et al. (1987). The disjoint models add to the semilattice frames an additional condition, where $x \leq y$ is defined as $x \sqcup y = y$,

(S5) $x \leq y \sqcup z$ only if there are $u, w \in P$ such that $u \sqcup w = x$, $u \leq y$, and $w \leq z$.

Two points x and y are disjoint, Jxy, iff for all $z \in P$, $z \leq x$ and $z \leq y$ only if $z = 0$. The verification condition for the conditional is then modified to use disjointness.

- $x \Vdash B \to C$ iff for all $y \in P$, if Jxy and $y \Vdash B$, then $x \sqcup y \Vdash C$.

Giambrone et al. (1987) show that the two classes of models, operational models without (S4) and disjoint models, validate the same logic.

While the semilattice logic outstrips R^+, a modification of the semilattice frames does agree with R^+. Humberstone (1988) adds an additional operation, $+$, to the frames, subject to some conditions. This operation is used to interpret disjunction, with the verification condition

- $x \Vdash B \vee C$ iff there are $y, z \in P$ such that $x = y + z$, $y \Vdash B$, and $z \Vdash C$.

The set of formulas valid on these frames is R^+, as shown by Humberstone.

Weiss (2019) showed that SL has the variable-sharing property. The argument uses the structure of prime numbers to SL enjoys the variable-sharing property, and Weiss (2020a) further exploits the structure of primes to obtain a characteristic frame for SL.

The semilattice logic has nice models. While the axiomatic systems provided by Fine and Charlwood include an unwieldy additional rule, the logic can be given well-behaved natural deduction and sequent systems. Charlwood (1978) provides an indexed natural deduction system for the logic and proves a Normalization theorem for it, which is crucial for his metatheoretic results.[22] One particularly pleasant feature of Charlwood's natural deduction system is that, unlike Anderson and Belnap's systems, no additional rule of distribution is required, as it follows from the usual rules for conjunction and disjunction. Giambrone and Urquhart (1987) and Kashima (2003) provide indexed sequent systems for the logic and prove cut elimination. Kashima's sequent system is used to provide an alternative axiomatic system for SL. Standefer (2022b) provides an overview of recent work on semilattice and operational models.

We will close this section by noting a recently obtained solution to an open problem. A logic is UNDECIDABLE when there is no effective procedure for determining whether a given formula is a theorem of the logic. Urquhart (1984) showed that R is undecidable, a result that extends to R^+. Urquhart (2016a) noted that the decidability of SL was then still open, but Knudstorp (2024) has shown that it is undecidable.

[22] See Prawitz (1965), Mancosu et al. (2021), or Restall and Standefer (2023b), among others, for details on techniques for proving Normalization.

3.9 Appendix: Fine Frames

Around the same time that the work of Sylvan and Meyer on ternary relational models was published, Fine (1974) presented different models for relevant logics. While there are clear connections between the two approaches, there are substantial differences. It will be useful to have these models in view because Fine (1988) presented a quantificational extension of them that permits a proof of Completeness and, due to Mares (1992c), there are extensions of those models with the machinery for interpreting identity.

DEFINITION 3.34 (Fine frame) *A Fine* B*-frame is a sextuple* $\langle T, S, 1, -, \cdot, \leq \rangle$ *such that the following conditions hold.*

(i) $T \neq \emptyset$, $S \subseteq T$, $1 \in T$, $\cdot : T \times T \mapsto T$, $- : S \mapsto S$, $\leq$ *is a reflexive, transitive, antisymmetric binary relation on* T.[23]
(ii) $\forall t,u,v \in T,\ u \leq t \Rightarrow (u \cdot v) \leq (t \cdot v)$.
(iii) $\forall a \in S, \forall t,u \in T((t \cdot u) \leq a \Rightarrow \exists b,c \in S(t \leq b \wedge u \leq c \wedge (b \cdot u) \leq a) \wedge (t \cdot c) \leq a)$.
(iv) $\forall t \in T,\ t \leq 1 \cdot t$.
(v) $\forall t \in T,\ 1 \cdot t \leq t$.
(vi) $\forall a \in S,\ --a = a$.
(vii) $\forall a,b \in S(a \leq b \Rightarrow -b \leq -a)$.

A Fine B*-model adds to a frame* F *a valuation* V: $\mathsf{At} \times T \mapsto \{0,1\}$ *such that*

(viii) $\forall t \in T \forall p \in \mathsf{At}(V(p,t) = 1$ *iff* $\forall a \in S(t \leq a \Rightarrow V(p,a) = 1))$.
Such a model is said to be built on F.

Some comments are in order. The set T is the set of points, which Fine calls *theories*, although these are not taken to be theories in the sense of Chapter 2. They are, rather, left as primitive, with Fine describing them as "sets of propositions closed under commitment."[24] The set S is the set of *saturated* theories, which settle disjunctive claims. The point 1 is the base point of the model. The $-$ operation is used to interpret negation. The relation $\leq$ is a containment relation, here taken as primitive rather than defined, as there is no ternary relation to use in defining it.

Fine's original presentation added the condition that for all $a \in S$, $1 \leq a \Rightarrow -a \leq a$. This condition validates excluded middle, yielding the logic here called BX, but the condition is not needed in general.

Following Fine, we will use s,t,u,v for points in T and a,b,c for points in S. Instead of writing $(t \cdot u)$, we will follow Fine in writing (tu). The verification conditions for the connectives are listed in Table 3.19.[25] Fine defines disjunction, $B \vee C$, as $\neg(\neg B \wedge \neg C)$. The verification condition for disjunction then comes out as a lemma, after

[23] Let $\leq$ be a binary relation on a set X. Then we can define the following properties of $\leq$.
- $\leq$ is reflexive iff for all $x \in X$, $x \leq x$.
- $\leq$ is transitive iff for all $x,y,z \in X$, if $x \leq y$ and $y \leq z$, then $x \leq z$.
- $\leq$ is antisymmetric iff for all $x,y \in X$, if $x \leq y$ and $y \leq x$, then $x = y$.

[24] Fine (1974, 348)

[25] I thank Shay Logan for suggesting the verification condition on fusion.

Table 3.19. Verification conditions

$t \Vdash p$	iff	$V(p,t) = 1$
$t \Vdash \neg B$	iff	$\forall a \in S(\text{if } t \leq a, \text{ then } -a \not\Vdash B)$
$t \Vdash B \wedge C$	iff	$t \Vdash B$ and $t \Vdash C$
$t \Vdash B \vee C$	iff	$\forall a \in S$, if $t \leq a$, then either $a \Vdash B$ or $a \Vdash C$
$t \Vdash B \rightarrow C$	iff	$\forall s \in T(s \Vdash B \Rightarrow ts \Vdash C)$
$t \Vdash B \circ C$	iff	$\forall a \in S \exists u, v \in T(t \leq a, uv \leq a, u \Vdash B$, and $v \Vdash C)$

Table 3.20. Conditions on Fine frames

	Axiom	Frame condition
(I1)	$A \wedge (A \rightarrow B) \rightarrow B$	$tt \leq t$
(I2)	$(A \rightarrow B) \wedge (B \rightarrow C) \rightarrow (A \rightarrow C)$	$t(tu) \leq tu$
(I3)	$(A \rightarrow B) \rightarrow ((C \rightarrow A) \rightarrow (C \rightarrow B))$	$t(uv) \leq (tu)v$
(I4)	$(A \rightarrow B) \rightarrow ((B \rightarrow C) \rightarrow (A \rightarrow C))$	$t(uv) \leq (ut)v$
(I5)	$A \rightarrow ((A \rightarrow B) \rightarrow B)$	$tu \leq ut$
(I7)	$(A \rightarrow (A \rightarrow B)) \rightarrow (A \rightarrow B)$	$(tu)u \leq tu$
(N1)	$(A \rightarrow \neg B) \rightarrow (B \rightarrow \neg A)$	$tb \leq a \Rightarrow t(-a) \leq (-b)$
(N2)	$(A \rightarrow \neg A) \rightarrow \neg A$	$a(-a) \leq a$
(N3)	$A \vee \neg A$	$-a \leq a$

unpacking the definitions. Nonetheless, we list it here along with the other conditions for illustrative purposes.

DEFINITION 3.35 (Holding, validity) *A formula A holds in a model* $\langle F, V \rangle$ *iff* $1 \Vdash A$.
A formula A is valid on a frame F iff A holds on all models built on F.
A formula A is valid in a class of frames $\mathcal{C}$ *iff A is valid on F for each* $F \in \mathcal{C}$.

Following Fine (1974), we note an important feature of these models to demonstrate, relating the sets S and T.

THEOREM 3.36 *Let M be a* B*-model. Then* $t \Vdash A$ *iff* $\forall a \in S(t \leq a \Rightarrow a \Vdash A)$.

Proof The proof is by induction on the construction of A. The details are left to the reader. □

The preceding theorem says that the saturated points are, in a sense, evenly distributed above the non-saturated points. Although Fine models contain many points that are not saturated, one can recover them from looking only at the saturated points containing them. This theorem is fundamental for the development of the theory of Fine models. We won't pursue the details of that theory here, for which the reader can consult Fine (1974), Bimbó and Dunn (2023), or Logan (2024).

As with ternary relational frames, we can give frames for stronger logics by imposing frame conditions. We will list some of the conditions for axioms in Table 3.20, following the convention that $a, b \in S$ and $t, u, v \in T$.

The reader may wonder about the relationship between Fine's frames and the ternary relational frames studied in most of this book. There are many apparent differences, some noted by Sylvan and Meyer as well as by Fine, such as the use of operations as opposed to relations and the use of non-prime points. We will note that one can render $uv \leq x$, in Fine's frames, as $Ruvx$ in the ternary relational setting. Further, it can be useful to consider all L-theories, not just the prime ones, in the development of the canonical frame construction for the ternary relational frames, which was done by Sylvan and Meyer in their early work.

Punčochár (2020) proved a duality result formally connecting classes of Fine frames with classes of ternary relational frames. Punčochár suggests that the Fine frames are more suited to an informational interpretation, while the ternary relational frames are more suited to an ontic interpretation. The philosophical interpretations and the formal connections between the two kinds of frames are further explored by Tedder (2023).

3.10 Exercises

(1) Use some of the frames from this chapter to create counterexamples to the following TV theorems.
  (i) $\neg(p \wedge \neg p)$
  (ii) $q \rightarrow (p \vee \neg p)$
  (iii) $\neg p \rightarrow (p \rightarrow r)$
  (iv) $p \rightarrow (q \rightarrow q)$
  (v) $p \rightarrow (q \rightarrow p)$
  (vi) $p \rightarrow (p \rightarrow p)$
  (vii) $(p \rightarrow p) \rightarrow \mathrm{t}$

(2) Use some of the frames from this chapter to create counterexamples to the following formulas, which are theorems of some relevant logics.
  (i) $(p \wedge (p \rightarrow q)) \rightarrow q$
  (ii) $((p \rightarrow p) \rightarrow q) \rightarrow q$
  (iii) $q \vee \neg q$
  (iv) $p \rightarrow ((p \rightarrow q) \rightarrow q)$
  (v) $((p \rightarrow q) \wedge (q \rightarrow r)) \rightarrow (p \rightarrow r)$

(3) Show that the following identities among propositions hold for any ternary relational model M and any formulas A and B.
  (i) $|\neg\neg A|_M = |A|_M$
  (ii) $|\neg(A \wedge B)|_M = |\neg A \vee \neg B|_M$
  (iii) $|\neg(A \vee B)|_M = |\neg A \wedge \neg B|_M$

(4) Show that C is closed under the disjunctive versions of (R1)–(R5). Conclude this holds for T, E, and R as well.

(5) Show that L(BX) is not closed under the disjunctive version of (R1), $C \vee A, C \vee (A \rightarrow B) \Rightarrow C \vee B$. Show directly that for any class of frames $\mathcal{C}$, L(r(C)) is closed under the disjunctive version of (R1).

(6) Given two frames $F = \langle K_1, N_1, R_1, {}^{*1} \rangle$ and $G = \langle K_2, N_2, R_2, {}^{*2} \rangle$, the PRODUCT $F \times G = \langle K, N, R, {}^* \rangle$ is defined as follows.

- $K = K_1 \times K_2$

- $N = N_1 \times N_2$
- $R\langle a_1, a_2\rangle\langle b_1, b_2\rangle\langle c_1, c_2\rangle$ iff $R_1 a_1 b_1 c_1$ and $R_2 a_2 b_2 c_2$
- $\langle a_1, a_2\rangle^* = \langle a_1^{*1}, a_2^{*2}\rangle$

  Prove that if F and G are B-frames, then so is the product, $F \times G$.

(7) Prove that if F and G both satisfy a frame condition among (FI1)–(FI8), (FPI1)–(FPI2), and (FN1)–(FN3), then so does $F \times G$.

(8) Define two frames F and G that both satisfy (FI9) but whose product $F \times G$ does not.

(9) Let $\mathcal{C}$ be a class of L-frames closed under products. Let $A \vee B$ be such that both $A \vee B \in \mathsf{L}(\mathcal{C})$, A and B do not share an propositional variables, and $A \vee B$ does not contain t. Prove that either $A \in \mathsf{L}(\mathcal{C})$ or $B \in \mathsf{L}(\mathcal{C})$.[26]

[26] This is Halldén completeness, also known as Halldén reasonableness. We will return to it in Section 7.11.

4 Metatheory: Soundness and Completeness

4.1 Introduction

We have a definition of L-provability and we have a definition of L-validity. In this chapter, we will show that for the logics we are considering, these two concepts coincide. In other words, we will show both

- if $\vdash_{\mathsf{L}} A$, then $\models_{\mathsf{L}} A$, and
- if $\models_{\mathsf{L}} A$, then $\vdash_{\mathsf{L}} A$.

The first is the Soundness of the logic with respect to validity over the class of L-frames. The second is the Completeness of the logic with respect to validity over the class of L-frames. We will prove both in this chapter. They are proved in more or less standard ways. Soundness is proved by an induction on the structure of proofs. Completeness is proved contrapositively by a canonical model construction, in the manner of Henkin.[1] Despite the techniques being fairly standard, going through the details in these cases will be instructive for seeing how the moving parts of the frames and the logics fit together. It will, additionally, simplify things when we move to less familiar settings later in the book, namely modal and quantified extensions of relevant logics.

We will begin in Section 4.2 with proving Soundness for B and then showing how to prove it for the extensions. Then, in Section 4.3, we move to proving Completeness. The structure will be the same, namely proving it for B and then showing how to adapt the proof for stronger logics. Finally, in Section 4.4, we briefly look at how the proof can be adapted for proving Completeness with respect to reduced frames to obtain a stronger result.

4.2 Soundness

In this section, we will show that if A is a theorem of L, then A is valid in all L-frames.

Soundness If $\vdash_{\mathsf{L}} A$, then $\models_{\mathsf{L}} A$.

[1] See Blackburn et al. (2002) for examples of canonical model constructions in (TV-based) modal logic.

Table 4.1. Axioms and rules for B

(A1)	$A \to A$	(R1)	$A, A \to B \Rightarrow B$
(A2)	$(A \wedge B) \to A, (A \wedge B) \to B$	(R2)	$A, B \Rightarrow A \wedge B$
(A3)	$((A \to B) \wedge (A \to C)) \to (A \to (B \wedge C))$	(R3)	$A \to \neg B \Rightarrow B \to \neg A$
(A4)	$A \to (A \vee B), A \to (B \vee A)$	(R4)	$A \to B \Rightarrow (B \to C) \to (A \to C)$
(A5)	$((B \to A) \wedge (C \to A)) \to ((B \vee C) \to A)$	(R5)	$A \to B \Rightarrow (C \to A) \to (C \to B)$
(A6)	$(A \wedge (B \vee C)) \to ((A \wedge B) \vee (A \wedge C))$	(R6)	$A \to (B \to C) \Rightarrow (A \circ B) \to C$
(A7)	$\neg\neg A \to A$	(R7)	$(A \circ B) \to C \Rightarrow A \to (B \to C)$
(A8)	t	(R8)	$A \Rightarrow \mathsf{t} \to A$

The proof is by induction on the construction of a proof of A. This induction breaks down into several cases. For the base cases, we will show that all the axioms are valid. The inductive steps show that all the rules preserve the property of being valid.

We will break the proof into cases. We will start with the axioms of B, and then cover the rules of B. Then we will consider the various axioms and their corresponding frame conditions. For ease of reference, Table 4.1 presents the axioms and rules for B.

LEMMA 4.1 *The axioms of* B *are valid in the class of* B*-frames.*

Proof Let M be an arbitrary B-model and $a \in K$ an arbitrary point. The proof breaks into several cases. Almost every case will use the Verification lemma, so we will not cite it in each case.

(A1) Suppose $a \Vdash A$. Then $a \Vdash A$, so $A \to A$ holds in M.

(A2) Suppose $a \Vdash A \wedge B$. Then $a \Vdash A$ and $a \Vdash B$. This establishes that both $(A \wedge B) \to A$ and $(A \wedge B) \to B$ hold in M.

(A3) Suppose $a \Vdash (A \to B) \wedge (A \to C)$. This implies $a \Vdash A \to B$ and $a \Vdash A \to C$. Suppose that $b, c \in K$ are such that $Rabc$ and $b \Vdash A$. Therefore, $c \Vdash B$ and $c \Vdash C$, so $c \Vdash B \wedge C$. So, $a \Vdash A \to (B \wedge C)$. This suffices to show that $((A \to B) \wedge (A \to C)) \to (A \to (B \wedge C))$ holds.

(A4) Suppose $a \Vdash A$. Then $a \Vdash A$ or $a \Vdash B$. So, $A \to (A \vee B)$ and $A \to (B \vee A)$ both hold in M.

(A5) Suppose $a \Vdash (B \to A) \wedge (C \to A)$. So, $a \Vdash B \to A$ and $a \Vdash C \to A$. Suppose $b, c \in K$ are such that $Rabc$ and $b \Vdash B \vee C$. Then either $b \Vdash B$ or $b \Vdash C$. If $b \Vdash B$, then $b \Vdash A$, and similarly if $b \Vdash C$. So, $c \Vdash A$. Therefore, $a \Vdash (B \vee C) \to A$. This suffices to show that $((B \to A) \wedge (C \to A)) \to ((B \vee C) \to A)$ holds in M.

(A6) Suppose $a \Vdash A \wedge (B \vee C)$. Then $a \Vdash A$ and $a \Vdash B \vee C$. So, $a \Vdash B$ or $a \Vdash C$. In the first case, $a \Vdash A \wedge B$, so $a \Vdash A \wedge B$ or $a \Vdash A \wedge C$. Therefore, $a \Vdash (A \wedge B) \vee (A \wedge C)$. In the second case, $a \Vdash A \wedge C$, so $a \Vdash A \wedge B$ or $a \Vdash A \wedge C$. Therefore, $a \Vdash (A \wedge B) \vee (A \wedge C)$. This suffices to establish that $(A \wedge (B \vee C)) \to ((A \wedge B) \vee (A \wedge C))$ holds in M.

(A7) Suppose $a \Vdash \neg\neg A$. Then $a^* \nVdash \neg A$, so $a^{**} \Vdash A$. As $a^{**} = a$, $a \Vdash A$. Thus, $\neg\neg A \to A$ holds in M.

(A8) This is the only case that does not use the Verification lemma. To show that t holds in M, one must show $a \Vdash \mathsf{t}$ for all $a \in N$. This is guaranteed by the verification condition for t. □

LEMMA 4.2 *If the premises of a rule of* B *are all valid, then the conclusion of the rule is valid.*

Proof (R1) Suppose $A \rightarrow B$ and A are both valid. Let M be an arbitrary model, and let $a \in N$ be arbitrary. By assumption, $a \Vdash A$ and $A \rightarrow B$ holds in M. So, by the Verification lemma, $a \Vdash B$, so B holds in M. Thus B is valid.

(R2) Suppose A and B are both valid. Let M be an arbitrary model, and let $a \in N$ be arbitrary. By assumption, $a \Vdash A$ and $a \Vdash B$. Therefore, $a \Vdash A \wedge B$, so $A \wedge B$ holds in M and so is valid.

(R3) Suppose $A \rightarrow \neg B$ is valid. Let M be an arbitrary model. Suppose $b \Vdash B$, so $b^* \nVdash \neg B$. Thus, $b^* \nVdash A$. So, $b \Vdash \neg A$. Thus, $B \rightarrow \neg A$ holds in M.

(R4) Suppose $A \rightarrow B$ is valid. Let M be an arbitrary model. Suppose $b \Vdash B \rightarrow C$. Suppose $c, d \in K$ are such that $Rbcd$ and $c \Vdash A$. From the assumption, $c \Vdash B$. So, $d \Vdash C$. Thus, $(B \rightarrow C) \rightarrow (A \rightarrow C)$ holds in M.

(R5) Suppose $A \rightarrow B$ is valid. Let M be an arbitrary model. Suppose $b \Vdash C \rightarrow A$. Suppose $c, d \in K$ are such that $Rbcd$ and $c \Vdash C$. From the assumptions, $d \Vdash A$, and then by the initial assumption, $d \Vdash B$. So, $b \Vdash C \rightarrow B$. Therefore, $(C \rightarrow A) \rightarrow (C \rightarrow B)$ holds in M.

(R6) Suppose $A \rightarrow (B \rightarrow C)$ is valid. Let M be an arbitrary model. Suppose $b \Vdash A \circ B$. Then there are points $c, d \in K$ such that $Rcdb, c \Vdash A$, and $d \Vdash B$. It follows that $c \Vdash B \rightarrow C$, whence $b \Vdash C$. Therefore, $(A \circ B) \rightarrow C$ holds in M.

(R7) Suppose $(A \circ B) \rightarrow C$ is valid. Let M be an arbitrary model. Suppose $b \Vdash A$. Let $c, d \in K$ be $Rbcd$ and $c \Vdash B$. From the assumptions, $d \Vdash A \circ B$. Therefore, $d \Vdash C$, so $A \rightarrow (B \rightarrow C)$ holds in M.

(R8) Suppose A is valid. Let M be an arbitrary model. Suppose $b \Vdash \mathsf{t}$. Then, $b \in N$. Since A is valid, $b \Vdash A$, so $\mathsf{t} \rightarrow A$ holds in M. □

Since we will discuss axioms and their corresponding frame conditions in what follows, it will be useful to have that information readily available. It can be found repeated in Table 4.2.[2]

The first place we will connect axioms and frame conditions is in the next Soundness lemma.

LEMMA 4.3 *The axiom* (Cn) *is valid in the class of frames obeying the condition* (FCn).

Proof For this proof, we will show that if the class of frames obeys condition (FCn), then axiom (Cn) is valid.

(I1) Let M be a model whose frame obeys condition (FI1). Suppose $a \Vdash A \wedge (A \rightarrow B)$. Therefore, $a \Vdash A$ and $a \Vdash A \rightarrow B$. As $Raaa$, $a \Vdash B$.

(I2) Let M be a model whose frame obeys condition (FI2). Suppose $a \Vdash (A \rightarrow B) \wedge (B \rightarrow C)$. So, $a \Vdash A \rightarrow B$ and $a \Vdash B \rightarrow C$. Suppose for $b, c \in K$, $Rabc$ and $b \Vdash A$. By (FI2), $Ra(ab)c$, so for some x, $Raxc$ and $Rabx$, so $x \Vdash B$. It follows that $c \Vdash C$, so $a \Vdash A \rightarrow C$.

[2] We will use "⇒" for the metalinguistic conditional in tables to save space.

Table 4.2. Axioms and their frame conditions

	Axioms		Conditions
(I1)	$(A \wedge (A \rightarrow B)) \rightarrow B$	(FI1)	$Raaa$
(I2)	$((A \rightarrow B) \wedge (B \rightarrow C)) \rightarrow (A \rightarrow C)$	(FI2)	$Rabc \Rightarrow Ra(ab)c$
(I3)	$(A \rightarrow B) \rightarrow ((C \rightarrow A) \rightarrow (C \rightarrow B))$	(FI3)	$Rabcd \Rightarrow Ra(bc)d$
(I4)	$(A \rightarrow B) \rightarrow ((B \rightarrow C) \rightarrow (A \rightarrow C))$	(FI4)	$Rabcd \Rightarrow Rb(ac)d$
(I5)	$A \rightarrow ((A \rightarrow B) \rightarrow B)$	(FI5)	$Rabc \Rightarrow Rbac$
(I6)	$(A \rightarrow (B \rightarrow C)) \rightarrow (B \rightarrow (A \rightarrow C))$	(FI6)	$Rabcd \Rightarrow Racbd$
(I7)	$(A \rightarrow (A \rightarrow B)) \rightarrow (A \rightarrow B)$	(FI7)	$Rabc \Rightarrow Rabbc$
(I8)	$(\mathrm{t} \rightarrow A) \rightarrow A$	(FI8)	$\exists x \in N, Raxa$
(I9)	$A \rightarrow (A \rightarrow A)$	(FI9)	$Rabc \Rightarrow a \leq c$ or $b \leq c$
(N1)	$(A \rightarrow \neg B) \rightarrow (B \rightarrow \neg A)$	(FN1)	$Rabc \Rightarrow Rac^*b^*$
(N2)	$(A \rightarrow \neg A) \rightarrow \neg A$	(FN2)	Raa^*a
(N3)	$A \vee \neg A$	(FN3)	$\forall a \in N, a^* \leq a$

(I3) Let M be a model whose frame obeys condition (FI3). Suppose $a \Vdash A \rightarrow B$. Let $b, e \in K$ be such that $Rabe$ and $b \Vdash C \rightarrow A$. Suppose $c, d \in K$ are such that $Recd$ and $c \Vdash C$. As $Rabcd$, by condition (FI3), $Ra(bc)d$, so for some x, $Raxd$ and $Rbcx$. From the assumptions, $x \Vdash A$, so $d \Vdash B$. Therefore, $e \Vdash C \rightarrow B$, so $a \Vdash (C \rightarrow A) \rightarrow (C \rightarrow B)$.

(I4) Let M be a model whose frame obeys condition (FI4). Suppose $a \Vdash A \rightarrow B$. Let $b, e \in K$ be such that $Rabe$ and $b \Vdash B \rightarrow C$. Suppose $c, d \in K$ are such that $Recd$ and $c \Vdash A$. As $Rabcd$, by condition (FI4), $Rb(ac)d$, so for some x, $Rbxd$ and $Racx$. From the assumptions, $x \Vdash B$, so $d \Vdash C$. Therefore, $e \Vdash A \rightarrow C$, so $a \Vdash (B \rightarrow C) \rightarrow (A \rightarrow C)$.

(I5) Let M be a model whose frame obeys condition (FI5). Suppose $a \Vdash A$. Let $b, c \in K$ be such that $Rabc$ and $b \Vdash A \rightarrow B$. By (FI5), $Rbac$, so $c \Vdash B$. Therefore, $a \Vdash (A \rightarrow B) \rightarrow B$.

(I6) Let M be a model whose frame obeys condition (FI6). Suppose $a \Vdash A \rightarrow (B \rightarrow C)$. Let $b, e \in K$ be such that $Rabe$ and $b \Vdash B$. Let $c, d \in K$ be such that $Recd$ and $c \Vdash A$. As $Rabcd$, by (FI6), $Racbd$, so there is some x such that $Racx$ and $Rxbd$. From the assumptions, $x \Vdash B \rightarrow C$, and so $d \Vdash C$. Therefore, $e \Vdash A \rightarrow C$, so $a \Vdash B \rightarrow (A \rightarrow C)$.

(I7) Let M be a model whose frame obeys condition (FI7). Suppose $a \Vdash A \rightarrow (A \rightarrow B)$ and let $b, c \in K$ be such that $Rabc$ and $b \Vdash A$. By (FI7), $Rabbc$, and so there is an x such that $Rabx$ and $Rxbc$. From the assumptions, $x \Vdash A \rightarrow B$, so $c \Vdash B$.

(I8) Let M be a model whose frame obeys condition (FI8). Suppose $a \Vdash \mathrm{t} \rightarrow A$. By the frame condition (FI8), there is an $x \in N$ such that $Raxa$. Since $x \in N$, $x \Vdash \mathrm{t}$. As $Raxa$, it follows that $a \Vdash A$, which we wanted to prove.

(I9) Let M be a model whose frame obeys condition (FI9). Suppose $a \Vdash A$. We want to show that $a \Vdash A \rightarrow A$, so suppose that $Rabc$ and $b \Vdash A$. By frame condition (FI9), either $a \leq c$ or $b \leq c$. In both cases, $c \Vdash A$, which suffices for what we wanted to show.

(N1) Let M be a model whose frame obeys condition (FN1). Suppose $a \Vdash A \rightarrow \neg B$. Let $b, c \in K$ be such that $Rabc$ and $b \Vdash B$. Then, by (FN1), Rac^*b^*. Suppose $c^* \Vdash A$. Then $b^* \Vdash \neg B$, so $b \nVdash B$, which contradicts an assumption. Therefore, $c^* \nVdash A$, so $c \Vdash \neg A$. Therefore, $a \Vdash B \rightarrow \neg A$.

(N2) Let M be a model whose frame obeys condition (FN2). Suppose $a \Vdash A \rightarrow \neg A$. Suppose $a^* \Vdash A$. By (FN2), Raa^*a, $a \Vdash \neg A$, so $a^* \nVdash A$, which contradicts our assumption. Therefore, $a^* \nVdash A$, so $a \Vdash \neg A$, as desired.

(N3) Let M be a model whose frame obeys condition (FN3). Suppose $a \in N$. Either $a \Vdash A$ or $a \nVdash A$. Suppose $a \nVdash A$, so $a^* \Vdash \neg A$. By the Heredity lemma and (FN3), $a \Vdash \neg A$, so $a \Vdash A \vee \neg A$. If $a \Vdash A$ then $a \Vdash A \vee \neg A$. So, $a \Vdash A \vee \neg A$.

□

The preceding lemmas suffice to establish the Soundness of L with respect to L-frames.

THEOREM 4.4 (Soundness) *If* $\vdash_{\mathsf{L}} A$, *then* $\models_{\mathsf{L}} A$.

We will now move to the next major metatheoretical result, Completeness.

4.3 Completeness

Completeness of a logic with respect to a class of frames says that whenever a formula is valid in that class, then it is provable in that logic. The strategy for proving Completeness is to proceed contrapositively. We assume that a formula is not provable and show how to construct a countermodel that is built on a frame in that class. This will be done via a canonical model construction.

There are many steps in the construction of the canonical model and in the verification that it works as intended. The details of the proof will be instructive, particularly regarding the role of theories in relevant logics, as well as the importance of primeness. Primeness is important as a key part of the Truth lemma below, showing that the canonical model supports the verification condition for disjunction. In the context of canonical model constructions for, say, modal logics over TV, maximal, consistent theories are used. The use of such theories ensures primeness, as TV has excluded middle as a theorem. In the context of relevant logics, excluded middle may not be a theorem, and consistency of a theory is not guaranteed. Therefore, a more direct route to the primeness of theories is used.

For the construction, we will work with the logic L and leave the 'L' subscript off throughout to reduce clutter.

DEFINITION 4.5 (Canonical L-frame) *Define the* CANONICAL L-FRAME $\langle K, N, R, ^* \rangle$ *as follows.*

- *K is the set of all prime* L*-theories.*
- *$a \in N$ iff* $\mathsf{t} \in a$.
- *$Rabc$ iff for all formulas $B, C \in \mathfrak{L}$, if $B \rightarrow C \in a$ and $B \in b$, then $C \in c$.*
- *$a^* = \{A \in \mathfrak{L} : \neg A \notin a\}$.*

The bullet point defining R can be rewritten, using the concepts of Chapter 2, as $a \bullet b \subseteq c$. We'll now prove that equivalence.

LEMMA 4.6 *Let $\langle K, N, R, {}^* \rangle$ be the canonical* L*-frame. Then, for all $a, b, c \in K$, $Rabc$ iff $a \bullet b \subseteq c$.*

Proof For the left to right direction, suppose $Rabc$. Suppose $C \in a \bullet b$. By theorem 2.30, $a \bullet b = a \star b$, so, for some B, $B \rightarrow C \in a$ and $B \in b$. From the definition of R, $C \in c$, so $a \bullet b \subseteq c$.

For the right to left direction, suppose $a \bullet b \subseteq c$. Suppose $B \rightarrow C \in a$ and $B \in b$. By theorem 2.30, it then follows that $C \in a \bullet b$, so it follows that $C \in c$, as desired. □

Note that neither the definition of R nor the proof of lemma 4.6 relies, in an essential way, on the primeness of the points. One could take the set K' of all L-theories and define a relation R' over that set and prove a corresponding lemma relating R' and $\bullet$. Indeed, some work on relevant logics proceeds to prove Completeness in this way. In some ways, it is more natural to consider the ternary relation over all L-theories, rather than just the prime ones, a fact also exploited by Fine in his operational frames. If one goes that route, using the Squeezing lemma appropriately permits one move from R' to R, as required for our canonical frame construction.

It will be useful to record a few facts about the star operation.

LEMMA 4.7 *The following hold of the canonical frame defined earlier for all $a \in K$.*

- *$A \in a$ iff $\neg A \notin a^*$.*
- *$\neg A \in a$ iff $A \notin a^*$.*

Proof For the first, suppose $A \in a$. Then, $\neg\neg A \in a$, so $\neg A \notin a^*$. Next, suppose $\neg A \notin a^*$. Then, by definition, $\neg\neg A \in a$, so $A \in a$.

For the second, suppose $\neg A \in a$. Then, $A \notin a^*$. Next, $A \notin a^*$. By definition, $\neg A \in a$. □

We must show that the canonical frame is, in fact, a frame.

LEMMA 4.8 (Heredity is inclusion) *In the canonical frame, $a \leq b$ iff $a \subseteq b$.*

Proof Suppose that $a \leq b$. Then, there is an $x \in N$ such that $Rxab$. Since $\mathsf{t} \in x$, $A \rightarrow A \in x$, for all $A \in a$. Therefore, $A \in b$, so $a \subseteq b$.

Suppose that $a \subseteq b$. First, we show that $\mathsf{L} \bullet a \subseteq b$. If a is the empty theory, then we are done, so we assume $B \in a$, for some B. Suppose not, so there is some $B \rightarrow C \in \mathsf{L}$ such that $B \in a$ but $C \notin b$. Since $a \subseteq b$, $B \in b$. As b is an L-theory, $C \in b$, contradicting an assumption. Therefore, $\mathsf{L} \bullet a \subseteq b$. As a and b are prime theories, we can apply lemma 2.39 to obtain a prime theory x such that $\mathsf{L} \subseteq x$ and $x \bullet a \subseteq b$. Since $\mathsf{L} \subseteq x$, $\mathsf{t} \in x$, so $x \in N$. Thus, $a \leq b$. □

Next, we will show that the canonical frame satisfies the conditions to be a B-frame.

LEMMA 4.9 *The canonical frame obeys the following for all* $a, b, c, d, e, f \in K$.

(i) $a \leq a$.
(ii) *If* $a \leq b$ *and* $b \leq c$ *then* $a \leq c$.
(iii) *If* $a \in N$ *and* $a \leq b$, *then* $b \in N$.
(iv) *If* $a \leq b$, *then* $b^* \leq a^*$.
(v) $a^{**} = a$.
(vi) *If* $Rabc, d \leq a, e \leq b$, *and* $c \leq f$, *then* $Rdef$.

Proof (i) Since $a \subseteq a$, $a \leq a$ by lemma 4.8.
(ii) Suppose $a \leq b$ and $b \leq c$. By lemma 4.8, $a \subseteq b$ and $b \subseteq c$, so $a \subseteq c$. Applying the lemma again yields $a \leq c$.
(iii) Suppose $a \in N$ and $a \leq b$. Then $\mathsf{t} \in a$. By lemma 4.8 and the second assumption, $\mathsf{t} \in b$, so $b \in N$.
(iv) Suppose $a \leq b$, so by lemma 4.8, $a \subseteq b$. Suppose $B \in b^*$. Then $\neg B \notin b$, so $\neg B \notin a$. Therefore $B \in a^*$.
(v) Suppose $B \in a$. This is equivalent to $\neg\neg B \in a$, which is the case iff $\neg B \notin a^*$. This, in turn, is equivalent to $B \in a^{**}$. Thus, $a^{**} = a$.
(vi) Suppose $Rabc, d \leq a, e \leq b$, and $c \leq f$. Using lemma 4.8 repeatedly with the definition of R yields $Rdef$, as desired. □

Thus, the canonical L-frame is a B-frame. For logics stronger than B, there is more yet to verify. For each of the axioms from Table 4.2 included in L, we need to verify that the L-frame obeys the corresponding condition.

LEMMA 4.10 *If the logic* L *contains the axiom* (Cn), *then the canonical* L*-frame obeys the condition* (FCn).

Proof There are several cases, one for each additional axiom.

(I1) Let L contain (I1). To show that $Raaa$, suppose $A \to B \in a$ and $A \in a$. Then $A \wedge (A \to B) \in a$, as a is a theory. Since L has (I1) as a theorem, $B \in a$, so $Raaa$.
(I2) Let L contain (I2). Suppose $Rabc$. To show $Ra(ab)c$, let $x = a \bullet b$. Suppose it is not the case that $a \bullet x \subseteq c$, so there are some formulas B and C such that $B \to C \in a$, $B \in x$, but $C \notin c$. Since $B \in x$, there is some formula D such that $D \to B \in a$ and $D \in b$. But then, $(D \to B) \wedge (B \to C) \in a$. From (I2), $D \to C \in a$, so $C \in c$, contradicting an assumption. So, $x \subseteq c$. Using lemma 2.39, there is a prime theory y such that $x \subseteq y$ and $a \bullet y \subseteq c$. By lemma 4.6, $Rayc$ and $Raby$.
(I3) Let L contain (I3). Suppose $Rabcd$. To show $Ra(bc)d$, let $x = b \bullet c$. Suppose it is not the case that $a \bullet x \subseteq d$. Then there are some formulas A and B such that $A \to B \in a$, $A \in x$, but $B \notin d$. By (I3) and the fact that a is an L-theory, $(C \to A) \to (C \to B) \in a$. Since $A \in x$, there is a formula C such that $C \to A \in b$ and $C \in c$. Since b is an L-theory, $C \to B \in z$, for some $z \in K$ such that $Rabz$ and $Rzcd$. Since $C \to B \in z$ and $C \in c$, $B \in d$, contradicting an assumption. Therefore $a \bullet x \subseteq d$. Using lemma 2.39, there is a prime theory y such that $x \subseteq y$ and $a \bullet x \subseteq d$. By lemma 4.6, it follows that $Rayd$ and $Rbcy$.

(I4) Let L contain (I4). Suppose $Rabcd$. To show $Rb(ac)d$, let $x = a \bullet c$. Suppose it is not the case that $b \bullet x \subseteq d$. Then there are some formulas B and C such that $B \to C \in b$, $B \in x$, but $C \notin d$. Since, $B \in x$, there is a formula A such that $A \to B \in a$ and $A \in c$. By (I4) and the fact that a is an L-theory, $(B \to C) \to (A \to C) \in a$. Since for some $z \in K$, $Rabz$, and $Rzcd$, $A \to C \in z$. As $A \in c$, $C \in d$, contradicting an assumption. Therefore, $b \bullet x \subseteq d$. As in the previous cases, we can use lemma 2.39 to obtain a prime theory y such that $x \subseteq y$ and $b \bullet x \subseteq d$, whence by lemma 4.6 we have that $Rbyd$ and $Racy$.

(I5) Let L contain (I5). Suppose $Rabc$. To show $Rbac$, suppose $A \to B \in b$ and $A \in a$. Since a is an L-theory, $(A \to B) \to B \in a$. As $Rabc$, $B \in c$, as desired.

(I6) Let L contain (I6). Suppose $Rabcd$. To show $Racbd$, let $x = a \bullet c$ and suppose it is not the case that $x \bullet b \subseteq d$. Then there are formulas B and C such that $B \to C \in x$, $B \in b$, but $C \notin d$. Since $B \to C \in x$, there is a formula A such that $A \to (B \to C) \in a$ and $A \in c$. Since a is an L-theory, $B \to (A \to C) \in a$. Since $Rabz$ and $Rzcd$ for some $z \in K$, $A \to C \in z$, and so $C \in d$, contradicting an assumption. Therefore, $x \bullet b \subseteq d$. As before, using lemma 2.39 we obtain a prime theory y such that $x \subseteq y$ and $x \bullet b \subseteq d$, and then by lemma 4.6 we have that $Racy$ and $Rybd$.

(I7) Let L contain (I7). Suppose $Rabc$. To show $Rabbc$, let $x = a \bullet b$. Suppose that it is not the case that $x \bullet b \subseteq c$. Then there are formulas A and B such that $A \to B \in x$, $A \in b$ but $B \notin c$. Since $A \to B \in x$, there is a formula C such that $C \to (A \to B) \in a$ and $C \in b$. Since b is an L-theory, $C \wedge A \in b$. Using lemma 2.38, we get $C \wedge A \to (C \wedge A \to B) \in a$, whence by axiom (I7), we get $C \wedge A \to B \in a$. Therefore, as $Rabc$, $B \in c$, contradicting an assumption. So, $x \bullet b \subseteq c$. As before, we use lemma 2.39 to obtain a prime theory y such that $y \bullet b \subseteq c$, and then we use lemma 4.6 to conclude $Raby$ and $Rybc$.

(I8) Let L contain (I8). Let $a \in K$. We want to show that $a \bullet \mathsf{L} \subseteq a$. Take $B \to C \in a$ where $B \in \mathsf{L}$. Since L contains (I8), it is closed under the rule $B \Rightarrow (B \to C) \to C$. Therefore $(B \to C) \to C \in \mathsf{L}$. Since a is a theory, it follows that $C \in a$, establishing $a \bullet \mathsf{L} \subseteq a$. We can then apply lemma 2.39 to obtain a prime theory $b \supseteq \mathsf{L}$ such that $Raba$. By construction, $b \in N$, so $\exists x \in N$, $Raxa$, as desired.

(I9) Let L contain (I9). Suppose that $Rabc$. Suppose that neither $a \leq c$ nor $b \leq c$. So there are formulas A and B such that $A \in a$, $A \notin c$, $B \in b$, and $B \notin c$. It follows that $A \circ B \in c$. It is a theorem of L that $(A \circ B) \to (A \vee B)$, as $(A \circ B) \to ((A \vee B) \circ (A \vee B))$ is a theorem of B and $(C \circ C) \to C$ is entailed by (I9). Since c is prime, either $A \in c$ or $B \in c$, contradicting the assumption. Therefore either $a \leq c$ or $b \leq c$.[3]

(N1) Let L contain (N1). Suppose $Rabc$. To show Rac^*b^*, suppose formulas A and B are such that $A \to B \in a$, $A \in c^*$, but $B \notin b^*$. Since $A \in c^*$, $\neg A \notin c$, and since $B \notin b^*$, $\neg B \in b$. As $A \to B \in a$, $A \to \neg\neg B \in a$ as well. Therefore, from axiom (N1), $\neg B \to \neg A \in a$, so $\neg A \in c$, which contradicts an assumption. Therefore, Rac^*b^*, as desired.

[3] This argument is due to Dunn (1979a).

(N2) Let L contain (N2). For this, we will use the equivalent axiom $(A \to B) \to (\neg A \vee B)$. To show Raa^*a, suppose $A \to B \in a$, $A \in a^*$, but $B \notin a$. Since $A \in a^*$, $\neg A \notin a^{**}$, so $\neg A \notin a$. From the equivalent form of (N2), $\neg A \vee B \in a$. Since a is prime, either $\neg A \in a$ or $B \in a$, so $B \in a$, contradicting an assumption. Therefore, Raa^*a.

(N3) Let L contain (N3). Suppose $a \in N$. From axiom (N3) and the assumption that $a \in N$, for all formulas A, $A \vee \neg A \in a$. Since a is prime, either $A \in a$ or $\neg A \in a$. Suppose there is some formula B such that $B \in a^*$ but $B \notin a$. Since $B \vee \neg B \in a$ and a is prime, it follows that $\neg B \in a$. So $B \notin a^*$, contradicting an assumption. Therefore, $a^* \subseteq a$, so by lemma 4.8, $a^* \leq a$. □

The preceding lemma together with lemma 4.3 supply the content to the claim that a particular frame condition corresponds with a particular axiom. The axiom (B) is valid in the class of frames obeying the condition (FB), and the canonical frame for a logic with (B) obeys the frame condition (FB). The significance of the latter result will not be fully established until we have shown that the canonical model is a model.

DEFINITION 4.11 (Canonical model) *Given the canonical* L*-frame* F_{L}*, the canonical valuation* V *is defined as* $V(p,a) = 1$ *iff* $p \in a$*, for all* $p \in$ At *and* $a \in K$.

The canonical L*-model is* $\langle F_{\mathsf{L}}, V\rangle$.

We must show that the canonical model is, in fact, a model. We must do this because we are showing that the logic L is complete with respect to the class of frames satisfying the conditions corresponding to the axioms of L, or, contrapositively, if a formula is not provable in L, then there is a counterexample in the appropriate class of frames. The canonical model needs to really be a model, in addition to satisfying the appropriate conditions, in order to provide the required counterexample. Verifying that the canonical model is, in fact, a model is done in lemma 4.13, which shows that the verification conditions for the connectives interact properly with the structure of theories in the canonical frame. In order to prove that lemma, we need the following lemma, which shows that there are always appropriate theories witnessing certain existential quantifiers used in the verification conditions.

LEMMA 4.12 (Witness lemma) *Let* $\langle F_{\mathsf{L}}, V\rangle$ *be the canonical* L*-model, and let* $a \in K$.

If $B \to C \notin a$*, then there are* $b, c \in K$ *such that* $Rabc$*,* $B \in b$*, and* $C \notin c$.

If $B \circ C \in a$*, then there are* $b, c \in K$ *such that* $Rbca$*,* $B \in b$*, and* $C \in c$.

Proof For the first claim, suppose $B \to C \notin a$. We construct c as follows. Let

$$z = \{D \in \mathfrak{L} : \ B \to D \in a\}.$$

Then $(z, \{C\})$ is disjoint. To see this, suppose that it is not, so there is $E \in \bigwedge z$ such that $\vdash_{\mathsf{L}} E \to C$. As $B \to E \in a$, by definition of z and the fact that a is a theory, $B \to C \in a$, contradicting an assumption. So, $(z, \{C\})$ is disjoint. We can then extend z to a prime theory c such that $z \subseteq c$ and $C \notin c$.

To construct b, let

$$y = \{D \in \mathfrak{L} : \ \exists E \notin c\, D \to E \in a\}.$$

The pair $(\{B\}, y)$ is disjoint. To see this, suppose it is not, so there is $F_1 \vee \cdots \vee F_n \in \bigvee y$ such that $\vdash_{\mathsf{L}} B \rightarrow (F_1 \vee \cdots \vee F_n)$. The disjunction in the consequent is of the form $F_1 \vee \cdots \vee F_n$ for some $n \geq 1$, so there are $G_1, \ldots, G_n \notin c$ such that $F_i \rightarrow G_i \in a$, from the definition of y. It follows that $\bigvee_i F_i \rightarrow \bigvee_i G_i \in a$. It follows from the definition of c that $\bigvee_i G_i \in c$. As c is prime, $G_j \in c$, for some $j \in [1, n]$, but this is a contradiction. Therefore, $(\{B\}, y)$ is a disjoint pair, and we can extend $\{B\}$ to a prime theory $b \supseteq \{B\}$ such that (b, y) is disjoint.

Next, we show that $Rabc$. Suppose $D \rightarrow E \in a$ and $D \in b$. Suppose that $E \notin c$. Then $D \in y$, which contradicts the fact that (b, y) is a disjoint pair. Therefore, $E \in c$.

For the second claim, suppose $B \circ C \in a$. Let $x = \theta_{\mathsf{L}}(\{B\})$ and $y = \theta_{\mathsf{L}}(\{C\})$. To show that $Rabc$, we argue by induction on the construction of $\theta_{\mathsf{L}}(x \odot y)$. Suppose $D \circ E \in x \odot y$. So, $\vdash_{\mathsf{L}} B \rightarrow D$ and $\vdash_{\mathsf{L}} C \rightarrow E$.[4] But then, as $B \circ C \in a$, $D \circ E \in a$. Next suppose $D \wedge E \in \theta_{\mathsf{L}}(x \odot y)$, so $D \in \theta_{\mathsf{L}}(x \odot y)$ and $E \in \theta_{\mathsf{L}}(x \odot y)$. By the inductive hypothesis, $D \in a$ and $E \in a$. As a is a theory, $D \wedge E \in a$. Finally, suppose $D \in \theta_{\mathsf{L}}(x \odot y)$ and $\vdash_{\mathsf{L}} D \rightarrow E$. By the inductive hypothesis, $D \in a$, so, as a is a theory, $E \in a$. Therefore, $Rabc$.

Using lemma 2.39, we can extend x to a prime theory b and y to a prime theory c such that $Rbca$, with $B \in b$ and $C \in c$. □

With the Witness lemma proven, we can proceed to show that in a canonical model, holding at a theory is equivalent to being a member of that theory.

LEMMA 4.13 (Truth lemma) *For all formulas A and all points $a \in K$, $A \in a$ iff $a \Vdash A$.*

Proof The proof is by induction on the construction of A.

- Base case: A is p. This holds by definition of V.
- Base case: A is t. $\mathsf{t} \in a$ iff $a \in N$ iff $a \Vdash \mathsf{t}$.
- Inductive case: A is $B \wedge C$. $B \wedge C \in a$ iff $B \in a$ and $C \in a$. By the inductive hypothesis, $a \Vdash B$ and $a \Vdash C$. Therefore, $a \Vdash B \wedge C$. The converse is similar.
- Inductive case: A is $B \vee C$. $B \vee C \in a$ only if $B \in a$ or $C \in a$, as a is prime. By the inductive hypothesis, $a \Vdash B$ or $a \Vdash C$, so $a \Vdash B \vee C$. The converse is similar.
- Inductive case: A is $\neg B$. But, $\neg B \in a$ iff $B \notin a^*$. By the inductive hypothesis, this is equivalent to $a^* \nVdash B$, which is equivalent to $a \Vdash \neg B$.
- Inductive case: A is $B \rightarrow C$. Suppose $B \rightarrow C \in a$. Suppose that $b, c \in K$ are such that $Rabc$ and $b \Vdash B$. By the inductive hypothesis, $B \in b$, so by definition of R, $C \in c$. By the inductive hypothesis, $c \Vdash C$, so $a \Vdash B \rightarrow C$.

  Next, suppose $B \rightarrow C \notin a$. Then by lemma 4.12, there are $b, c \in K$ such that $Rabc$, $B \in b$, and $C \notin c$. By the inductive hypothesis, $b \Vdash B$ and $c \nVdash C$, so $a \nVdash B \rightarrow C$.

[4] See definition 2.29 for the definition of $x \odot y$.

- Inductive case: A is $B \circ C$. Suppose $B \circ C \in a$. By lemma 4.12, there are $b, c \in K$ such that $Rbca$, $B \in b$, and $C \in c$. By the inductive hypothesis, $b \Vdash B$ and $c \Vdash C$. Therefore, $a \Vdash B \circ C$.

  Suppose $a \Vdash B \circ C$. Then there are $b, c \in K$ such that $Rbca$, $b \Vdash B$, and $c \Vdash C$. By the inductive hypothesis, $B \in b$ and $C \in c$. As $B \rightarrow (C \rightarrow (B \circ C))$ is a theorem, $C \rightarrow (B \circ C) \in b$. As $Rbca$, $B \circ C \in a$, as desired. □

As one can see from the details of the proof, lemma 4.12 was used in two places, the subcase where an implication is not in a theory and the subcase where a fusion is in a theory. The corresponding verification conditions for these cases use existential quantifiers, and lemma 4.12 ensures that there are appropriate prime theories witnessing those quantifiers.

Next, we need to show that there are sufficiently many L-regular theories. Each non-theorem of L is excluded from some L-regular theory. The L-canonical model will serve as a counterexample for each non-theorem of L, provided there are appropriate theories to act as counterexamples.

LEMMA 4.14 (Existence lemma) *Let $\langle F_{\mathsf{L}}, V \rangle$ be the canonical* L*-model. Let A be a non-theorem of* L*. Then there is some $a \in N$ such that $A \notin a$.*

Proof Let A be a non-theorem of L. This means that $(\mathsf{L}, \{A\})$ is a disjoint pair. Therefore, we can extend L to a prime theory a such that $A \notin a$. □

While not every logic is prime, some are. The logic R is an example of a logic that is not prime, and B is an example of a prime logic, a fact that we will establish in Chapter 9. If the logic is prime, the logic itself can serve as the L-regular prime theory excluding A. In fact, the logic can serve as the desired theory for any non-theorem. For a non-prime logic L, such as R, different non-theorems may require different L-regular theories to witness their failure.

Now that we have all the pieces on the table, we can prove Completeness.

THEOREM 4.15 (Completeness) *If $\models_{\mathsf{L}} A$, then $\vdash_{\mathsf{L}} A$.*

Proof Suppose it is not the case that $\vdash_{\mathsf{L}} A$. Then by lemma 4.14, there is, in the canonical L-model $\langle F_{\mathsf{L}}, V \rangle$, $a \in N$ such that $A \notin a$. By lemma 4.13, $a \nVdash A$. Therefore, it is not the case that $\models_{\mathsf{L}} A$. Contraposing establishes that if $\models_{\mathsf{L}} A$, then $\vdash_{\mathsf{L}} A$. □

Thus, for a wide range of logics, their non-theorems have a counterexample in an appropriate class of frames. In later chapters, we will see how to extend Completeness results to modal and quantificational extensions of relevant logics. Before we proceed to the next chapter, we will look at what needs to change in the completeness proof in order to obtain Completeness with respect to *reduced frames*. This material is not required for the later chapters.

4.4 Completeness for Reduced Frames

In Chapter 3, we defined reduced frames as frames whose set of regular points is generated by a single point. The Soundness of L with respect to the class of L-frames suffices for the Soundness with respect to the class of reduced L-frames. Completeness, however, is another matter. We saw that there are logics that are complete with respect to their frames but not with respect to their reduced frames.

The techniques for proving Completeness with respect to reduced frames are interesting and worth examining in detail. Routley et al. (1982, 328ff.) showed how to prove Completeness for many logics that are not prime, the weakest of which is the logic C. Slaney (1987a) extended this reduced modeling to weaker logics, logics that are required to be prime. Slaney uses the technique of metavaluations, about which we will say more in Chapter 9. For present purposes, it will suffice to leave those details as a black box, relying only on the fact that they show that some logics are prime.

For the remainder of this section, we will drop t and $\circ$ from the language, so the associated axioms and rules will also be dropped. This section will follow the approach of Routley et al. (1982, 306ff.) and restrict attention to logics extending C, that is, those logics containing (I1), (I3), (I4), and (N1).

Rather than simply use L-theories, we will use T-L-theories, as defined in definition 2.42. Recall that T-L-theories differ from L-theories in requiring closure under not just the provable implications of L but also the implications contained in T. The definitions of special kinds of L-theories, such as regular or prime, carry over to the T-L-theories.

First, we will prove that a given L-theory T is a T-L-theory, provided L extends C.

LEMMA 4.16 *Let T be an* L*-theory, where* L *extends* C*. Then T is a T-*L*-theory.*

Proof Closure under conjunction follows from the fact that T is an L-theory. For the second part, suppose that $A \in T$ and $A \rightarrow B \in T$. By closure under conjunction, $A \wedge (A \rightarrow B) \in T$. Since (I1) is in the logic and the fact that T is an L-theory, $B \in T$. □

The canonical model construction proceeds similarly to the case for non-reduced frames. There are a few differences, which we will highlight below. The first difference is that with reduced frames, it is not the case that a single canonical frame provides counterexamples to every non-theorem. Different canonical frames will be used for counterexamples for different invalid formulas. A canonical reduced frame is defined as follows.

DEFINITION 4.17 (Canonical reduced frame) *Let T be a regular, prime, T-*L*-theory. Then the canonical T-*L*-frame is defined as follows.*

- *K_T is the set of prime T-*L*-theories.*
- *$N_T = \{X \in K_T : T \subseteq X\}$.*
- *$R_T abc$ iff for all $A, B \in \mathfrak{L}$, if $A \rightarrow B \in a$ and $A \in b$, then $B \in c$.*
- *$a^* = \{A \in \mathfrak{L} : \neg A \notin a\}$.*

Note that since t is being dropped from the language, a regular L-theory will be an L-theory X such that $\mathsf{L} \subseteq X$. There are two major differences between the canonical reduced frames and the canonical frame of definition 4.5. The first is that the former uses T-L-theories, whereas the latter uses L-theories. The second is that the former obeys the reduction condition, with T being the minimal point in N, while the latter need not obey the reduction condition. For some logics, the canonical L-frame defined in definition 4.5 does obey the reduction condition, namely the cases where L is itself already prime.

We next need to show that for prime T-L-theories, the Routley star operation produces prime T-L-theories.

LEMMA 4.18 *Let $a \in K_T$. Then, $a^* \in K_T$.*

Proof Suppose $a \in K_T$. First, to show that a^* is closed under T-implication, suppose $B \in a^*$ and $B \rightarrow C \in T$. So, $\neg B \notin a$. By (N1), $\neg C \rightarrow \neg B \in T$, so $\neg C \notin a$. It then follows that $C \in a^*$.

The remaining conditions are proved as in the corresponding proofs for the canonical L-frame. □

To verify that a canonical T-L-frame satisfies the various conditions, we need the fact that theorem 2.27 and lemma 2.39 can be extended to yield appropriate prime T-L-theories. For those extensions, the definition of a disjoint pair will need a generalization. Say that B is L-DERIVABLE from A iff B can be obtained from A using the rules of L together with the provable implications of L. Say that a set Y of formulas is L-DERIVABLE from X iff for some formulas $B_1, \ldots, B_n \in Y$ and some for formulas $A_1, \ldots, A_m \in X$, $B_1 \vee \cdots \vee B_n$ is L-derivable from $A_1 \wedge \cdots \wedge A_m$. The pair (X, Y) of sets of formulas X and Y is an L-DISJOINT PAIR iff Y is not L-derivable from X. Using this definition, the proof of the prime extension theorem goes through with minor modifications. With those pieces in place, the arguments that a canonical T-L-frame satisfies all the required conditions go through as before.

The canonical T-L-model adds to the canonical T-L-frame a valuation, V_T, defined similarly to the non-reduced canonical valuations:

- For $a \in K_T$, $V_T(p,a) = 1$ iff $p \in a$.

While much of the argument that the canonical L-model is a model still goes through, there is one lemma, namely the Witness lemma, that has a notable amendment.

LEMMA 4.19 (Witness lemma) *Let $\langle K_T, N_T, R_T, ^* \rangle$ be a canonical T-L-frame. If $B \rightarrow C \notin a$, then there are $b, c \in K_T$ such that $R_T abc$, $B \in b$, and $C \notin c$.*

Proof Suppose $B \rightarrow C \notin a$. Then, let

$$y = \{D \in \mathfrak{L} : B \rightarrow D \in T\}.$$

It follows that y is a T-L-theory. To see this, suppose that $D \rightarrow E \in T$ and $D \in y$. Since (I4) is in L and $B \rightarrow D \in T$, it follows that $(D \rightarrow E) \rightarrow (B \rightarrow E) \in T$. It then follows that $D \rightarrow E \in T$, and this implies that $B \rightarrow E \in T$, so $E \in y$, as desired.

Closure under adjunction is shown in the usual way. Note that since $B \rightarrow B \in T$, $B \in y$.

Next, let

$$z = \{D \in \mathfrak{L} : \exists B \in \mathfrak{L}, B \rightarrow D \in a \text{ and } B \in y\}.$$

It follows that z is a T-L-theory. To show this, suppose that $F \rightarrow E \in T$ and $F \in z$. Since $F \in z$, there is a D such that $D \rightarrow F \in a$ and $D \in y$. As $F \rightarrow E \in T$ and (I3) is a theorem, $(D \rightarrow F) \rightarrow (D \rightarrow E) \in T$. Since a is a T-L-theory, it follows that $D \rightarrow E \in a$, so $E \in z$, as desired. Closure under adjunction is shown in the usual way.

We can then show that for all $D, E \in \mathfrak{L}$, if $D \rightarrow E \in a$ and $D \in y$, then $E \in z$, so we can extend y and z to appropriate prime T-L-theories b and c, respectively, such that $R_T abc$. The rest of the proof is similar to the proof of lemma 4.12, with "T-L-theory" in place of "L-theory," so we omit it. □

With those changes to the Witness lemma, the Truth lemma for the T-L-canonical model goes through as before.

LEMMA 4.20 (Truth lemma) *For all formulas A and all points $a \in K_T$, $A \in a$ iff $a \Vdash A$.*

The Existence lemma is provable, as before, so one can prove Completeness for reduced frames.

THEOREM 4.21 *Let* L *be a logic extending* C *with axioms from Table 4.2, and let $\mathcal{C}$ be the class of* L*-frames. If $\models_{r(\mathcal{C})} A$, then $\vdash_L A$.*

Although we have not gone into the details, the techniques used here actually permit something stronger to be proved, where proofs are permitted to have assumptions.

THEOREM 4.22 *Let* L *be a logic extending* C *with axioms from Table 4.2, and let $\mathcal{C}$ be the class of* L*-frames. If $X \models_{r(\mathcal{C})} A$, then $X \vdash_{\mathsf{L}} A$.*

This second theorem is known as Strong Completeness, and it permits proofs to use assumptions, rather than just involving logical truths. For the proof of Strong Completeness, we pick the prime theory T such that $X \subseteq T$. Then T verifies every member of X while falsifying A.

In the section so far, we have restricted attention to logics L extending C, following the treatment of reduced frames offered by Routley et al. (1982). One might wonder where the characteristic additional axioms of C are used. Axiom (I1) was used in lemma 4.16, axiom (I3) in lemma 4.19, and axiom (N1) in lemma 4.18. Slaney (1987a) has shown that inclusion of these axioms is a sufficient but not necessary condition on a logic being complete with respect to its reduced frames. The trick is to use a theory that obeys certain closure conditions, defined by Slaney.

DEFINITION 4.23 (Varieties of theories) *Let T be an* L*-theory.*

T is DETACHED iff if $A \rightarrow B \in T$ and $A \in T$, then $B \in T$.

T is AFFIXED *iff if $A \rightarrow B \in T$, then both $(C \rightarrow A) \rightarrow (C \rightarrow B) \in T$ and $(B \rightarrow C) \rightarrow (A \rightarrow C) \in T$.*

T is TRANSPOSITIVE *iff if $A \rightarrow B \in T$, then $\neg B \rightarrow \neg A \in T$.*

T is STABLE *iff it is detached, affixed, and transpositive.*

For some of the logics weaker than C, choice of a stable, regular theory T permits the earlier construction to go through, even in the absence of the axioms (I1), (I3), and (N1). The proofs do not work for all logics, as there are some hurdles to obtaining appropriate prime theories in the absence of contraction principles, such as (I1). Obtaining a prime T-L-theory will require the technique of metavaluations, which will be introduced in Chapter 9.

4.5 Exercises

(1) Prove that rules (R1)–(R5) all preserve verification across normal points. In other words, show that for each rule, if the premises of the rule hold at a normal point, then the conclusion of the rule does as well. As an example, for (R1), for $a \in N$, if $a \Vdash A \rightarrow B$ and $a \Vdash A$, then $a \Vdash B$.[5]

(2) Show that R7 preserves verification across normal points.

(3) Using Read's E-frame from Table 3.14, show that (R8) does not preserve verification across all normal points.

(4) Using the DW-frame in Table 3.5, show that (R8) does not preserve verification at normal points in reduced frames.

(5) Prove that (R8) preserves verification at the 0 point of any reduced frame.

(6) Using Read's E-frame from Table 3.14, show that (R6) does not preserve verification across normal points. Hint: Use the instance of the rule with $A = B = C = p$.

(7) Using the DW-frame in Table 3.5, show that (R6) does not preserve verification at normal points in reduced frames.

(8) Prove that (R6) preserves verification at the 0 point of any reduced frame.

(9) Prove that the rule (γ), $A, \neg A \vee B \Rightarrow B$, is sound for frames satisfying the condition that for all $a \in N, a \leq a^*$.

[5] I would like to thank Tore Fjetland Øgaard for suggesting the first eight exercises.

5 Matrices

5.1 Introduction

Models are a comparatively recent addition to the toolkit for studying relevant logics. Although at this point they have been around for a while, with much of the early work on them having been published in the 1970s, algebraic and matrix methods have been used in the study of relevant logics for even longer. There are important connections between matrices for relevant logics and ternary relational frames.

Matrices are often useful for proving things about relevant logics. In particular, they are extremely useful for showing that a certain formula is not a theorem of a particular logic. The usual way to do this is in two steps. First, one shows by an inductive argument that all axioms are designated on all valuations on a particular matrix and that the rules preserve the property of being designated or being valid. Next, one provides a valuation that results in the target non-theorem being undesignated. As all theorems are designated on all valuations, the target formula is thus not a theorem.

We will present enough background on algebraic concepts to make this chapter self-contained, but we cannot provide a general introduction to algebraic methods in philosophical logic. There is much more to say about algebraic methods in philosophical logic, and there is more to say about the use of algebraic and matricial techniques in the study of relevant logics in particular. For the reader interested in getting more background in the area, we recommend consulting the excellent books by Dunn and Hardegree (2001), Galatos et al. (2007), Font (2016), Ono (2019), and Cintula and Noguera (2021).

The plan for the chapter is to start in Section 5.2 by presenting the basic algebraic structures and matrices for relevant logics. In Section 5.3, we will present a connection between the matrices and frames, showing how to construct matrices from frames and frames from matrices. Section 5.4 will present several matrices that have been used in the relevant logic literature in some way. Then, we will turn in Section 5.5 to one of the most important applications of the matrix techniques of this chapter, namely showing that relevant logics have the variable-sharing property, as well as some prominent variations. Finally, in Section 5.6 we will briefly consider issues of interpolation and look at Urquhart's surprising result that many well known relevant logics lack interpolation.

5.2 Basic Algebras

Matrix models for relevant logics consist of an ordered, non-empty set of values, an upwardly closed subset of those values, and algebraic operations that are used to interpret the connectives of the logic. The matrix models we will consider are all based on distributive lattices, which we will now define. We will use the convention that boldface symbols, such as $\boldsymbol{a}$ and $\boldsymbol{0}$, are used for values of an algebra.

DEFINITION 5.1 (Distributive lattice) *A distributive lattice L is a triple* $\langle K, \wedge, \vee \rangle$ *where K is a non-empty set, and both* $\wedge$ *and* $\vee$ *are binary operations obeying the following conditions.*

(L1) $\boldsymbol{a} \wedge (\boldsymbol{b} \wedge \boldsymbol{c}) = (\boldsymbol{a} \wedge \boldsymbol{b}) \wedge \boldsymbol{c}$	(L6) $\boldsymbol{a} \vee \boldsymbol{a} = \boldsymbol{a}$
(L2) $\boldsymbol{a} \wedge \boldsymbol{b} = \boldsymbol{b} \wedge \boldsymbol{a}$	(L7) $\boldsymbol{a} \wedge (\boldsymbol{a} \vee \boldsymbol{b}) = \boldsymbol{a}$
(L3) $\boldsymbol{a} \wedge \boldsymbol{a} = \boldsymbol{a}$	(L8) $\boldsymbol{a} \vee (\boldsymbol{a} \wedge \boldsymbol{b}) = \boldsymbol{a}$
(L4) $\boldsymbol{a} \vee (\boldsymbol{b} \vee \boldsymbol{c}) = (\boldsymbol{a} \vee \boldsymbol{b}) \vee \boldsymbol{c}$	(L9) $\boldsymbol{a} \wedge (\boldsymbol{b} \vee \boldsymbol{c}) = (\boldsymbol{a} \wedge \boldsymbol{b}) \vee (\boldsymbol{a} \wedge \boldsymbol{c})$
(L5) $\boldsymbol{a} \vee \boldsymbol{b} = \boldsymbol{b} \vee \boldsymbol{a}$	(L10) $\boldsymbol{a} \vee (\boldsymbol{b} \wedge \boldsymbol{c}) = (\boldsymbol{a} \vee \boldsymbol{b}) \wedge (\boldsymbol{a} \vee \boldsymbol{c})$

In a distributive lattice, say that $\boldsymbol{a} \leq \boldsymbol{b}$ *iff* $\boldsymbol{a} \wedge \boldsymbol{b} = \boldsymbol{a}$.

This presentation of distributive lattices is nice for some purposes, but it will make it easier to prove things to have a few facts and rules proved as lemmas.

LEMMA 5.2 *Let L be a distributive lattice. Then the following hold.*

(i) $\boldsymbol{a} \wedge \boldsymbol{b} \leq \boldsymbol{a}, \boldsymbol{a} \wedge \boldsymbol{b} \leq \boldsymbol{b}$.
(ii) $\boldsymbol{a} \leq \boldsymbol{a} \vee \boldsymbol{b}, \boldsymbol{a} \leq \boldsymbol{b} \vee \boldsymbol{a}$.
(iii) $\boldsymbol{a} \wedge \boldsymbol{b} = \boldsymbol{a}$ *iff* $\boldsymbol{a} \vee \boldsymbol{b} = \boldsymbol{b}$.
(iv) *If* $\boldsymbol{a} \leq \boldsymbol{b}$ *and* $\boldsymbol{a} \leq \boldsymbol{c}$, *then* $\boldsymbol{a} \leq \boldsymbol{b} \wedge \boldsymbol{c}$.
(v) *If* $\boldsymbol{b} \leq \boldsymbol{a}$ *and* $\boldsymbol{c} \leq \boldsymbol{a}$, *then* $\boldsymbol{b} \vee \boldsymbol{c} \leq \boldsymbol{a}$.
(vi) $\boldsymbol{a} \leq \boldsymbol{a}$.
(vii) *If* $\boldsymbol{a} \leq \boldsymbol{b}$ *and* $\boldsymbol{b} \leq \boldsymbol{c}$, *then* $\boldsymbol{a} \leq \boldsymbol{c}$.

Proof The proofs are left to the reader. □

The operations $\wedge$ and $\vee$ are called, respectively, "meet" and "join." It is useful to have some additional terminology for lattices.

DEFINITION 5.3 (Bounds) *Let L be a lattice with* $\boldsymbol{a}, \boldsymbol{b} \in K$.

An element $\boldsymbol{x}$ *is a* LOWER BOUND *of* $\{\boldsymbol{a}, \boldsymbol{b}\}$ *iff* $\boldsymbol{x} \leq \boldsymbol{a}$ *and* $\boldsymbol{x} \leq \boldsymbol{b}$.

An element $\boldsymbol{x}$ *is an* UPPER BOUND *of* $\{\boldsymbol{a}, \boldsymbol{b}\}$ *iff* $\boldsymbol{a} \leq \boldsymbol{x}$ *and* $\boldsymbol{b} \leq \boldsymbol{x}$.

An element $\boldsymbol{x}$ *is the* GREATEST LOWER BOUND *of* $\{\boldsymbol{a}, \boldsymbol{b}\}$ *iff both (i)* $\boldsymbol{x}$ *is a lower bound of* $\{\boldsymbol{a}, \boldsymbol{b}\}$ *and (ii) for all lower bounds* $\boldsymbol{y}$ *of* $\{\boldsymbol{a}, \boldsymbol{b}\}$, $\boldsymbol{y} \leq \boldsymbol{x}$.

An element $\boldsymbol{x}$ *is the* LEAST UPPER BOUND *of* $\{\boldsymbol{a}, \boldsymbol{b}\}$ *iff both (i)* $\boldsymbol{x}$ *is an upper bound of* $\{\boldsymbol{a}, \boldsymbol{b}\}$ *and (ii) for all upper bounds* $\boldsymbol{y}$ *of* $\{\boldsymbol{a}, \boldsymbol{b}\}$, $\boldsymbol{x} \leq \boldsymbol{y}$.

The meet and join of a pair of elements are, respectively, the greatest lower bound and least upper bound of that set.

From the definitions, nothing requires infinite sets of elements to have a greatest lower bound or a least upper bound. There is a special class of lattices that does, and it will be useful to define that class.

DEFINITION 5.4 (Complete lattice) *A lattice $L = \langle L, \wedge, \vee \rangle$ is* COMPLETE *if for any set $X \subseteq L$,*

- *there is an element $\bigwedge X \in L$ such that $y \leq \bigwedge X$ iff $y \leq x$, for each $x \in X$, and*
- *there is an element $\bigvee X \in L$ such that $\bigvee X \leq y$ iff $x \leq y$, for each $x \in X$.*

$\bigwedge X$ is the GREATEST LOWER BOUND *of X and $\bigvee X$ is the least upper bound of X.*

Distributive lattices underpin the matrices for relevant logics. They interpret the so-called extensional, or additive, connectives. We can add *operations* to these structures to interpret the other connectives, where n-ary operations are functions from n-tuples of K to K. We first extend them with a De Morgan negation as follows.

DEFINITION 5.5 (De Morgan algebra) *A* DE MORGAN ALGEBRA *L is a quadruple $\langle K, \wedge, \vee, - \rangle$ where $\langle K, \wedge, \vee \rangle$ is a distributive lattice and $-$ is a unary operation on K satisfying the following.*

(DM1) $--\boldsymbol{a} = \boldsymbol{a}$.
(DM2) $-(\boldsymbol{a} \vee \boldsymbol{b}) = -\boldsymbol{a} \wedge -\boldsymbol{b}$.
(DM3) $-(\boldsymbol{a} \wedge \boldsymbol{b}) = -\boldsymbol{a} \vee -\boldsymbol{b}$.
(DM4) $\boldsymbol{a} \leq \boldsymbol{b} \Rightarrow -\boldsymbol{b} \leq -\boldsymbol{a}$.

Interpreting the other connectives requires different algebraic machinery. In particular, we need to add operations to interpret the implication and fusion, and we need something to interpret the Ackermann constant. This last will move us to focus on matrices.

DEFINITION 5.6 (Matrix) *A* MATRIX *M is an algebra together with a non-empty set $D \subseteq K$ of designated values.*

DEFINITION 5.7 (Ackermann groupoid) *An* ACKERMANN GROUPOID *G is a septuple $\langle K, \cdot, \rightarrow, \wedge, \vee, -, \mathbb{1} \rangle$ where $\langle K, \wedge, \vee, - \rangle$ is a De Morgan algebra, $\cdot$ and $\rightarrow$ are binary operations on K, and $\mathbb{1} \in K$, obeying the following conditions. We will write $\boldsymbol{ab}$ for $\boldsymbol{a} \cdot \boldsymbol{b}$.*

(G1) $\mathbb{1}\boldsymbol{a} = \boldsymbol{a}$.
(G2) *If $\boldsymbol{a} \leq \boldsymbol{b}$, then $\boldsymbol{ac} \leq \boldsymbol{bc}$ and $\boldsymbol{ca} \leq \boldsymbol{cb}$.*
(G3) *$\boldsymbol{ab} \leq \boldsymbol{c}$ iff $\boldsymbol{a} \leq \boldsymbol{b} \rightarrow \boldsymbol{c}$.*
(G4) $\boldsymbol{a}(\boldsymbol{b} \vee \boldsymbol{c}) = (\boldsymbol{ab}) \vee (\boldsymbol{ac})$.
(G5) $(\boldsymbol{b} \vee \boldsymbol{c})\boldsymbol{a} = (\boldsymbol{ba}) \vee (\boldsymbol{ca})$.

Table 5.1. Conditions on groupoids

	Axiom		Condition
(I1)	$A \wedge (A \to B) \to B$	(CI1)	$\boldsymbol{a \leq aa}$
(I2)	$(A \to B) \wedge (B \to C) \to (A \to C)$	(CI2)	$\boldsymbol{ab \leq a(ab)}$
(I3)	$(A \to B) \to ((C \to A) \to (C \to B))$	(CI3)	$\boldsymbol{(ab)c \leq a(bc)}$
(I4)	$(A \to B) \to ((B \to C) \to (A \to C))$	(CI4)	$\boldsymbol{(ab)c \leq b(ac)}$
(I5)	$A \to ((A \to B) \to B)$	(CI5)	$\boldsymbol{ab \leq ba}$
(I6)	$(A \to (B \to C)) \to (B \to (A \to C))$	(CI6)	$\boldsymbol{(ab)c \leq (ac)b}$
(I7)	$(A \to (A \to B)) \to (A \to B)$	(CI7)	$\boldsymbol{ab \leq (ab)b}$
(N1)	$(A \to \neg B) \to (B \to \neg A)$	(CN1)	$\boldsymbol{ab \leq c \Rightarrow a(-c) \leq -b}$
(N2)	$(A \to \neg A) \to \neg A$	(CN2)	$\boldsymbol{ab \leq -b \Rightarrow a \leq -b}$
(N3)	$A \vee \neg A$	(CN3)	$\mathbb{1} \leq \boldsymbol{-a \vee a}$

Ackermann groupoids will also be called B-groupoids. One can obtain restricted classes of groupoids, appropriate, in a sense to be explained, to stronger logics L, by imposing conditions on the groupoids. Axioms and their associated conditions are listed in Table 5.1.

In an Ackermann groupoid, the set $\{\boldsymbol{x} \in K : \mathbb{1} \leq \boldsymbol{x}\}$ is the set of designated values. This set plays a crucial role in the definitions of satisfaction and validity.

DEFINITION 5.8 (Valuation, satisfaction, holding, validity) *Given an Ackermann groupoid G, a* VALUATION *v on G is a function from* At *to K such that*

- $v(\mathrm{t}) = \mathbb{1}$.
- $v(\neg A) = -v(A)$.
- $v(A \wedge B) = v(A) \wedge v(B)$.
- $v(A \vee B) = v(A) \vee v(B)$.
- $v(A \to B) = v(A) \to v(B)$.
- $v(A \circ B) = v(A) \cdot v(B)$.

A valuation v SATISFIES *a formula A on G iff* $\mathbb{1} \leq v(A)$.

A valuation v is a COUNTEREXAMPLE *to A on G iff* $\mathbb{1} \not\leq v(A)$.

A formula A HOLDS *on G iff there is no counterexample to A on G.*

A formula A is VALID *in a class C of Ackermann groupoids iff for all* $G \in \mathcal{C}$, *for all valuations v on G, v satisfies A.*

LEMMA 5.9 *The following hold in all Ackermann groupoids.*

(i) $\boldsymbol{a \leq b \to (ab)}$.
(ii) $\boldsymbol{(a \to b)a \leq b}$.
(iii) $\boldsymbol{a \leq b}$ *iff* $\mathbb{1} \leq \boldsymbol{a \to b}$.

Proof For (i), $\boldsymbol{ab \leq ab}$, by reflexivity, so by (G3), $\boldsymbol{a \leq b \to (ab)}$.

For (ii), $\boldsymbol{a \to b \leq a \to b}$, by (G3), $\boldsymbol{(a \to b)a \leq b}$.

For (iii), suppose $\boldsymbol{a \leq b}$. By (G1), $\mathbb{1}\boldsymbol{a \leq b}$, so by (G3), $\mathbb{1} \leq \boldsymbol{a \to b}$. The converse is similar. □

Part (iii) of the lemma justifies us in showing that a conditional axiom $A \rightarrow B$ is valid by showing that $v(A) \leq v(B)$. For this section, we will adopt the convention that $v(A) = \boldsymbol{a}$, $v(B) = \boldsymbol{b}$, etc.

LEMMA 5.10 (Algebraic soundness for axioms of B) *All axioms of* B *are valid in the class of Ackermann groupoids.*

Proof Let G be an arbitrary Ackermann groupoid. There are several cases.

(A1) $A \rightarrow A$. This is immediate as $\boldsymbol{a} \leq \boldsymbol{a}$ by reflexivity.

(A2) $(A \wedge B) \rightarrow A$. This case follows from lemma 5.2. The other subcase, $(A \wedge B) \rightarrow B$, is similar.

(A3) $((A \rightarrow B) \wedge (A \rightarrow C)) \rightarrow (A \rightarrow (B \wedge C))$. From lemma 5.9, $(\boldsymbol{a} \rightarrow \boldsymbol{b})\boldsymbol{a} \leq \boldsymbol{b}$ and $(\boldsymbol{a} \rightarrow \boldsymbol{c})\boldsymbol{a} \leq \boldsymbol{c}$. Therefore, $((\boldsymbol{a} \rightarrow \boldsymbol{b}) \wedge (\boldsymbol{a} \rightarrow \boldsymbol{c}))\boldsymbol{a} \leq \boldsymbol{b}$ and $((\boldsymbol{a} \rightarrow \boldsymbol{b}) \wedge (\boldsymbol{a} \rightarrow \boldsymbol{c}))\boldsymbol{a} \leq \boldsymbol{c}$, so $((\boldsymbol{a} \rightarrow \boldsymbol{b}) \wedge (\boldsymbol{a} \rightarrow \boldsymbol{c}))\boldsymbol{a} \leq \boldsymbol{b} \wedge \boldsymbol{c}$. By (G3), $(\boldsymbol{a} \rightarrow \boldsymbol{b}) \wedge (\boldsymbol{a} \rightarrow \boldsymbol{c}) \leq \boldsymbol{a} \rightarrow (\boldsymbol{b} \wedge \boldsymbol{c})$.

(A4) $A \rightarrow (A \vee B)$, $A \rightarrow (B \vee A)$. This case is handled by lemma 5.2.

(A5) $((B \rightarrow A) \wedge (C \rightarrow A)) \rightarrow ((B \vee C) \rightarrow A)$. This is similar to case (A3).

(A6) $(A \wedge (B \vee C)) \rightarrow ((A \wedge B) \vee (A \wedge C))$. This is handled by postulate (L9) of definition 5.1.

(A7) $\neg\neg A \rightarrow A$. This case is handled by postulate (DM1) of the definition 5.5.

(A8) t. As $v(\mathrm{t}) = \mathbb{1}$, $\mathbb{1} \leq v(\mathrm{t})$. □

LEMMA 5.11 *If the premises of a rule of* B *are satisfied, then the conclusion is as well.*

Proof (R1) Suppose $\mathbb{1} \leq \boldsymbol{a}$ and $\mathbb{1} \leq \boldsymbol{a} \rightarrow \boldsymbol{b}$. Then $\boldsymbol{a} \leq \boldsymbol{b}$. By lemma 5.2, $\mathbb{1} \leq \boldsymbol{b}$, so B is satisfied.

(R2) Suppose $\mathbb{1} \leq \boldsymbol{a}$ and $\mathbb{1} \leq \boldsymbol{b}$. By lemma 5.2, $\mathbb{1} \leq \boldsymbol{a} \wedge \boldsymbol{b}$, so $A \wedge B$ is satisfied.

(R3) Suppose $\mathbb{1} \leq \boldsymbol{a} \rightarrow -\boldsymbol{b}$. Then, $\boldsymbol{a} \leq -\boldsymbol{b}$. By (DM4), $--\boldsymbol{b} \leq -\boldsymbol{a}$, so $\boldsymbol{b} \leq -\boldsymbol{a}$. Therefore, $\mathbb{1} \leq \boldsymbol{b} \rightarrow -\boldsymbol{a}$, so $B \rightarrow \neg A$ is satisfied.

(R4) Suppose $\mathbb{1} \leq \boldsymbol{a} \rightarrow \boldsymbol{b}$, so $\boldsymbol{a} \leq \boldsymbol{b}$. By (G2), $(\boldsymbol{b} \rightarrow \boldsymbol{c})\boldsymbol{a} \leq (\boldsymbol{b} \rightarrow \boldsymbol{c})\boldsymbol{b}$. From lemma 5.9, $(\boldsymbol{b} \rightarrow \boldsymbol{c})\boldsymbol{b} \leq \boldsymbol{c}$, so $(\boldsymbol{b} \rightarrow \boldsymbol{c})\boldsymbol{a} \leq \boldsymbol{c}$. Therefore, $\boldsymbol{b} \rightarrow \boldsymbol{c} \leq \boldsymbol{a} \rightarrow \boldsymbol{c}$, so $(B \rightarrow C) \rightarrow (A \rightarrow C)$ is satisfied.

(R5) Suppose $\mathbb{1} \leq \boldsymbol{a} \rightarrow \boldsymbol{b}$, so $\boldsymbol{a} \leq \boldsymbol{b}$. By lemma 5.9, $(\boldsymbol{c} \rightarrow \boldsymbol{a})\boldsymbol{c} \leq \boldsymbol{a}$, so by lemma 5.2, $(\boldsymbol{c} \rightarrow \boldsymbol{a})\boldsymbol{c} \leq \boldsymbol{b}$. By (G3), $\boldsymbol{c} \rightarrow \boldsymbol{a} \leq \boldsymbol{c} \rightarrow \boldsymbol{b}$, so $(C \rightarrow A) \rightarrow (C \rightarrow B)$ is satisfied.

(R6) Suppose $\mathbb{1} \leq \boldsymbol{a} \rightarrow (\boldsymbol{b} \rightarrow \boldsymbol{c})$, so $\boldsymbol{a} \leq \boldsymbol{b} \rightarrow \boldsymbol{c}$. By (G3), $\boldsymbol{ab} \leq \boldsymbol{c}$. Therefore, $(A \circ B) \rightarrow C$ is satisfied.

(R7) Suppose $\mathbb{1} \leq (\boldsymbol{ab}) \rightarrow \boldsymbol{c}$, so $\boldsymbol{ab} \leq \boldsymbol{c}$. Then by (G3), $\boldsymbol{a} \leq \boldsymbol{b} \rightarrow \boldsymbol{c}$, so $A \rightarrow (B \rightarrow C)$ is satisfied, as desired.

(R8) Suppose $\mathbb{1} \leq \boldsymbol{a}$. Then by (G1), $\mathbb{1}\mathbb{1} \leq \boldsymbol{a}$, so by (G3), $\mathbb{1} \leq \mathbb{1} \rightarrow \boldsymbol{a}$, and so $\mathrm{t} \rightarrow A$ is satisfied. □

LEMMA 5.12 *Let G be an Ackermann groupoid obeying the condition* (CDn) *corresponding to the axiom* (Dn) *in Table 5.1. Then* (Dn) *holds on G.*

Proof Let G be an Ackermann groupoid. There are then several cases.

(I1) Let G obey condition (CI1). To show that (I1) holds, we start by showing $a \wedge b \leq ab$. By lemma 5.9, $a \wedge b \leq b \rightarrow (ab)$, so $(a \wedge b)(a \wedge b) \leq ab$. By (CI1), $a \wedge b \leq (a \wedge b)(a \wedge b)$, so $a \wedge b \leq ab$. As $a \wedge (a \rightarrow b) \leq (a \rightarrow b) \wedge a$, $a \wedge (a \rightarrow b) \leq (a \rightarrow b)a$, and by lemma 5.9, $(a \rightarrow b)a \leq b$, $a \wedge (a \rightarrow b) \leq b$.

(I2) Let G obey condition (CI2). By lemma 5.9, $(a \rightarrow b)a \leq b$ and so $(b \rightarrow c)((a \rightarrow b)a) \leq c$. From (G2), $((a \rightarrow b) \wedge (b \rightarrow c))(((a \rightarrow b) \wedge (b \rightarrow c))a) \leq c$. By (CI2), $((a \rightarrow b) \wedge (b \rightarrow c))a \leq c$, so $(a \rightarrow b) \wedge (b \rightarrow c) \leq a \rightarrow c$.

(I3) Let G obey condition (CI3). By lemma 5.9, $(a \rightarrow b)((c \rightarrow a)c) \leq b$. By (CI3), $((a \rightarrow b)(c \rightarrow a))c \leq b$, so $(a \rightarrow b)(c \rightarrow a) \leq c \rightarrow b$, and so $a \rightarrow b \leq (c \rightarrow a) \rightarrow (c \rightarrow b)$.

(I4) Let G obey condition (CI4). From lemma 5.9 we have $(b \rightarrow c)((a \rightarrow b)a) \leq c$. From (CI4), we have $((a \rightarrow b)(b \rightarrow c))a \leq c$. Thus by (G3), $(a \rightarrow b)(b \rightarrow c) \leq a \rightarrow c$ and so $a \rightarrow b \leq (b \rightarrow c) \rightarrow (a \rightarrow c)$.

(I5) Let G obey condition (CI5). By lemma 5.9, $(a \rightarrow b)a \leq b$, so by (CI5), $a(a \rightarrow b) \leq b$. Therefore, $a \leq (a \rightarrow b) \rightarrow b$.

(I6) Let G obey condition (CI6). By lemma 5.9, $((a \rightarrow (b \rightarrow c))a)b \leq c$. By (CI6), $((a \rightarrow (b \rightarrow c))b)a \leq c$, so $a \rightarrow (b \rightarrow c) \leq b \rightarrow (a \rightarrow c)$, as desired.

(I7) Let G obey condition (CI7). By lemma 5.9, $((a \rightarrow (a \rightarrow b))a)a \leq b$, so by (CI7), $(a \rightarrow (a \rightarrow b))a \leq b$. Thus $a \rightarrow (a \rightarrow b) \leq a \rightarrow b$.

(N1) Let G obey condition (CN1). By lemma 5.9, $(a \rightarrow -b)a \leq -b$, so by (CN1), $(a \rightarrow -b) - -b \leq -a$. Therefore, $(a \rightarrow -b)b \leq -a$, so $a \rightarrow -b \leq b \rightarrow -a$.

(N2) Let G obey condition (CN2). By lemma 5.9, $(a \rightarrow -a)a \leq -a$, so by (CN2), $a \rightarrow -a \leq -a$, as desired.

(N3) This case is immediate from condition (CN3). □

We will say that the L-groupoids are the Ackermann groupoids that obey the additional conditions corresponding to the axioms of L. Some L-groupoids have been studied under other names. One class of such groupoids is the class of DUNN MONOIDS.[1] These are R-groupoids, in our terminology. They are named after Jon Michael Dunn, who first defined them as algebraic models for R.[2]

The preceding lemmas permit us to conclude with an algebraic Soundness theorem.

THEOREM 5.13 (Algebraic Soundness) *If* $\vdash_{\mathsf{L}} A$ *then* A *is valid in the class of* L-*groupoids.*

While one can prove an algebraic Completeness theorem, it will not be as useful to us as the Soundness theorem.[3] The Soundness theorem permits us to conclude that a particular formula A is not a theorem of L if we can find a counterexample on an L-algebra. In practice, appeals to Soundness seem to be more common. This is not to say that the Completeness results are uninteresting or unuseful. Our focus is, primarily,

[1] Meyer and Routley (1972).

[2] See Dunn (1966, ch. X), although Dunn used the terminology "De Morgan monoid," which remains common in the literature.

[3] Font (2016) gives a good presentation of Completeness using algebraic techniques.

on the logics, obtained via the proof systems or the frames, and not so much on the classes of groupoids and other algebraic structures.

5.3 Connections to Frames

Matrices are connected to ternary relational frames in an important way. One can generate a matrix from a frame and a frame from a matrix. This section will demonstrate that connection.

First, we will recall some definitions from Chapter 3. Given a frame $F = \langle K, N, R, ^*\rangle$, the set $\mathcal{H}_F$ is defined as $\{X \subseteq K : \text{ if } a \leq b \text{ and } a \in X, \text{ then } b \in X\}$. The elements of $\mathcal{H}_F$ are PROPOSITIONS on the frame F. Given propositions X and Y, we can define the following operations that yield propositions.

(i) $X \Rightarrow Y = \{a \in K : \forall b, c \in K(\text{if } Rabc \text{ and } b \in X, \text{ then } c \in Y)\}$.
(ii) $X \otimes Y = \{a \in K : \exists b, c \in K(Rbca \text{ and } b \in X \text{ and } c \in Y)\}$.
(iii) $-X = \{a \in K : a^* \notin X\}$.

In addition, the intersection and union of two propositions is a proposition, as shown by lemma 7.7.

LEMMA 5.14 *Let F be a* B*-frame. Then $\langle \mathcal{H}_F, \otimes, \Rightarrow, \cap, \cup, -, N\rangle$ is an Ackermann groupoid.*

Proof Let F be a B-frame and let G be defined as $\langle \mathcal{H}_F, \otimes, \Rightarrow, \cap, \cup, -, N\rangle$. We will show that G is an Ackermann groupoid.

That $\langle \mathcal{H}_F, \cap, \cup\rangle$ is a distributive lattice follows from the definitions of set-theoretic intersection and union. It follows that $X \leq Y$ iff $X \subseteq Y$. If $X \subseteq Y$, then $X \cap Y = X$. Suppose $X \cap Y = X$ and $a \in X$ but $a \notin Y$. Then $a \notin X \cap Y$, contradicting an assumption. Thus, $X \subseteq Y$.

We will show that $\langle \mathcal{H}_F, \cap, \cup, -\rangle$ is a De Morgan algebra.

(DM1) We argue as follows.

$$\begin{aligned} --X &= \{Y \in \mathcal{H}_F : Y^* \notin -X\} = \{Y \in \mathcal{H}_F : Y^{**} \in X\} \\ &= \{Y \in \mathcal{H}_F : Y \in X\} = X. \end{aligned}$$

(DM2) The following equations establish what we want.

$$\begin{aligned} -(X \cup Y) &= \{Z \in \mathcal{H}_F : Z^* \notin X \cup Y\} \\ &= \{Z \in \mathcal{H}_F : Z^* \notin X\} \cap \{Z \in \mathcal{H}_F : Z^* \notin Y\} = -X \cap -Y. \end{aligned}$$

(DM3) This is similar to the preceding case.
(DM4) Suppose $X \leq Y$. So, $X \subseteq Y$. Suppose $a \in -Y$, so $a^* \notin Y$. Then $a^* \notin X$, so $a \in -X$, as desired.

Finally, we have to show that G obeys the conditions on the other components.

(G1) Suppose $a \in N \otimes X$. Then there are $b, c \in K$ such that $Rbca$ and $b \in N$ and $c \in X$. By definition, $c \leq a$, so $a \in X$.

Suppose $a \in X$. For all $a \in K$, there is a $b \in N$ such that $Rbaa$, so $a \in N \otimes X$.

(G2) Suppose $X \leq Y$. Suppose $a \in X \otimes Z$. Then there are $b, c \in K$ such that $Rbca$, $b \in X$, and $c \in Z$. By assumption, $b \in Y$, so $a \in Y \otimes Z$, which suffices for $X \otimes Z \leq Y \otimes Z$ and the other subcase, mutatis mutandis.

(G3) Suppose $X \otimes Y \leq Z$. Suppose $a \in X$. To show that $a \in Y \Rrightarrow Z$, suppose $b, c \in K$ are such that $Rabc$ and $b \in Y$. Then $c \in X \otimes Y$, so by assumption, $c \in Z$. Thus, $a \in Y \Rrightarrow Z$.

Suppose $X \leq Y \Rrightarrow Z$. Suppose $a \in X \otimes Y$. Then there are $b, c \in K$ such that $Rbca$, $b \in X$, and $c \in Y$. So, $b \in Y \Rrightarrow Z$. By definition, if $Rbca$ and $c \in Y$, then $a \in Z$. From the assumptions, we have $Rbca$ and $c \in Y$, so $a \in Z$, as desired.

(G4) Suppose $a \in X \otimes (Y \cup Z)$. Then there are $b, c \in K$ such that $Rbca$, $b \in X$ and $c \in Y \cup Z$. So, $c \in Y$ or $c \in Z$. Thus, $a \in X \otimes Y$ or $a \in X \otimes Z$, so $a \in (X \otimes Y) \cup (X \otimes Z)$.

Suppose $a \in (X \otimes Y) \cup (X \otimes Z)$. So, $a \in X \otimes Y$ or $a \in X \otimes Z$. So, there are $b, c \in K$ such that $Rbca$, $b \in X$, and $c \in Y$, or there are $d, e \in K$ such that $Rdea$, $d \in X$, and $e \in Z$. So, there are $b, c \in K$ such that $Rbca$, $b \in X$, and $c \in Y \cup Z$, which establishes that $a \in X \otimes (Y \cup Z)$.

(G5) This is similar to the preceding case. □

The groupoid defined in the way laid out in the preceding lemma from a frame F will be denoted $\Gamma(F)$. B-frames generate B-groupoids. The result extends to frames with conditions.

LEMMA 5.15 *Let F be an* L*-frame. Then* $\langle \mathcal{H}_F, \otimes, \Rrightarrow, \cap, \cup, -, N\rangle$ *is an* L*-groupoid.*

Proof Let F be an L-frame. From the preceding lemma, we know that G, defined as $\langle \mathcal{H}_F, \otimes, \Rrightarrow, \cap, \cup, -, N\rangle$, is a B-groupoid. It remains to check that G obeys the appropriate conditions to be an L-groupoid. There are several cases, one for each frame condition. For this proof, letters from the start of the alphabet will be used for points in F and (bolded) letters from the end of the alphabet will be used for points from the associated groupoid.

(I1) Let F obey condition (FI1). To show that (CI1) holds, we assume $a \in \mathbf{x}$. To show that $a \in \mathbf{xx}$, we need to show that there are $b, c \in K$ such that $Rabc$, $b \in \mathbf{x}$, and $c \in \mathbf{x}$. By (FI1), $Raaa$, so $a \in \mathbf{xx}$.

(I2) Let F obey condition (FI2). To show that (CI2) holds, suppose $a \in \mathbf{xy}$. So, there are $b, c \in K$ such that $Rbca$, $b \in \mathbf{x}$ and $c \in \mathbf{y}$. It follows that $Rb(bc)a$, so there is a $d \in K$ such that $Rbda$ and $Rbcd$. Then, $d \in \mathbf{xy}$, so $a \in \mathbf{x}(\mathbf{xy})$, as desired.

(I3) Let F obey condition (FI3). To show that (CI3) holds, suppose $a \in (\mathbf{xy})\mathbf{z}$. Unpacking this yields that there are $b, c, d, e \in K$ such that $Rbca$, $Rdeb$, $b \in \mathbf{xy}$, $c \in \mathbf{z}$, $d \in \mathbf{x}$, and $e \in \mathbf{y}$. From condition (FI3), $Rd(ec)a$, which suffices for $a \in \mathbf{x}(\mathbf{yz})$.

(I4) Let F obey condition (FI4). Suppose that $a \in (\boldsymbol{x}\boldsymbol{y})\boldsymbol{z}$. This yields that there are $b,c,d,e \in K$ such that $Rbca$, $Rdeb$, $b \in \boldsymbol{x}\boldsymbol{y}$, $c \in \boldsymbol{z}$, $d \in \boldsymbol{x}$, and $e \in \boldsymbol{y}$. From condition (FI4), $Re(dc)a$, which suffices for $a \in \boldsymbol{y}(\boldsymbol{x}\boldsymbol{z})$.

(I5) Let F obey condition (FI5). Suppose $a \in \boldsymbol{x}\boldsymbol{y}$. So there are $b,c \in K$ such that $Rbca$, $b \in \boldsymbol{x}$, and $c \in \boldsymbol{y}$. From condition (FI5), $Rcba$, so $a \in \boldsymbol{y}\boldsymbol{x}$, as desired.

(I6) Let F obey condition (FI6). Suppose $a \in (\boldsymbol{x}\boldsymbol{y})\boldsymbol{z}$. This yields that there are $b,c,d,e \in K$ such that $Rbca$, $Rdeb$, $b \in \boldsymbol{x}\boldsymbol{y}$, $c \in \boldsymbol{z}$, $d \in \boldsymbol{x}$, and $e \in \boldsymbol{y}$. From condition (FI6) we have $Rdcea$, which suffices for $a \in (\boldsymbol{x}\boldsymbol{z})\boldsymbol{y}$.

(I7) Let F obey condition (FI7). Suppose $a \in \boldsymbol{x}\boldsymbol{y}$. So there are $b,c \in K$ such that $Rbca$, $b \in \boldsymbol{x}$, and $c \in \boldsymbol{y}$. By condition (FI7), $Rbcca$, so there is $d \in K$ such that $Rbcd$ and $Rdca$. It follows that $d \in \boldsymbol{x}\boldsymbol{y}$, so $a \in (\boldsymbol{x}\boldsymbol{y})\boldsymbol{y}$, as desired.

(N1) Let F obey condition (FN1). Suppose $\boldsymbol{x}\boldsymbol{y} \leq \boldsymbol{z}$ and $a \in \boldsymbol{x}(-\boldsymbol{z})$. So there are $b,c \in K$ such that $Rbca$, $b \in \boldsymbol{x}$, and $c \in -\boldsymbol{z}$. From the definition of $-\boldsymbol{z}$, $c^* \notin \boldsymbol{z}$. Suppose that $a^* \in \boldsymbol{y}$. From condition (FN1), Rba^*c^*, so $c^* \in \boldsymbol{x}\boldsymbol{y}$. From the initial assumptions, $c^* \in \boldsymbol{z}$, which is a contradiction. Thus, $a^* \notin \boldsymbol{y}$, so $a \in -\boldsymbol{y}$, as desired.

(N2) Let F obey condition (FN2). Suppose $\boldsymbol{x}\boldsymbol{y} \leq -\boldsymbol{y}$ and $a \in \boldsymbol{x}$. Suppose that $a \notin -\boldsymbol{y}$, so $a^* \in \boldsymbol{y}$. By (FN2), Raa^*a, so $a \in \boldsymbol{x}\boldsymbol{y}$. From the assumptions, $a \in -\boldsymbol{y}$, which is a contradiction. Therefore, $a \in -\boldsymbol{y}$, as desired.

(N3) Let F obey condition (FN3). Suppose $a \in \mathbb{1}$. Suppose for reductio that $a \notin -\boldsymbol{x} \cup \boldsymbol{x}$. Then $a \notin -\boldsymbol{x}$ and $a \notin \boldsymbol{x}$. The former implies $a^* \in \boldsymbol{x}$. Since $a \in \mathbb{1}$, $a \in N$, which implies, by condition (FN3), that $a^* \leq a$. Since $\boldsymbol{x}$ is a proposition, $a \in \boldsymbol{x}$, which is a contradiction. Therefore, $a \in -\boldsymbol{x} \cup \boldsymbol{x}$, as desired. □

Let us recall the definition of a propositional L-frame from Chapter 3.

DEFINITION 5.16 (Propositional frames) *Let $F = \langle K, N, R, {}^* \rangle$ be an* L*-frame and let* Prop *be a model-like set of propositions on F. Then the pair $\langle F, \mathsf{Prop} \rangle$ is a propositional* L*-frame.*

The set Prop can be used to define a restricted groupoid.

LEMMA 5.17 *Let $\langle F, \mathsf{Prop} \rangle$ be a propositional* L*-frame. Let H be defined as $\langle \mathcal{H}_F, \otimes, \Rrightarrow, \cap, \cup, -, N \rangle$ and let G be defined as $\langle \mathsf{Prop}, \otimes_G, \Rrightarrow_G, \cap, \cup, -_G, N \rangle$. Then G is an* L*-groupoid, with $\otimes_G$, $\Rrightarrow_G$, and $-_G$ identical to $\otimes$, $\Rrightarrow$, and $-$, respectively, restricted to* Prop.

Proof Let H and G be defined as earlier. H is an L-groupoid, by the previous lemmas.

Seeing that the operations of H are simply restrictions to Prop of the corresponding operations on G is a matter of grinding through the definitions, which we will leave to the interested reader. □

Let us turn to getting frames from matrices. For this, we need to begin with some definitions.

DEFINITION 5.18 (Filters, prime filters, $\mathbb{1}$-filters) *Let $L = \langle K, \wedge, \vee \rangle$ be a distributive lattice with a defined partial order $\leq$ and $\mathbb{1} \in K$ a distinguished element.*

X is a FILTER on K iff both

- *if $\boldsymbol{x} \in X$ and $\boldsymbol{y} \in X$, then $\boldsymbol{x} \wedge \boldsymbol{y} \in X$, and*
- *if $\boldsymbol{x} \in X$ and $\boldsymbol{x} \leq \boldsymbol{y}$, then $\boldsymbol{y} \in X$.*

A filter X is PRIME iff $\boldsymbol{x} \vee \boldsymbol{y} \in X$ implies $\boldsymbol{x} \in X$ or $\boldsymbol{y} \in X$.

A filter X is a $\mathbb{1}$-FILTER iff $\mathbb{1} \in X$.

The filter $[\mathbb{1}) = \{\boldsymbol{x} \in K : \mathbb{1} \leq \boldsymbol{x}\}$ is the PRINCIPAL FILTER on L.

Let G be an L*-groupoid. Then G is PRIME iff $[\mathbb{1})$ is prime.*

DEFINITION 5.19 (Matrix-generated frame) *Let $G = \langle J, \cdot, \rightarrow, \cap, \cup, -, \mathbb{1} \rangle$ be a prime* L*-groupoid.*

Let X and Y be filters on G. Define $X + Y = \{\boldsymbol{y} \in J : \exists \boldsymbol{x} \in J(\boldsymbol{x} \rightarrow \boldsymbol{y} \in X$ and $\boldsymbol{x} \in Y)\}$

Define $\Phi(G)$ as $\langle K, N, R, ^* \rangle$ as follows.

- *K is the set of prime filters on G.*
- *For $a, b, c \in K$, $Rabc$ iff for all $\boldsymbol{x}, \boldsymbol{y} \in G$, if $\boldsymbol{x} \rightarrow \boldsymbol{y} \in a$ and $\boldsymbol{x} \in b$, then $\boldsymbol{y} \in c$.*
- *$N = \{a \in K : [\mathbb{1}) \subseteq a\}$.*
- *$a^* = \{\boldsymbol{x} \in G : -\boldsymbol{x} \notin a\}$.*

We will prove some lemmas on the road to proving that for a prime L-groupoid G, $\Phi(G)$ is a L-frame.

LEMMA 5.20 *Let G be an* L*-groupoid. If X is a filter on G, $\mathbb{1} \leq \boldsymbol{x} \rightarrow \boldsymbol{y}$, and $\boldsymbol{x} \in X$, then $\boldsymbol{y} \in X$.*

Proof Let G be an L-groupoid. Suppose X is a filter on G, $\mathbb{1} \leq \boldsymbol{x} \rightarrow \boldsymbol{y}$, and $\boldsymbol{x} \in X$. Since $\mathbb{1} \leq \boldsymbol{x} \rightarrow \boldsymbol{y}$, $\boldsymbol{x} \leq \boldsymbol{y}$. As X is a filter, $\boldsymbol{y} \in X$. □

LEMMA 5.21 *Let $G = \langle J, \cdot, \rightarrow, \cap, \cup, -, \mathbb{1} \rangle$ be a prime* L*-groupoid, and define $\Phi(G)$ as in definition 5.19. Then the following are true.*

- *$N \neq \emptyset$.*
- *For $a, b \in K$, $a \leq b$ iff $a \subseteq b$.*

Proof Let G and $\Phi(G)$ be as in the statement of the lemma.

For the first claim, $[\mathbb{1})$ is prime, by assumption, so $[\mathbb{1}) \in K$.

For the second, suppose $a \leq b$. Then for some $d \in N$, $Rdab$. Since $[\mathbb{1}) \subseteq d$ and $\boldsymbol{x} \rightarrow \boldsymbol{x} \in [\mathbb{1})$, $\boldsymbol{x} \rightarrow \boldsymbol{x} \in d$. Suppose $\boldsymbol{x} \in a$, By definition, $\boldsymbol{x} \in b$, so $a \subseteq b$. Suppose $a \subseteq b$. Let $\boldsymbol{x}$ and $\boldsymbol{y}$ be arbitrary, and suppose $\boldsymbol{x} \rightarrow \boldsymbol{y} \in [\mathbb{1})$ and $\boldsymbol{x} \in a$. By lemma 5.20, $\boldsymbol{x} \leq \boldsymbol{y}$, so $\boldsymbol{y} \in a$. As $a \subseteq b$, $\boldsymbol{y} \in b$, so $a \leq b$, as desired. □

LEMMA 5.22 *Let $G = \langle J, \cdot, \rightarrow, \cap, \cup, -, \mathbb{1} \rangle$ be a prime* L*-groupoid, and define $\Phi(G)$ as in definition 5.19. Then the following are true.*

- *For $a \in K$, $a \leq a$.*
- *For $a, b, c \in K$, if $a \leq b$ and $b \leq c$, then $a \leq c$.*

- *For $a, b \in K$, if $a \in N$ and $a \leq b$, then $b \in N$.*
- *For $a, b, c, d, e, f \in K$, if $Rabc$, $d \leq a$, $e \leq b$, and $c \leq f$, then $Rdef$.*

Proof Let G and $\Phi(G)$ be as in the statement of the lemma.

For the first claim, $a \subseteq a$, so by lemma 5.21, $a \leq a$.

The second claim follows from lemma 5.21 together with the transitivity of the subset relation.

For the third claim, suppose $a \in N$ and $a \leq b$. By lemma 5.21, $a \subseteq b$. Since $a \in N$, $[\mathbb{1}) \subseteq a$, so by the transitivity of the subset relation, $[\mathbb{1}) \subseteq b$. Thus $b \in N$.

For the fourth claim, suppose $Rabc$, $d \leq a$, $e \leq b$, and $c \leq f$. By lemma 5.21, $d \subseteq a$, $e \subseteq b$, and $c \subseteq f$. Suppose $\boldsymbol{x} \rightarrow \boldsymbol{y} \in d$ and $\boldsymbol{x} \in e$. Then $\boldsymbol{x} \rightarrow \boldsymbol{y} \in a$ and $\boldsymbol{x} \in b$, so $\boldsymbol{y} \in c$. Therefore, $\boldsymbol{y} \in f$, and so $Rdef$, as desired. □

LEMMA 5.23 *Let $G = \langle J, \cdot, \rightarrow, \cap, \cup, -, \mathbb{1} \rangle$ be a prime* L*-groupoid, and define $\Phi(G)$ as in definition 5.19. Then the following are true.*

- *For $a, b \in K$, $a \leq b$ only if $b^* \leq a^*$.*
- *For $a \in K$, $a^{**} = a$.*

Proof Let G and $\Phi(G)$ be as in the statement of the lemma.

For the first, suppose $a \leq b$. By lemma 5.21, $a \subseteq b$. Suppose $\boldsymbol{x} \in b^*$. So, $-\boldsymbol{x} \notin b$. Therefore, $-\boldsymbol{x} \notin a$, so $\boldsymbol{x} \in a^*$.

For the second, we argue as follows.

$$a^{**} = \{\boldsymbol{x} \in J : -\boldsymbol{x} \notin a^*\} = \{\boldsymbol{x} \in J : --\boldsymbol{x} \in a\} = \{\boldsymbol{x} \in J : \boldsymbol{x} \in a\} = a. \quad \square$$

Before stating and proving the desired theorem, we will need to state two mathematical results to which we will appeal, namely Zorn's lemma and the Prime filter separation principle. Both of these are important results whose proofs would take us too far afield. Rather than presenting proofs, we will provide references for the reader to consult, treating the results as black boxes for the time being. For Zorn's lemma, a definition will be needed.

DEFINITION 5.24 (Chain) *A set C of sets $\{X_i : i \in I\}$ is a* CHAIN *iff for every $X, Y \in C$, $X \subseteq Y$ or $Y \subseteq X$.*

LEMMA 5.25 (Zorn's lemma) *If E is a non-empty set of sets, and if the union $\bigcup C \in E$, for every non-empty chain C from E, then E has a maximal member.*

See Dunn and Hardegree (2001, 30) for consequences of Zorn's lemma.

LEMMA 5.26 (Prime filter separation principle) *Let L be a distributive lattice with $\boldsymbol{a}, \boldsymbol{b} \in L$ such that $\boldsymbol{a} \nleq \boldsymbol{b}$. Then there is a prime filter $X \subseteq L$ such that $\boldsymbol{a} \in X$ and $\boldsymbol{b} \notin X$.*

Proof See Dunn and Hardegree (2001, 298). □

The Pair extension lemma can be seen as a logical counterpart of the the Prime filter separation principle.

THEOREM 5.27 *Let* $G = \langle J, \cdot, \rightarrow, \cap, \cup, -, \mathbb{1}\rangle$ *be a prime* B*-groupoid. Then* $\Phi(G)$ *is a* B*-frame and* G *is isomorphic to a subalgebra of* $\Gamma(\Phi(G))$.

Proof Verification that $\Phi(G)$ obeys the six conditions on B-frames was presented in lemmas 5.22 and 5.23.

To define the isomorphism from G to $\Phi(G)$, for $\boldsymbol{i} \in J$, set $h(\boldsymbol{i}) = \{a \in K : \boldsymbol{i} \in a\}$. The isomoprhism is extended to the operations as follows.

(i) $h(-\boldsymbol{i}) = -h(\boldsymbol{i})$.
(ii) $h(\boldsymbol{i} \wedge \boldsymbol{j}) = h(\boldsymbol{i}) \cap h(\boldsymbol{j})$.
(iii) $h(\boldsymbol{i} \vee \boldsymbol{j}) = h(\boldsymbol{i}) \cup h(\boldsymbol{j})$.
(iv) $h(\boldsymbol{i} \cdot \boldsymbol{j}) = h(\boldsymbol{i}) \otimes h(\boldsymbol{j})$.
(v) $h(\boldsymbol{i} \rightarrow \boldsymbol{j}) = h(\boldsymbol{i}) \Rrightarrow h(\boldsymbol{j})$.

We need to verify that these identities hold.

(i) $a \in h(-\boldsymbol{i})$ iff $-\boldsymbol{i} \in a$ iff $\boldsymbol{i} \notin a^*$ iff $a^* \notin h(\boldsymbol{i})$ iff $a \in -h(\boldsymbol{i})$.
(ii) $a \in h(\boldsymbol{i} \wedge \boldsymbol{j})$ iff $\boldsymbol{i} \wedge \boldsymbol{j} \in a$ iff $\boldsymbol{i} \in a$ and $\boldsymbol{j} \in a$ iff $a \in h(\boldsymbol{i})$ and $a \in h(\boldsymbol{j})$ iff $a \in h(\boldsymbol{i}) \cap h(\boldsymbol{j})$.
(iii) $a \in h(\boldsymbol{i} \vee \boldsymbol{j})$ iff $\boldsymbol{i} \vee \boldsymbol{j} \in a$ iff $\boldsymbol{i} \in a$ or $\boldsymbol{j} \in a$, as a is prime, iff $a \in h(\boldsymbol{i})$ or $a \in h(\boldsymbol{j})$ iff $a \in h(\boldsymbol{i}) \cup h(\boldsymbol{j})$.
(iv) $a \in h(\boldsymbol{i} \cdot \boldsymbol{j})$ iff $\boldsymbol{i} \cdot \boldsymbol{j} \in a$. One direction of the desired equivalence is straightforward. Suppose for some $b, c \in K$, $\boldsymbol{i} \in b$, $\boldsymbol{j} \in c$, and $Rbca$. From the definition of R and the fact that $\boldsymbol{j} \rightarrow (\boldsymbol{i} \cdot \boldsymbol{j}) \in b$, it then follows that $\boldsymbol{i} \cdot \boldsymbol{j} \in a$. So, $b \in h(\boldsymbol{i})$, $c \in h(\boldsymbol{j})$, and $Rbca$. So, $a \in h(\boldsymbol{i}) \otimes h(\boldsymbol{j})$.

For the other direction, suppose $\boldsymbol{i} \cdot \boldsymbol{j} \in a$. Next, consider $[\boldsymbol{i})$ and $[\boldsymbol{j})$. Suppose $\boldsymbol{m} \rightarrow \boldsymbol{n} \in [\boldsymbol{i})$ and $\boldsymbol{m} \in [\boldsymbol{j})$, so $\boldsymbol{i} \leq \boldsymbol{m} \rightarrow \boldsymbol{n}$ and $\boldsymbol{j} \leq \boldsymbol{m}$. It follows that $\boldsymbol{i} \cdot \boldsymbol{j} \leq \boldsymbol{n}$, so $\boldsymbol{n} \in a$, as a is a filter. By appeal to Zorn's lemma, we can extend a to a maximal filter, and maximal filters are prime. By similar appeal to Zorn's lemma, there are prime filters b and c such that $[\boldsymbol{i}) \subseteq b$ and $[\boldsymbol{j}) \subseteq c$ and $Rbca$. Therefore, $b \in h(\boldsymbol{i})$and $c \in h(\boldsymbol{j})$, so $a \in h(\boldsymbol{i}) \otimes h(\boldsymbol{j})$.
(v) $a \in h(\boldsymbol{i} \rightarrow \boldsymbol{j})$ iff $\boldsymbol{i} \rightarrow \boldsymbol{j} \in a$. As in the previous case, one direction of the desired equivalence is straightforward. Suppose for $b, c \in K$, $Rabc$, and $b \in h(\boldsymbol{i})$. It follows that $\boldsymbol{i} \in b$, so by definition of R, $\boldsymbol{j} \in c$. Therefore, $c \in h(\boldsymbol{j})$, so $a \in h(\boldsymbol{i}) \Rrightarrow h(\boldsymbol{j})$.

Suppose $\boldsymbol{i} \rightarrow \boldsymbol{j} \notin a$. It is the case that $[\boldsymbol{i})$ and $a + [\boldsymbol{i})$ are filters. Further, for all $\boldsymbol{m}, \boldsymbol{n} \in G$, if $\boldsymbol{m} \rightarrow \boldsymbol{n} \in a$ and $\boldsymbol{m} \in [\boldsymbol{i})$, then $\boldsymbol{n} \in a + [\boldsymbol{i})$. Suppose that $\boldsymbol{j} \in a + [\boldsymbol{i})$. Then there is some $\boldsymbol{m} \in G$ such that $\boldsymbol{m} \rightarrow \boldsymbol{j} \in a$ and $\boldsymbol{m} \in [\boldsymbol{i})$. But then, $\boldsymbol{i} \leq \boldsymbol{m}$, so $\boldsymbol{m} \rightarrow \boldsymbol{j} \leq \boldsymbol{i} \rightarrow \boldsymbol{j}$, whence $\boldsymbol{i} \rightarrow \boldsymbol{j} \in a$, contradicting an assumption. So, $\boldsymbol{j} \notin a + [\boldsymbol{i})$. Using Zorn's lemma, we can extend $[\boldsymbol{i})$ and $a + [\boldsymbol{i})$ to prime filters b and c such that $[\boldsymbol{i}) \subseteq b$, $a + [\boldsymbol{i}) \subseteq c$, $\boldsymbol{j} \notin c$, and $Rabc$. Thus, as $b \in h(\boldsymbol{i})$ and $c \notin h(\boldsymbol{j})$, $a \notin h(\boldsymbol{i}) \Rrightarrow h(\boldsymbol{j})$.

To establish that h is an isomorphism on its range, we must show that if $\boldsymbol{i} \neq \boldsymbol{j}$, then $h(\boldsymbol{i}) \neq h(\boldsymbol{j})$. Suppose $\boldsymbol{i} \neq \boldsymbol{j}$. We can assume $\boldsymbol{i} \not\leq \boldsymbol{j}$. By the Prime filter separation principle, there is a prime filter a on $\Phi(G)$ such that $\boldsymbol{i} \in a$ and $\boldsymbol{j} \notin a$.[4] Therefore, $h(\boldsymbol{i}) \neq h(\boldsymbol{j})$. □

We can say a bit more about the frame $\Phi(G)$. By assumption, $[\mathbb{1})$ is a prime filter, so $[\mathbb{1}) \in K$. Suppose $a \in N$. Then $[\mathbb{1}) \subseteq a$, so $[\mathbb{1}) \leq a$. This means that $[\mathbb{1})$ is the least point in N, so $\Phi(G)$ is a reduced frame.

The preceding theorem says that G is isomorphic to a subalgebra of $\Gamma(\Phi(G))$. There's more information about what that subalgebra looks like. Suppose for $a, b \in K$, $a \leq b$. By definition, $h(\boldsymbol{i})$ is the set of prime filters $c \in K$ such that $\boldsymbol{i} \in c$. Suppose $a \in h(\boldsymbol{i})$; then $\boldsymbol{i} \in a$. By assumption, $a \leq b$, so $a \subseteq b$. Therefore, $\boldsymbol{i} \in b$, so $b \in h(\boldsymbol{i})$. Thus, $h(\boldsymbol{i}) \in \mathcal{H}_{\phi(G)}$, which is to say that $h(\boldsymbol{i})$ is a proposition.

The carrier set of that algebra is

$$\{X \subseteq K : X \in \mathcal{H}_{\Phi(G)} \text{ and } X = h(a), \text{ for some } a \in G\},$$

so that set is a set of propositions, which we can call Prop. From the proof of the theorem, Prop is closed under the operations $\Rrightarrow$, $\otimes$, $\cap$, $\cup$, and $-$. Further, as $[\mathbb{1})$ is a prime filter, $[\mathbb{1}) \in K$ and $h(\mathbb{1}) = N$. So, $N \in$ Prop. Thus, Prop is model-like.

5.4 Notable Matrices

In this section, we will provide a small collection of useful matrices for investigating relevant logics.[5] We will comment on what they have been used for and what logics are valid on these matrices. Given the importance of prime filters in the constructions of the previous section, we will identify the prime filters. We will give the tables for the arrow and negation operators of each matrix. The conjunction and disjunction tables can be read off the Hasse diagrams. The fusion operators can be recovered via a definition: $\boldsymbol{a} \cdot \boldsymbol{b} = glb(\{\boldsymbol{x} \in V : \boldsymbol{a} \leq \boldsymbol{b} \to \boldsymbol{x}\})$. For matrices that satisfy both (CI5), $\boldsymbol{ab} \leq \boldsymbol{ba}$, and (CN1), if $\boldsymbol{ab} \leq \boldsymbol{c}$, then $\boldsymbol{a}(-\boldsymbol{c}) \leq -\boldsymbol{b}$, fusion can be defined as $\boldsymbol{a} \cdot \boldsymbol{b} = \neg(\boldsymbol{b} \to \neg\boldsymbol{a})$.

The first matrix is the $RM3$ matrix, presented in Table 5.2. This is a three-element matrix that is a model of RM.[6] Since RM is valid on this matrix, R is also valid. This matrix has been used for many purposes. One notable application is showing that relevant arithmetic has a finitary consistency proof.[7]

We can see that the axiom of weakening, $p \to (q \to p)$, is invalid on this matrix. Set $v(p) = \frac{1}{2}$ and $v(q) = \mathbf{1}$. Then we compute

$$v(p \to (q \to p)) = v(p) \to (v(q) \to v(p)) = \frac{\mathbf{1}}{\mathbf{2}} \to \left(\mathbf{1} \to \frac{\mathbf{1}}{\mathbf{2}}\right) = \frac{\mathbf{1}}{\mathbf{2}} \to \mathbf{0} = \mathbf{0}.$$

Not all non-theorems of R are invalidated on this matrix. For example, $A \to (A \to A)$ is valid on this algebra while being a non-theorem of R.

4 Dunn and Hardegree (2001, 298).

5 For more examples of matrices, see, for example, Tokarz (1980).

6 See Anderson and Belnap (1975, §29.3) and Avron (1992) for more on RM.

7 See Meyer and Mortensen (1984) or Meyer (2021c).

Table 5.2. Matrix *RM*3, $\mathbb{1} = \frac{1}{2}$

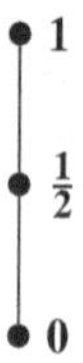

$\rightarrow$	0	$\frac{1}{2}$	1	$\neg$
0	1	1	1	1
$\frac{1}{2}$	0	$\frac{1}{2}$	1	$\frac{1}{2}$
1	0	0	1	0

Table 5.3. Ternary relation and star for frame generated by *RM*3

R	a	b	c	d	$*$
a	abc	abc	c	$abcd$	b
b	abc	bc	c	$abcd$	a
c	c	c	c	$abcd$	d
d	$abcd$	$abcd$	$abcd$	$abcd$	c

The prime filters of $RM3$ are easy to identify. The ordering is linear, and all filters on linear orders are prime. To see, this, note that $\boldsymbol{x} \leq \boldsymbol{y}$ iff $\boldsymbol{x} \vee \boldsymbol{y} = \boldsymbol{y}$. Linearity means that for all $\boldsymbol{x}$ and $\boldsymbol{y}$, either $\boldsymbol{x} \leq \boldsymbol{y}$ or $\boldsymbol{y} \leq \boldsymbol{x}$, so $\boldsymbol{x} \vee \boldsymbol{y} = \boldsymbol{x}$ or $\boldsymbol{x} \vee \boldsymbol{y} = \boldsymbol{y}$, which suffices for primeness. So, all upward-closed sets are prime filters. That makes the prime filters $\emptyset$, $\{\mathbf{1}\}$, $\{\mathbf{1}, \frac{1}{2}\}$, and $\{\mathbf{1}, \frac{1}{2}, \mathbf{0}\}$. Some of the algebras to be discussed below are not linear, and identifying the prime filters on those algebras requires more work.

As an example of the general construction of the preceding section, we will present the frame that is generated by $RM3$. The set $K = \{a, b, c, d\}$, where

- $a = \{\mathbf{1}\}$,
- $b = \{\mathbf{1}, \frac{1}{2}\}$,
- $c = \{\mathbf{1}, \frac{1}{2}, \mathbf{0}\}$, and
- $d = \emptyset$.

The set $N = \{b, c\}$. The star operation and ternary relation are defined as in Table 5.3. This frame is an R-frame as can be verified by checking the frame conditions.

The frame can be turned into an algebra using the construction of the preceding section. The set of points $J = \{\boldsymbol{x}, \boldsymbol{y}, \boldsymbol{z}, \boldsymbol{v}, \boldsymbol{w}\}$, where the points are defined as

- $\boldsymbol{x} = \emptyset$,
- $\boldsymbol{y} = \{c\}$,
- $\boldsymbol{z} = \{b, c\}$,
- $\boldsymbol{v} = \{a, b, c\}$, and
- $\boldsymbol{w} = \{a, b, c, d\}$.

The points are ordered linearly as $\boldsymbol{x} \leq \boldsymbol{y} \leq \boldsymbol{z} \leq \boldsymbol{v} \leq \boldsymbol{w}$ and $\mathbb{1} = \boldsymbol{z}$. The implication and negation operations are defined as in Table 5.4.

Table 5.4. Implication and negation operations for the generated algebra

$\to$	x	y	z	v	w	$\neg$
x	w	w	w	w	w	w
y	x	v	v	v	w	v
z	x	y	z	v	w	z
v	x	y	y	v	w	y
w	x	x	x	x	w	x

The matrix $RM3$ is a subalgebra of the generated matrix. The points $\mathbf{0}, \frac{1}{2}$, and $\mathbf{1}$ correspond, respectively, to $\boldsymbol{x}, z$, and $\boldsymbol{w}$. Restricting the operations in Table 5.4 to the set $\{\boldsymbol{x}, z, \boldsymbol{w}\}$ yields the $RM3$ operations.

The Belnap–Dunn matrix M_4 has played an important role in the development of relevant logics.[8] It is the matrix characteristic of the logic FDE, which was presented in the appendix of Chapter 2. M_4 is often studied without an implication table, so the focus is on the vocabulary $\{\wedge, \vee, \neg\}$, which is sometimes known as the TRUTH-FUNCTIONAL VOCABULARY. Thus, M_4 is often studied as just a De Morgan algebra.

Following Belnap (1977a; 1977b), we can think of the values of M_4 as representing the status of claims in a database. The database receives information about different basic claims. If it receives information saying that a certain claim holds, then it records a T, and if it receives information that the claim doesn't hold, then it records an F. If it receives conflicting information, it will record both T and F, and if it receives no information about a claim, it will record nothing. The values of the algebra can be viewed as representing a set of "truth values," T and F, with $\mathbf{0}$ representing $\{F\}$, $\boldsymbol{b}$ (both) representing $\{T, F\}$, $\mathbf{1}$ representing $\{T\}$, and $\boldsymbol{n}$ (neither) representing $\emptyset$. The designated values then represent sets containing T. The tables for the connectives come naturally given this interpretation of the values.

A given De Morgan algebra may support more than one implication table, as is the case with M_4. We have presented two different implication tables here. The matrix in Table 5.5 is known as $BN4$.[9] The matrix in Table 5.6 is known as $E4$. As can be seen, the implication tables are quite similar, differing only on some entries where $\boldsymbol{n}$ is an input.

Where A and B are truth-functional, an entailment $A \models_{\mathsf{FDE}} B$ is valid in FDE iff for all valuations v, $v(A) \leq v(B)$, which is equivalent to $\mathbf{1} \leq v(A \to B)$. A feature of the valid formulas of FDE is that they have a form of the variable-sharing property. In particular, if $A \to B$ is valid in FDE, then A and B share a propositional variable.

[8] See Dunn (1966), Dunn (1976), and Anderson et al. (1992, §81).

[9] There is an intuitive idea behind the $BN4$ implication table. As Restall (2000, 171–172) describes the idea, we can view each of the different elements as representing a set of truth values, T and F, as above. We can compute the table by using the following biconditionals.

- $T \in \boldsymbol{x} \to \boldsymbol{y}$ iff both if $T \in \boldsymbol{x}$ then $T \in \boldsymbol{y}$ and if $F \in \boldsymbol{y}$ then $F \in \boldsymbol{x}$.
- $F \in \boldsymbol{x} \to \boldsymbol{y}$ iff $T \in \boldsymbol{x}$ and $F \in \boldsymbol{y}$.

Brady (1982) showed that this idea can be used to generate a two-element ternary relational frame that agrees with the matrix on valid formulas.

Table 5.5. Belnap–Dunn matrix M_4 with the *BN*4 implication table, $\mathbb{1} = \boldsymbol{b}$.

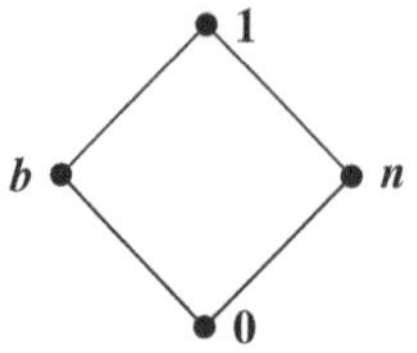

→	**0**	*b*	*n*	**1**	¬
0	**1**	**1**	**1**	**1**	**1**
b	**0**	*b*	*n*	**1**	*b*
n	*n*	*n*	**1**	**1**	*n*
1	**0**	**0**	*n*	**1**	**0**

Table 5.6. Belnap–Dunn matrix M_4 with the *E*4 implication table, $\mathbb{1} = \boldsymbol{b}$

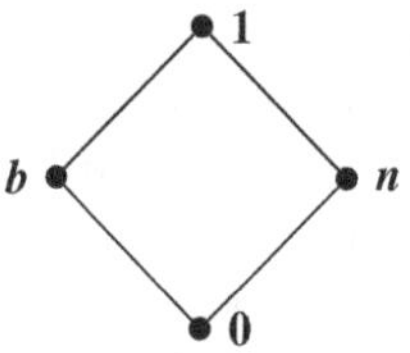

→	**0**	*b*	*n*	**1**	¬
0	**1**	**1**	**1**	**1**	**1**
b	**0**	*b*	**0**	**1**	*b*
n	**0**	**0**	*b*	**1**	*n*
1	**0**	**0**	**0**	**1**	**0**

To prove this, suppose that A and B do not share a propositional variable. Then set $v(p) = \boldsymbol{b}$ for all atoms p in A and $v(q) = \boldsymbol{n}$ for all atoms q in B. We can prove by an induction on the construction of A and of B, that $v(A) = \boldsymbol{b}$ and that $v(B) = \boldsymbol{n}$, so $v(A) \not\leq v(B)$.

As $M4$ is not linear, it is not the case that every filter is a prime filter. An example of a non-prime filter is $\{\mathbf{1}\}$. Since $\boldsymbol{b} \vee \boldsymbol{n} = \mathbf{1}$ but $\boldsymbol{b} \notin \{\mathbf{1}\}$ and $\boldsymbol{n} \notin \{\mathbf{1}\}$, $\{\mathbf{1}\}$ is not a prime filter. Extending $\{\mathbf{1}\}$ with the addition of either $\boldsymbol{b}$ or $\boldsymbol{n}$ results in a prime filter. The prime filters are $\emptyset, \{\boldsymbol{b}, \mathbf{1}\}$, $\{\boldsymbol{n}, \mathbf{1}\}$, and $\{\mathbf{0}, \boldsymbol{b}, \boldsymbol{n}, \mathbf{1}\}$.

Finally, we will comment on what logics these matrices validate. As Robles and Méndez (2016) show, $BN4$ validates an extension of RW.[10] It invalidates some R-theorems, such as (N2). Set $v(p) = \boldsymbol{n}$, and then compute

$$v((p \to \neg p) \to \neg p) = (v(p) \to \neg v(p)) \to \neg v(p) = (\boldsymbol{n} \to \neg \boldsymbol{n}) \to \neg \boldsymbol{n} = \mathbf{1} \to \boldsymbol{n} = \mathbf{0},$$

which demonstrates the invalidity of (N2).

The logic validated by $E4$ is close to E, although it is not an extension of E, as shown by Robles and Méndez (2016). Of the logics we have named, TW is the strongest of which it is a proper extension. This matrix also validates (I1) and (I7). It does not validate (N2). Set $v(p) = \boldsymbol{n}$. Then we compute

$$v((p \to \neg p) \to \neg p) = (v(p) \to \neg v(p)) \to \neg v(p) = (\boldsymbol{n} \to \neg \boldsymbol{n}) \to \neg \boldsymbol{n} = \boldsymbol{b} \to \boldsymbol{n} = \mathbf{0},$$

so (N2) is not valid according to $E4$.

[10] Brady (1982) axiomatized this logic first. Robles and Méndez (2016) provided an alternative axiomatization.

Table 5.7. Matrix $C4$, $\mathbb{1} = 1$

$\to$	0	1	2	3	$\neg$
0	3	3	3	3	0
1	0	1	2	3	2
2	0	0	1	3	1
3	0	0	0	3	3

One of the axioms over which $BN4$ and $E4$ differ is (I5). The axiom is valid in the former but not in the latter. To see that it is invalid, set $v(p) = v(q) = \boldsymbol{n}$ and compute

$$v(p \to ((p \to q) \to q)) = \boldsymbol{n} \to ((\boldsymbol{n} \to \boldsymbol{n}) \to \boldsymbol{n}) = \boldsymbol{n} \to (\boldsymbol{b} \to \boldsymbol{n}) = \boldsymbol{n} \to \mathbf{0} = \mathbf{0}.$$

Another axiom over which they differ is (I1), which is not valid in $BN4$. For the invalidity, set $v(p) = \boldsymbol{n}$ and $v(q) = \mathbf{0}$ and compute

$$v((p \wedge (p \to q)) \to q) = (\boldsymbol{n} \wedge (\boldsymbol{n} \to \mathbf{0})) \to \mathbf{0} = (\boldsymbol{n} \wedge \boldsymbol{n}) \to \mathbf{0} = \boldsymbol{n} \to \mathbf{0} = \boldsymbol{n}.$$

Thus, (I1) is not valid in $BN4$, although it is valid in $E4$.

The next matrix is the four-valued matrix $C4$, presented in Table 5.7. $C4$ has values that are linearly ordered and it validates R. A notable feature of $C4$ is that the $v(\neg \mathrm{t}) = \mathbf{2}$ while $v(\mathrm{t}) = \mathbf{1}$. Some contradictory formulas can be assigned greater values than some logical truths. For example, if $v(p) = \mathbf{1}$, then $v(p \to p) = \mathbf{1}$, while $v(\neg(p \to p)) = \mathbf{2}$. This is, in a sense, a *deeply* non-classical matrix. Slaney (2022) shows that $C4$ has a special place among the finite models of relevant arithmetic, a topic to which we will return in Chapter 13.

The next matrix is Belnap's M_8, displayed in Table 5.8.[11] This matrix was used to show that E has the variable-sharing property. In fact, R is valid on the matrix, and the technique can be used to show that R has the variable-sharing property.

Table 5.8. Belnap's matrix M_8, $\mathbb{1} = +\mathbf{0}$

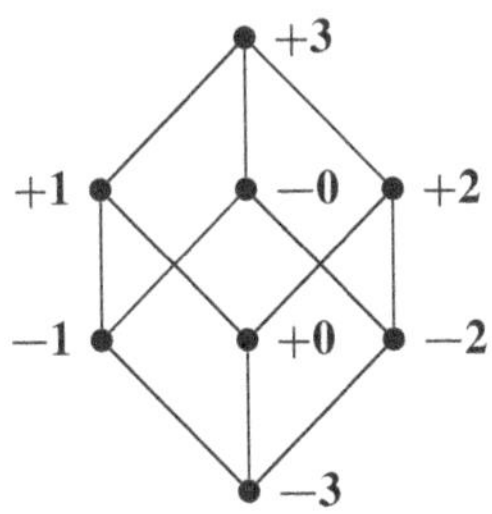

$\to$	−3	−2	−1	−0	+0	+1	+2	+3	$\neg$
−3	+3	+3	+3	+3	+3	+3	+3	+3	+3
−2	−3	+2	−3	+2	−3	−3	+2	+3	+2
−1	−3	−3	+1	+1	−3	+1	−3	+3	+1
−0	−3	−3	−3	+0	−3	−3	−3	+3	+0
+0	−3	−2	−1	−0	+0	+1	+2	+3	−0
+1	−3	−3	−1	−1	−3	+1	−3	+3	−1
+2	−3	−2	−3	−2	−3	−3	+2	+3	−2
+3	−3	−3	−3	−3	−3	−3	−3	+3	−3

[11] Anderson and Belnap (1975, 252–254). Note that Belnap called it "M_0." A different name is used here for some uniformity with the names of other matrices.

To show that R has the variable-sharing property, let $A \rightarrow B$ be valid, and suppose A and B share no propositional atoms and do not contain t. For all atoms p in A, set $v(p) \in \{\mathbf{+1}, \mathbf{-1}\}$, and for all atoms q in B, set $v(q) \in \{\mathbf{+2}, \mathbf{-2}\}$. As can be shown by induction on the construction of A, $v(A) \in \{\mathbf{+1}, \mathbf{-1}\}$, and similarly $v(B) \in \{\mathbf{+2}, \mathbf{-2}\}$. But then, $v(A) \rightarrow v(B) = \mathbf{-3}$, so $A \rightarrow B$ is not valid on the algebra. As all the theorems of R are valid on the algebra, $A \rightarrow B$ is not a theorem of R, contradicting an assumption. So, A and B share an atom and thus R has the variable-sharing property. Strengthenings of the variable-sharing property will be discussed in Section 5.5.

The short sketch of a proof of the variable-sharing property in the preceding paragraph suggests some general features of a matrix that are sufficient for using it to prove that a logic has the variable-sharing property. There need to be two disjoint sets of values $\mathfrak{X}$ and $\mathfrak{Y}$ that are closed under the operations of the matrix, and if $\boldsymbol{x} \in \mathfrak{X}$ and $\boldsymbol{y} \in \mathfrak{Y}$, then $\mathbb{1} \not\leq \boldsymbol{x} \rightarrow \boldsymbol{y}$. As the proof sketch shows, $\{\mathbf{+1}, \mathbf{-1}\}$ and $\{\mathbf{+2}, \mathbf{-2}\}$ are the two desired sets of values. Robles and Méndez (2011a; 2012) identified this sufficient condition, and they explore matrices for variable sharing in a more general context.[12]

As with M_4, M_8 is not linear, and many filters fail to be prime on this algebra. The prime filters are $\emptyset$, $\{\mathbf{-1}, \mathbf{+1}, \mathbf{-0}, \mathbf{+3}\}$, $\{\mathbf{-2}, \mathbf{+2}, \mathbf{-0}, \mathbf{+3}\}$, $\{\mathbf{+0}, \mathbf{+1}, \mathbf{+2}, \mathbf{+3}\}$, and $\{\mathbf{-0}, \mathbf{+0}, \mathbf{-1}, \mathbf{+1}, \mathbf{-2}, \mathbf{+2}, \mathbf{-3}, \mathbf{+3}\}$.

The logic of M_8 is the set of formulas that hold in every valuation on M_8. This logic has been axiomatized, by Brady (2003, 128), but its axiomatization uses the rule (γ).[13] This logic is one of the two maximal extensions of R with the variable-sharing property.[14] The other such logic will be described shortly.

The next matrix we will consider, in Table 5.9, is known as the CRYSTAL LATTICE, CL, although "M_6" would fit better with our naming conventions. It has been used in automated theorem proving by Thistlewaite et al. (1988). Routley et al. (1982, §3.6) use it to show that an extension of R has the variable-sharing property. Urquhart uses it to show the failure of an interpolation property for many strong relevant logics, a result we will look at in Section 5.6. Overall, it is quite a useful structure.[15]

As can be seen by inspection of Table 5.9, the sets $\{\boldsymbol{c}\}$ and $\{\boldsymbol{d}\}$ are both closed under the operations of the matrix. Further, $\boldsymbol{c} \rightarrow \boldsymbol{d} = \boldsymbol{f}$, so this matrix can be used to show that a logic has the variable-sharing property. We can use this fact to show that one of

[12] For more on variable sharing using algebraic techniques, see Robles et al. (2010; 2011) and Méndez et al. (2012).

[13] The axiomatization adds the following axioms to R, with the rule (γ), where $A \supset B$ is $\neg A \vee B$. In this notation, (γ) is the rule $A, A \supset B \Rightarrow B$.

- $(A \wedge \neg A \wedge B) \rightarrow (A \rightarrow B)$
- $A \supset ((\neg A \wedge B) \rightarrow (A \vee (A \rightarrow B)))$
- $(A \wedge B) \supset ((\neg A \rightarrow A) \vee (A \rightarrow B))$
- $(\neg A \wedge (A \rightarrow B)) \supset (\neg A \rightarrow (A \vee (A \rightarrow B)))$
- $((\neg A \rightarrow A) \wedge (A \rightarrow B)) \supset ((B \rightarrow A) \vee (C \rightarrow B))$
- $A \supset ((\neg A \rightarrow A) \vee ((A \wedge \neg A) \rightarrow B))$

We will discuss (γ) more in Section 9.8. As far as we know, there is no published alternative axiomatization of the logic that does not use that rule.

[14] See Swirydowicz (1999).

[15] Brady (2003, 222–230).

Table 5.9. Crystal lattice CL, $\mathbb{1} = e$

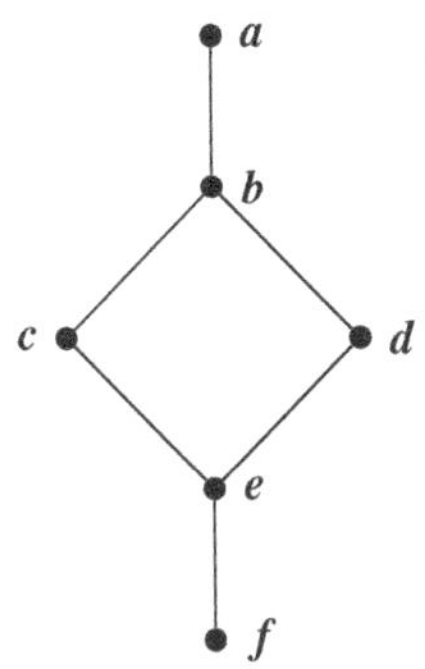

$\to$	f	e	c	d	b	a	$\neg$
f	a	a	a	a	a	a	a
e	f	e	c	d	b	a	b
c	f	f	c	f	c	a	c
d	f	f	f	d	d	a	d
b	f	f	f	f	e	a	e
a	f	f	f	f	f	a	f

Table 5.10. *M*5, Meyer's matrix, $\mathbb{1} = N$

$\to$	0	N	A	P	1	$\neg$
0	N	N	N	N	N	1
N	0	N	N	N	N	P
A	0	0	N	N	N	A
P	0	0	0	N	N	N
1	0	0	0	0	N	0

the paradoxes of implication is not a theorem of R. Let $v(q) = \boldsymbol{c}$ and $v(p) = \boldsymbol{d}$. Then $v(p \to p) = \boldsymbol{d}$, so $v(q \to (p \to p)) = \boldsymbol{c} \to \boldsymbol{d} = \boldsymbol{f}$.

Much like $E4$ and M_8, CL is not linearly ordered. The prime filters on CL are $\emptyset$, $\{\boldsymbol{a}\}$, $\{\boldsymbol{c}, \boldsymbol{b}, \boldsymbol{a}\}$, $\{\boldsymbol{d}, \boldsymbol{b}, \boldsymbol{a}\}$, $\{\boldsymbol{e}, \boldsymbol{c}, \boldsymbol{d}, \boldsymbol{b}, \boldsymbol{a}\}$, and $\{\boldsymbol{f}, \boldsymbol{e}, \boldsymbol{c}, \boldsymbol{d}, \boldsymbol{b}, \boldsymbol{a}\}$.

The logic of CL, which is the set of formulas that holds in every valuation on CL, can be axiomatized by adding the following two axioms to R.[16]

(CL1) $(\neg A \wedge B) \to ((\neg A \to A) \vee (A \to B))$.
(CL2) $A \vee (A \to B)$.

This is the second of the two maximal extensions of R with the variable-sharing property, as shown by Swirydowicz (1999).

The next matrix we will consider, displayed in Table 5.10, is the five-element $M5$. This algebra validates E but not R. One striking feature of this algebra is that implications are only ever "barely satisfied." They take at most the value N, which one might gloss as saying that they can be true, but only just. One would expect that R would not be valid on this algebra, as every formula in R is equivalent to an implication, but not every formula must take the value N.

[16] Brady (2003, 114).

For the next point, let us work in the language without fusion. The matrix is notable for being used in a quick proof that E has the ACKERMANN PROPERTY, which says that no formula of the form $B \to (C \to D)$, where B does not contain implication or t, is a theorem of E.[17] The algebra and the proof are due to Meyer. For the proof, let $v(p) = \boldsymbol{A}$ for all atoms p in B. Then $v(B) = \boldsymbol{A}$. No matter what values are assigned to atoms in C and D, $v(C \to D) \in \{\boldsymbol{0}, \boldsymbol{N}\}$. But then, $v(B \to (C \to D)) = \boldsymbol{0}$, as $\boldsymbol{A} \to \boldsymbol{N} = \boldsymbol{A} \to \boldsymbol{0} = \boldsymbol{0}$. Thus E has no theorems of the described form.

Since $M5$ validates E but not R, it will be worthwhile to see an example theorem of R that is invalid here. The formula $p \to ((p \to q) \to q)$ is a theorem of R and has a form described in the proof sketch earlier. Set $v(p) = \boldsymbol{A}$ and $v(q) = \boldsymbol{A}$. Then $v(p \to q) = v((p \to q) \to q) = \boldsymbol{N}$, so $v(p \to ((p \to q) \to q)) = \boldsymbol{0}$. For a slightly more involved example, consider $(p \to (q \to r)) \to (q \to (p \to r))$. Set $v(p) = \boldsymbol{N}$ and $v(q) = v(r) = \boldsymbol{P}$. Then

$$v(p \to (q \to r)) = \boldsymbol{N} \to \boldsymbol{N} = \boldsymbol{N}$$

and

$$v(q \to (p \to r)) = \boldsymbol{P} \to \boldsymbol{0} = \boldsymbol{0},$$

so $v((p \to (q \to r)) \to (q \to (p \to r))) = \boldsymbol{0}$.

The prime filters are straightforward to identify. As $M5$ is linearly ordered, all upwardly closed subsets are prime filters. That makes the prime filters $\emptyset$, $\{\boldsymbol{1}\}$, $\{\boldsymbol{1}, \boldsymbol{P}\}$, $\{\boldsymbol{1}, \boldsymbol{P}, \boldsymbol{A}\}$, $\{\boldsymbol{1}, \boldsymbol{P}, \boldsymbol{A}, \boldsymbol{N}\}$, and $\{\boldsymbol{1}, \boldsymbol{P}, \boldsymbol{A}, \boldsymbol{N}, \boldsymbol{0}\}$.

The example matrices so far have all been finite. Nothing in the definitions or the techniques discussed requires limiting oneself to finite matrices. There are infinite matrices, and a notable one was presented by Meyer (1970).[18] It will take a bit more work to describe than the finite matrices. The set of points J is $\mathbb{Z} \cup \{\omega\}$, the set of integers together with an additional point, ω. Let $\mathbb{Z}^-$ and $\mathbb{Z}^+$ be the sets, respectively, of negative integers and positive integers, and for $\boldsymbol{y}, \boldsymbol{z} \in \mathbb{Z}^+$, let $\gcd(\boldsymbol{y}, \boldsymbol{z})$ be the greatest common divisor of $\boldsymbol{y}$ and $\boldsymbol{z}$, and $\mathrm{lcm}(\boldsymbol{y}, \boldsymbol{z})$ be the least common multiple of $\boldsymbol{y}$ and $\boldsymbol{z}$.

The negation operation, which will be written $\neg$, is defined, for $\boldsymbol{x} \in J$, as follows.

$$\neg \boldsymbol{x} = \begin{cases} \omega & \text{if } \boldsymbol{x} = \boldsymbol{0}, \\ \boldsymbol{0} & \text{if } \boldsymbol{x} = \omega, \\ -\boldsymbol{y} & \text{if } \boldsymbol{x} = +\boldsymbol{y}, \\ +\boldsymbol{y} & \text{if } \boldsymbol{x} = -\boldsymbol{y}. \end{cases}$$

The lattice operations are defined as follows.

- For $\boldsymbol{x} \in J$, $\boldsymbol{x} \wedge \boldsymbol{0} = \boldsymbol{0} \wedge \boldsymbol{x} = \boldsymbol{0}$ and $\boldsymbol{x} \vee \boldsymbol{0} = \boldsymbol{0} \vee \boldsymbol{x} = \boldsymbol{x}$.
- For $\boldsymbol{x} \in J$, $\boldsymbol{x} \wedge \omega = \omega \wedge \boldsymbol{x} = \boldsymbol{x}$ and $\boldsymbol{x} \vee \omega = \omega \vee \boldsymbol{x} = \omega$.

[17] See Anderson and Belnap (1975, 243) for details, along with discussion of the more general Maksimova property.

[18] Note, though, that there was an error in Meyer's presentation regarding the lattice operations when one input is positive and the other is negative.

- For $y, z \in \mathbb{Z}^+$, $y \wedge z = \gcd(y, z)$ and $y \vee z = \mathrm{lcm}(y, z)$.
- For $u, v \in \mathbb{Z}^-$, $u \wedge v = \neg(\neg u \vee \neg v)$ and $u \vee v = \neg(\neg u \wedge \neg v)$.
- For $u \in \mathbb{Z}^-$ and $y \in \mathbb{Z}^+$, $u \wedge y = y \wedge u = y$ and $u \vee y = y \vee u = u$.

The resulting lattice order, part of which is displayed in Table 5.11, is not linear, with $\mathbf{0} < y < u < \omega$, for $y \in \mathbb{Z}^+$ and $u \in \mathbb{Z}^-$. For $y, z \in \mathbb{Z}^+$, $y \leq z$ iff z is divisible by y, and for $u, v \in \mathbb{Z}^-$, $u \leq v$ iff $\neg u$ is divisible by $\neg v$, which is to say $\neg v \leq \neg u$.

Implication is defined as follows.

- For $x \in J$, $\mathbf{0} \to x = \omega$ and $x \to \omega = \omega$.
- For $x \in J$, if $x \neq \mathbf{0}$, then $x \to \mathbf{0} = \mathbf{0}$.
- For $x \in J$, if $x \neq \omega$, then $\omega \to x = \mathbf{0}$.
- For $y, z \in \mathbb{Z}^+$,

$$y \to z = \begin{cases} z/y & \text{if } z \text{ is divisible by } y, \\ \mathbf{0} & \text{otherwise.} \end{cases}$$

- For $u, v \in \mathbb{Z}^-$, $u \to v = \neg v \to \neg u$.
- For $u \in \mathbb{Z}^-$ and $y \in \mathbb{Z}^+$, $u \to y = \mathbf{0}$ and $y \to u = y \times u$.

Finally, $\mathbb{1} = +\mathbf{1}$. Thus, the only way to get a counterexample to a formula A in this algebra is to find a valuation v such that $v(A) = \mathbf{0}$.

The matrix affords a simple definition of the fusion operation on the positive integers. The general definition of fusion still holds, $x \cdot y = \neg(y \to \neg x)$; however, in the present context, it simplifies to multiplication, so $y \cdot z = y \times z$.

Let us call this matrix the DIVISION GROUPOID. The division groupoid has a feature not shared by any of the other lattices we have surveyed, for which we will need a definition. The division groupoid is the first example of a lattice that is not complete, as all finite lattices are complete. The set $\{\mathbf{2}^n \in \mathbb{Z}^+ : n \in \mathbb{Z}^+\}$ does not have a least upper bound. Every negative integer is an upper bound of each member of X. No negative integer, however, is the *least* upper bound. Thus the division groupoid is not complete.

Let us look at an example. We will show that $p \to (q \to q)$ has a counterexample in the division groupoid. Set $v(q) = v(p) = -\mathbf{1}$. Then $v(q \to q) = +\mathbf{1}$, and $v(p \to (q \to q)) = \mathbf{0}$, as $v(p) \in \mathbb{Z}^-$ and $v(q \to q) \in \mathbb{Z}^+$. Thus, $p \to (q \to q)$ is not a theorem of R. This example also works to show that $p \to (p \to p)$ is not a theorem of R.

Let us look at one more example since it took so much work to set out the matrix. We will show how to invalidate $(p \wedge \neg p) \to q$. Set $v(p) = -\mathbf{3}$ and $v(q) = +\mathbf{2}$. Then $v(p \wedge \neg p) = +\mathbf{3}$, and, as $+\mathbf{3} \to +\mathbf{2} = \mathbf{0}$, $v((p \wedge \neg p) \to q) = \mathbf{0}$.

The next example of an infinite matrix we will present is the SUGIHARA MATRIX.[19] It is related to the matrix $RM3$, as we will see. The set of elements J is $\mathbb{Z}$, the integers. The lattice operations are defined as $x \wedge y = \min(x, y)$ and $x \vee y = \max(x, y)$. The

[19] Anderson and Belnap (1975, 336–337).

Table 5.11. A partial diagram for the division groupoid, $\mathbb{1} = +1$

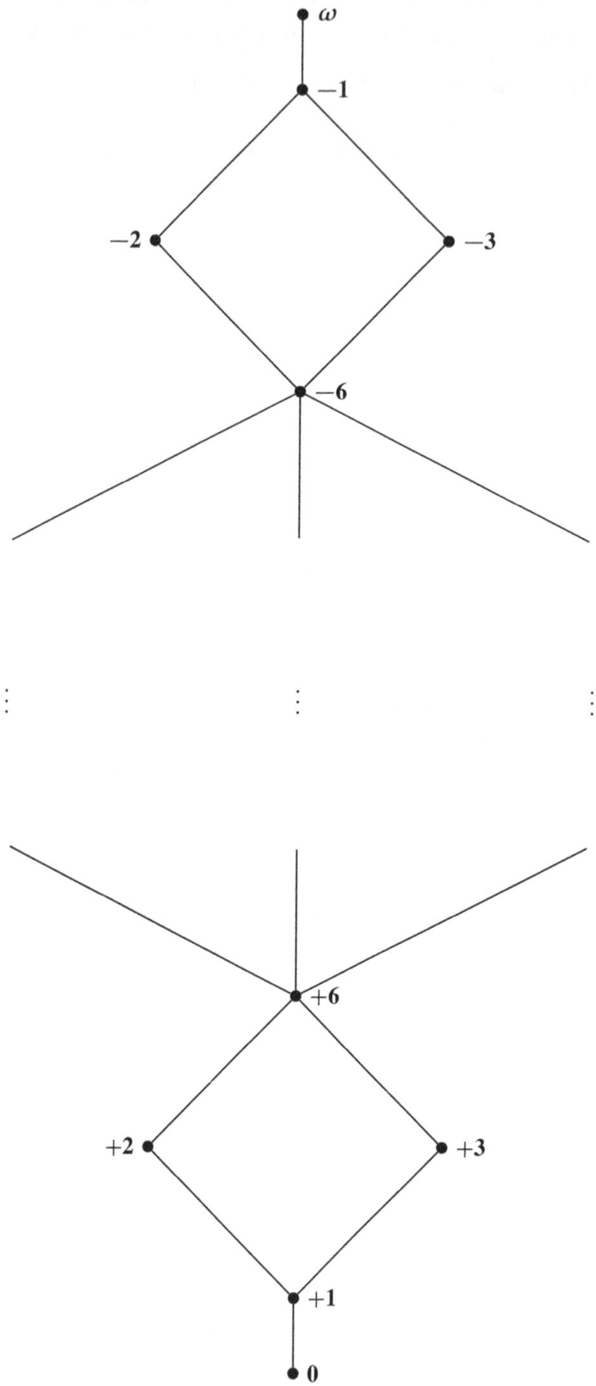

negation operation, $\neg \boldsymbol{x}$, flips the sign for $\boldsymbol{x} \neq \mathbf{0}$ and $\neg \mathbf{0} = \mathbf{0}$. The conditional operation is defined as

$$\boldsymbol{x} \to \boldsymbol{y} = \begin{cases} (\neg \boldsymbol{x}) \vee \boldsymbol{y} & \text{if } \boldsymbol{x} \leq \boldsymbol{y}, \\ (\neg \boldsymbol{x}) \wedge \boldsymbol{y} & \text{if } \boldsymbol{x} > \boldsymbol{y}. \end{cases}$$

Finally, $\mathbb{1} = \mathbf{0}$. It follows that a counterexample to a formula is a valuation that assigns the formula a negative number value.

The Sugihara matrix is related to $RM3$ in the following way. If we rewrite $RM3$ using $+\mathbf{1}, \mathbf{0}, -\mathbf{1}$, respectively, for the elements $\mathbf{1}, \frac{\mathbf{1}}{\mathbf{2}}, \mathbf{0}$, then there is an obvious embedding of $RM3$ into the Sugihara matrix. We can define another matrix, which we might call $RM5$, where $J = \{+\mathbf{2}, +\mathbf{1}, \mathbf{0}, -\mathbf{1}, -\mathbf{2}\}$ with the operations defined as earlier. We can continue to define RMn, for all odd $n \in \omega$. That would make the Sugihara matrix $RM\omega$, a natural limit of the process. This matrix is not complete, as $\bigvee J$ is not defined. It can be extended with a top and bottom element, $+\infty$ and $-\infty$ respectively, such that $-\infty < \boldsymbol{x} < +\infty$ for all $\boldsymbol{x} \in J$. The extended matrix is complete, although we have not yet specified the extended values for the negation and conditional operators.

The Sugihara matrix is notable for being characteristic for the logic RM, which is to say that a formula is a theorem of RM just in case it is valid in the matrix. As a consequence, R is valid in the matrix, as is $A \to (A \to A)$. To see why this formula is valid, note that $v(A \to A) = \max(v(A), -v(A))$, which is also the same as $v(A \to (A \to A))$. This value is guaranteed to be at least $\mathbf{0}$, and so there are no counterexamples on this matrix. As an example of how the matrix invalidates a non-theorem, consider $p \to (q \to p)$. Let $v(p) = -\mathbf{1}$ and $v(q) = +\mathbf{2}$. Then

$$v(q \to p) = (\neg +\mathbf{2}) \wedge -\mathbf{1} = -\mathbf{2} \wedge -\mathbf{1} = -\mathbf{2}.$$

So

$$v(p \to (q \to p)) = (\neg -\mathbf{1}) \wedge -\mathbf{2} = +\mathbf{1} \wedge -\mathbf{2} = -\mathbf{2}.$$

Thus, $p \to (q \to p)$ is invalid.

There are many other noteworthy matrices. Rather than continue to present examples, we will mention one invaluable tool for discovering matrices for counterexamples, John Slaney's program MaGIC, Matrix Generator for Implication Connectives.[20] Given a specification of a logic, MaGIC generates finite matrices that satisfy the logic. The user can input a formula to be falsified, the bad guy, and MaGIC will search for a matrix that satisfies the logic but provides a counterexample to the bad guy. If a counterexample is found, it will present the values, the lattice ordering, the value of $\mathbb{1}$, the negation table, and the implication table along with the counterexample valuation. It is an immensely useful program to whose outputs we will appeal in later chapters.[21]

20 See http://users.cecs.anu.edu.au/~jks/magic.html for more information. Slaney (1992) presents some technical details concerning the program.

21 For a short tutorial on MaGIC, see Logan (2024, 15–20).

5.5 Variable Sharing and Depth Relevance

In Section 5.4, we noted that Robles and Méndez have identified some sufficient features for a matrix to be usable to show that a logic has the variable-sharing property. Adapting their terminology, let us adopt the following definition.

DEFINITION 5.28 (Basic relevance matrix) *A matrix is a* BASIC RELEVANCE MATRIX *iff both*

- *there are two disjoint sets of values $\mathfrak{X} \subseteq K$ and $\mathfrak{Y} \subseteq K$ that are each closed under the operations of the matrix, and*
- *if $\boldsymbol{x} \in \mathfrak{X}$ and $\boldsymbol{y} \in \mathfrak{Y}$, then $\mathbb{1} \not\leq \boldsymbol{x} \rightarrow \boldsymbol{y}$.*

They show the following.

THEOREM 5.29 *If M is a basic relevance matrix and* L *is valid on M, then* L *has the variable-sharing property.*

Proof Suppose L is valid on a basic relevant matrix M. Suppose that A and B do not contain t and have no propositional atoms in common. Let v be such that $v(p) \in \mathfrak{X}$, for all atoms p occurring in A, and $v(q) \in \mathfrak{Y}$, for all atoms q occurring in $\mathfrak{Y}$. From the definition of a basic relevant matrix, it follows that $v(A) \in \mathfrak{X}$ and $v(B) \in \mathfrak{Y}$. It then follows that $\mathbb{1} \not\leq v(A \rightarrow B)$, so $\not\vdash_L A \rightarrow B$. □

Tedder (2025) shows that the existence of two such subsets is also necessary, in addition to being sufficient, for demonstrating variable sharing.

As the name of the matrices suggests, this form of variable sharing is relatively weak. While the property is weak, the weakness is, we think, a philosophical strength. The variable-sharing property is simple to state, and it is exceedingly straightforward to understand. It seems to capture a basic type of formal relevance, and that makes it suitable for defining the class of relevant logics, as discussed by Standefer (2025b). The class of logics defined in terms of just the variable-sharing property goes far beyond the logics that are the focus of this book. The logics discussed in this book have a lot going for them, but they do not exhaust the class of relevant logics.

While the basic form of variable sharing is relatively weak, there are strengthened forms available.[22] For presenting these forms, we will drop fusion and t from the language. This is in part a simplification and in part because some of the forms do not appear to work well with them.

The first strengthening involves assigning each formula occurrence a polarity, antecedent part (ap), or consequent part (cp). This is defined recursively as follows.

DEFINITION 5.30 (Polarity) *Given a formula A, the polarity of the occurrences of subformulas of A is defined recursively as in Table 5.12, where B/C means both B and C. A is a cp of itself.*

22 There are also related principles, such as the proscriptive principle of the logics of analytic implication, that we will not have space to discuss. Ferguson (2017) provides an excellent discussion of the proscriptive principle. Ferguson (2025) discusses the relationship between logics of analytic implication and relevant logics.

Table 5.12. Definition of ap and cp

If ...	is a ...	of A, then ...	is a/an... of A
$\neg B$	cp	B	ap
$\neg B$	ap	B	cp
$B \wedge C$	cp	B/C	cp
$B \wedge C$	ap	B/C	ap
$B \vee C$	cp	B/C	cp
$B \vee C$	ap	B/C	ap
$B \to C$	cp	C	cp
$B \to C$	cp	B	ap
$B \to C$	ap	C	ap
$B \to C$	ap	B	cp

This is used to define the strong variable-sharing property.

DEFINITION 5.31 (Strong variable-sharing property) *A logic* L *has the* STRONG VARIABLE-SHARING PROPERTY *iff whenever* $\vdash_{\mathsf{L}} A \to B$*, then* A *and* B *share an atom with the same polarity.*

The strong variable-sharing property says that implicational theorems must have an atom shared between antecedent and consequent that occurs with the same polarity in A and in B. In the formula $A \to B$, the atom will occur with differing polarities, since A and B occur with differing polarities.

Belnap (1960) showed that R and all of its sublogics have the strong variable-sharing property.[23] The matrix M_8 can be used to verify this. The proof is similar to the proof that a logic has the variable-sharing property. We assume that A and B share no atoms with the same polarity and show that $A \to B$ is not designated on some valuation. As with the variable-sharing property, we have two disjoint sets of values $\mathfrak{X}$ and $\mathfrak{Y}$ closed under the operations of the algebra. Unlike the proof of the variable-sharing property, we do not assign arbitrary values in $\mathfrak{X}$ to atoms in A and arbitrary values in $\mathfrak{Y}$ to atoms in B. Rather, the polarities of the atoms are factored into the assignment of values.

THEOREM 5.32 *All sublogics of the logic of* M_8 *have the strong variable-sharing property.*

Proof See Anderson and Belnap (1975, 253–254) for details of the proof, although we will note the role of the polarities. Define a valuation v as follows. For atoms p that occur in both A and B but with differing polarities, $v(p) = \mathbf{+3}$ if p occurs as cp in A, and $v(p) = \mathbf{-3}$ if p occurs as ap in A $v(p) = \mathbf{-3}$. For atoms p that are only in A, $v(p) = \mathbf{-1}$ if p is cp and $v(p) = \mathbf{+1}$ if p is ap. For atoms that are only in B, $v(p) = \mathbf{-2}$ if p is cp and $v(p) = \mathbf{+2}$ if p is ap. An inductive argument then shows that A takes a value in $\{\mathbf{+1}, \mathbf{-1}, \mathbf{+3}\}$, and B takes a value in $\{\mathbf{+2}, \mathbf{-2}, \mathbf{-3}\}$, so $\mathbb{1} \not\leq v(A \to B)$. □

[23] Belnap's original argument was for E, but the matrix he used works for R.

Robles and Mendéz (2012) provide sufficient conditions for a matrix to be usable to verify that a logic has the strong variable-sharing property. Further, they provide a logic, in fact a class of logics, not contained in RM with the variable-sharing property but not the strong variable-sharing property. While the variable-sharing property and strong variable-sharing property differ, within the scope of logics focused on in this book, they coincide.

Another variable-sharing property was defined by Brady (1984a), called DEPTH RELEVANCE. Depth relevance adds to the variable-sharing property the requirement that the shared atom occur at the same depth of nested implications. To make this precise, we define depth.

DEFINITION 5.33 (Depth) *The DEPTH of an occurrence of the formula D in A, $d(D, A)$ is defined as follows.*

- $d(A, A) = 0$.
- *If* $d(\neg B, A) = n$, *then* $d(B, A) = n$.
- *If* $d(B \wedge C, A) = n$, *then* $d(B, A) = n$ *and* $d(C, A) = n$.
- *If* $d(B \vee C, A) = n$, *then* $d(B, A) = n$ *and* $d(C, A) = n$.
- *If* $d(B \rightarrow C, A) = n$, *then* $d(B, A) = n + 1$ *and* $d(C, A) = n + 1$.

Intuitively, the depth of an occurrence D in a formula is the number of nested implications it occurs under. For example, in the formula

$$(\underset{1}{p} \wedge (\underset{2}{p} \rightarrow \underset{2}{q})) \rightarrow \underset{1}{q},$$

we've put the depth of each occurrence of an atom under it. So, while this formula does have an atom shared between antecedent and consequent, the occurrences of the atom are not at the same depth.

Depth considerations can be used to define another variable-sharing property. For this, we will use the common abuse of notation of using formulas for occurrences of formulas.

DEFINITION 5.34 (Depth-relevance property) *A logic* L *has the DEPTH-RELEVANCE PROPERTY iff whenever* $\vdash_{\mathsf{L}} A \rightarrow B$, *there is an atom p that occurs in both A and B and* $d(p, A) = d(p, B)$

If L *has the depth relevance property, then* L *is a DEPTH-RELEVANT LOGIC.*

Following Robles and Méndez (2014c), we can specify a sufficient condition for matrices to verify depth relevance. We start with a basic relevance matrix $M = (K, \wedge_M, \vee_M, \neg_M, \rightarrow_M, \mathbb{1}_M)$. We will define a new function V_ω on M via a sequence of a valuations.[24] Let $v_0, v_1, \ldots, v_n, \ldots$ be an ω-sequence of valuations on M. For $i < \omega$, define V_i, from formulas to K, as follows.

- $V_i(p) = v_i(p)$.
- $V_i(\neg B) = \neg_M V_i(B)$.

[24] For the reader familiar with the literature on paradoxes, the construction bears some similarities to the inductive constructions of Gupta and Belnap (1993), Brady (2006b), and Field (2008).

- $V_i(B \wedge C) = V_i(B) \wedge_M V_i(C)$.
- $V_i(B \vee C) = V_i(B) \vee_M V_i(C)$.
- $V_0(B \to C) \geq_M \mathbb{1}_M$.
- $V_{i+1}(B \to C) = V_i(B) \to_M V_i(C)$.

Finally, V_ω is a function defined on formulas such that $V_\omega(A) \geq \mathbb{1}_M$ iff for all $i \geq 0$, $V_i(A) \geq \mathbb{1}_M$. Let us say that any such V_ω defined on a basic relevance matrix M is a DEPTH VALUATION on M. A formula A is DEPTH VALID on M iff for all depth valuations V_ω on M, $V_\omega(A) \geq \mathbb{1}_M$. A set of formulas is depth valid iff each formula in the set is depth valid.

Let's say that the maximum depth of a formula is

$$md(A) = \max\{n \in \omega : \text{ for some } p, d(p, A) = n\}.$$

The V_is have the feature that if $i < md(A) - 1$, implications are not guaranteed to be evaluated using the operation $\to_M$. Instead, the implications may be treated as atoms, according to the penultimate clause of the definition of the V_is. As an example, consider the formula $p \to (p \to p)$, which has maximum depth of two. Any V_0 will assign this a designated value. $V_1(p \to (q \to r)) = V_0(p) \to_M \boldsymbol{x}$, for some designated value $\boldsymbol{x}$. Finally, $V_2(p \to (q \to r)) = V_1(p) \to_M (V_2(q) \to_M V_2(r))$.

Let us turn to the connection between depth validity and depth relevant logics.

THEOREM 5.35 *If M is a basic relevance matrix and L is depth valid on M, then L has the depth-relevance property.*

Proof Let M be a basic relevance matrix and suppose L is depth valid on M. Suppose $\vdash_{\mathsf{L}} A \to B$ but A and B do not share an atom at the same depth. Let $n = d(A, A \to B) = d(B, A \to B)$. Define the sequence of valuations v_i as follows for atoms p.

- For $i \leq n$, $v_i(p) \in \mathfrak{X}$, if p occurs with depth i in A.
- For $i \leq n$, $v_i(p) \in \mathfrak{Y}$, if p occurs with depth i in B.
- For $i > n$, $v_i(p)$ can be arbitrary.
- In the case where p does not occur in A or B, $v_i(p)$ can be arbitrary.

These valuations are well defined, since no atom occurs in A and in B at the same depth.

We can then show that, $V_n(A) \in \mathfrak{X}$ and $V_n(B) \in \mathfrak{Y}$. This is proved via induction on the structure of subformulas C of A and B. Given this, $V_{n+1}(A \to B) = V_n(A) \to_M V_n(B) \not\geq \mathbb{1}_M$. Therefore $V_\omega(A \to B) \not\geq \mathbb{1}_M$. This contradicts the assumption that L was depth valid on M. □

There is a feature of the definition of depth valuations that is worth remarking on. The definition provided matches that of Logan (2021), where $V_\omega(A)$ is merely constrained by whether the $V_i(A)$s are all designated values. Brady (1984a) and Robles and Méndez (2014c), in contrast, define V_ω to act more like a valuation, respecting the operations for meet, join, and negation, so that $V_\omega(A \wedge B) = V_\omega(A) \wedge V_\omega(B)$, for example. This has subtle consequences for the resulting logics.

The depth-relevance property carves out a proper subset of the relevant logics covered in this book. Brady (1984a) shows that the logic DJ extended with the rule $A \Rightarrow \neg(A \rightarrow \neg A)$ has the depth-relevance property, as do all of its sublogics.[25] Many of the other relevant logics that we discuss are not depth-relevant logics. In fact, any logic that contains (I1), (I3), (I4), (I5), (I6), (I7), (I9), or (N2) will fail to be a depth-relevant logic. Most of the logics considered explicitly in this book are not depth-relevant logics. There are, however, many depth-relevant logics when one is willing to consider less common axioms and rules, as demonstrated by Robles and Méndez (2014c).

The depth-relevant logics are interesting, in part, because of their connection to the set-theoretic paradoxes. As we will see in Chapter 13, the addition of the axioms of naive set theory to many relevant logics results in triviality. As shown by Brady (1989b), Robles and Méndez (2014a; 2014b) and Salto et al. (2018), depth-relevant logics can handle the axioms of naive set theory in a non-trivial way, providing a way to avoid Curry's paradox.

One can combine polarity and depth considerations. Logan (2021) defines the STRONG DEPTH-RELEVANCE PROPERTY as requiring that if $A \rightarrow B$ is a theorem of L, then A and B share an atom at the same depth with the same polarity at that depth. Logan shows that some relevant logics have the strong depth-relevance property using using M_8 in the earlier construction, paying careful attention to the polarities at different depths.

Another notable feature of depth-relevant logics is that they are hyperformal. Following Logan (2022), let us define a depth substitution as follows.

DEFINITION 5.36 (Depth substitution) *A* DEPTH SUBSTITUTION δ *is a function* $\mathfrak{L} \times \omega \mapsto \mathfrak{L}$ *such that*

- $\delta(p,n) \in \mathfrak{L}$.
- $\delta(\neg B,n) = \neg\delta(B,n)$.
- $\delta(B \wedge C,n) = \delta(B,n) \wedge \delta(C,n)$.
- $\delta(B \vee C,n) = \delta(B,n) \vee \delta(C,n)$.
- $\delta(B \rightarrow C,n) = \delta(B,n+1) \rightarrow \delta(C,n+1)$.

A logic L is HYPERFORMAL iff for all depth substitutions δ, all $n \in \omega$, and all formulas A, if $A \in$ L, then $\sigma(A,n) \in$ L. Logan (2022) shows that the sublogics of R with the strong depth-relevance property are hyperformal. In a hyperformal logic, atoms occurring at different depths are, from the point of view of the logic, different atoms.

5.6 Interpolation

To finish this chapter, we will consider a property of logics that has a feeling similar to that of variable sharing, namely INTERPOLATION. Interpolation properties say that

[25] Brady's result actually shows that that logic further extended by disjunctive rules also has the depth-relevance property.

whenever an implication $A \rightarrow B$ is valid in a logic, then there is a formula C from the common language of A and B that acts as an intermediary between them. There are different varieties of interpolation, but we will focus on a simple one for this section: basic Craig interpolation.

DEFINITION 5.37 (Interpolation) *A logic* L *has the* INTERPOLATION PROPERTY *iff whenever* $\vdash_{\mathsf{L}} A \rightarrow B$*, then there is a formula* C *such that* $\vdash_{\mathsf{L}} A \rightarrow C$*,* $\vdash_{\mathsf{L}} C \rightarrow B$*, and the atoms occurring in* C *occur in both* A *and* B*.*

Interpolation seems like a property that a relevant logic should have. After all, on an intuitive level, it seems like C is relevant to both A and B, since by definition atoms are shared between all three. Indeed, interpolation seems antecedently plausible and desirable for relevant logics.

As noted by Anderson and Belnap (1975), FDE has the interpolation property. In fact, FDE has what they call the "perfected interpolation property." That name comes from the fact that in TV, the standard form of interpolation adds to the earlier definition clauses saying that either A is the negation of a logical truth or B is a logical truth, that is, either (i) there is an interpolant C, (ii) $\vdash_{\mathsf{TV}} \neg A$, or (iii) $\vdash_{\mathsf{TV}} B$. Interpolation for FDE does not require the latter two clauses.

Perhaps surprisingly, many relevant logics lack interpolation, as shown by Urquhart (1993; 1999; 2003).

THEOREM 5.38 (Urquhart (2003)) *Interpolation fails for all logics between* TW *and* R*.*

The proof of this theorem is surprising, in that the failure of interpolation requires neither the Ackermann constant nor negation. The proof has two parts. One part shows that the target formula is a theorem of TW, and so of all stronger logics. The second part of the proof shows that the target formula has no interpolant in R, and so no interpolant in any weaker logics.

For the first part of the proof, we will follow Urquhart in defining some notation. Let p, q, r, s, u be distinct atoms. Then we define the following for arbitrary formulas A and B.

- ΔA is $u \rightarrow A$.[26]
- $-A$ is $(A \rightarrow s)$.
- $A \otimes B$ is $-(A \rightarrow -B)$.

To get the target formula, we need some more definitions.

- $A_0 =_{Df} (\Delta q \rightarrow q) \wedge (-(p \vee \Delta q))$.
- $A_1 =_{Df} ((q \wedge p) \rightarrow u) \wedge u$.
- $B_0 =_{Df} \Delta u \wedge (s \rightarrow (p \vee r))$.
- $B_1 =_{Df} \Delta r \wedge \Delta(r \rightarrow r)$.

[26] Urquhart uses "□" rather than "Δ," but we are adjusting the notation to avoid a clash with the defined necessity of E.

Given these definitions, $A_0 \rightarrow (B_0 \rightarrow (B_1 \rightarrow (A_1 \rightarrow (\Delta q \rightarrow r))))$ is a theorem of TW. The proof is left to the reader. From this, one can obtain the target theorem:

$$(A_0 \otimes \Delta\Delta(A_1 \otimes \Delta q)) \rightarrow -- (B_0 \rightarrow \Delta(B_1 \rightarrow \Delta -- r)).$$

This is the target formula $A \rightarrow B$. It is a theorem of TW, and it is one that lacks an interpolant. The atoms of any interpolant can only come from the set $\{p, s, u\}$.

To show that there is no possible interpolant, suppose C is an interpolant for $A \rightarrow B$. Urquhart uses CL to show that there is no such C. Consider a valuation v such that $v(p) = \boldsymbol{c}$, $v(q) = v(r) = \boldsymbol{d}$, $v(s) = \boldsymbol{b}$, and $v(u) = \boldsymbol{e}$. On this valuation, $v(E) = v(F) = \boldsymbol{d}$. It follows then that $v(C) = \boldsymbol{d}$. This is because $\boldsymbol{e} \leq v(A \rightarrow C)$, so $v(A) \leq v(C)$, and similarly, $\boldsymbol{e} \leq v(C \rightarrow F)$, so $v(C) \leq v(F)$. The set of values $\{\boldsymbol{a}, \boldsymbol{b}, \boldsymbol{c}, \boldsymbol{e}, \boldsymbol{f}\}$ is closed under the operations of the matrix. Since $v(p), v(s)$, and $v(u)$ are all in the set, so is the value of any formula D built from just those atoms. However, C is assumed to be such a formula, but $v(C)$ is not in the set. That is a contradiction, so there is no such interpolant. Since CL is a matrix for which R is sound, it follows that there is no interpolant for $A \rightarrow B$ in any logic contained in R.

Urquhart offers additional proofs, using geometric methods, that extend to logics stronger than R. Some of these logics, however, lack variable sharing, and, more importantly, the proof methods would require a lot more background than we have space for, so we will not go into those details.

It is worth noting that the derivability of the target theorem, $A \rightarrow B$, above uses the prefixing and suffixing axioms of TW in an important way. Therefore, the specific choice of the valid implication $A \rightarrow B$ above will not work for weaker relevant logics that lack TW's prefixing and suffixing axioms. For example, it need not extend to DJ and DW. As far as we are aware, it is not currently known whether any logics between B and R enjoy the interpolation property.

As mentioned earlier, there are other interpolation properties and other ways to define interpolation. Galatos and Ono (2006) give a definition of interpolation for consequence relations. Using algebraic methods, they show that many substructural logics, including R, have interpolation in that sense. Brotherston and Goré (2011) use display logic proof methods to obtain interpolation results for consecution presentations of some relevant logics. They note that their methods do not extend to logics that contain contraction principles, such as R.

What are we to make of the failures of interpolation in relevant logics? Urquhart's example formula that lacks an interpolant only involves three connectives: implication, conjunction, and disjunction. The failure of interpolation depends on the interplay between these three.[27] The interplay between conjunction and disjunction on the one hand, and implication, on the other, is subtle and difficult. It is, plausibly, behind many of the difficulties in constructing proof systems for relevant logics. In informal terms, to introduce a conjunction, the conjuncts have to depend on the same antecedents, and there is a similar sort of dependency involved in eliminating a disjunction. It is this issue that causes problems for the derivation of the distribution principle, which has

[27] For this discussion, familiarity with alternative proof systems for relevant logics would be useful.

to be added as a separate axiom in our axiom systems. To eliminate an implication, however, there need not be any such shared dependence.

At the start of this section, we said that interpolation has a feeling similar to that of variable sharing. Whereas variable sharing requires that all (t-free) implicational theorems have a shared atom, interpolation requires that all implicational theorems be "splittable" with a formula containing only the shared atoms. Such a split appears to break the important logical connections required for establishing an implicational theorem. This can be seen in Urquhart's formula. It may seem prima facie strange that, in $A \rightarrow B$, the atoms occurring in A but not in B could play an important role in the theoremhood of the implication. Notice, however, that, one of the subformulas of A is $(q \wedge p) \rightarrow u$. While p and u are in the set of atoms shared between A and B, q is not. Yet, if we think proof-theoretically, one would need to obtain q in order to conjoin it with p to detach and get u. Further, q and p would need to depend on the same antecedents.

The conjunctions (and disjunctions) require that there are certain logical connections among the subformulas, and the implications require that there be other logical connections. These connections are not adequately respected in the restriction to shared atoms for the interpolant, as illustrated by Urquhart's proof.

While it is hard to see failure of interpolation as a virtue, one can get into the mindset that it is not disastrous. Suppose that one must use an antecedent in obtaining the consequent, as suggested by Anderson and Belnap (1975). It is plausible that some important part of the antecedent A used in obtaining B is lost in obtaining a putative interpolant C, which no longer has sufficient logical power to obtain B. As has been said before, the implications of relevant logics are demanding. At least for some of the stronger relevant logics, the implications are sufficiently demanding that they cannot, in general, be "split" in a way required by interpolation.

5.7 Exercises

(1) Provide proofs for lemma 5.2.

(2) Test the validity of the following formulas in the finite matrices of Section 5.4.

- $A \rightarrow ((A \rightarrow B) \rightarrow B)$
- $(A \rightarrow (A \rightarrow B)) \rightarrow (A \rightarrow B)$
- $(A \wedge (A \rightarrow B)) \rightarrow B$
- $(A \rightarrow B) \rightarrow ((B \rightarrow C) \rightarrow (A \rightarrow C))$
- $A \vee \neg A$
- $(A \rightarrow \neg A) \rightarrow \neg A$

(3) Which of the following relevantly unacceptable formulas has counterexamples in the matrices of Section 5.4?

- $A \rightarrow (B \rightarrow A)$
- $A \rightarrow (B \rightarrow B)$
- $A \rightarrow (B \rightarrow (A \wedge B))$
- $A \rightarrow (B \vee \neg B)$

- $(((A \rightarrow A) \wedge B) \rightarrow C) \rightarrow (B \rightarrow C)$
- $((A \wedge B) \rightarrow C) \rightarrow ((A \wedge \neg C) \rightarrow \neg B)$

(4) Compute the prime filter frame for CL using the construction of Section 5.3.

(5) Prove that the axioms of R are valid on M_8 and that the rules of R preserve validity.

(6) Axiomatize the logic of M_8 without using the rule (γ). (Warning: This may be hard.)

(7) Prove that any logic that contains any of the following will fail to be a depth-relevant logic: (I1), (I3), (I4), (I5), (I6), (I7), (I9), or (N2).

(8) Prove that in the full language with fusion, no logic extending B has the Ackermann property.

Part II

Modality

6 Modal Extensions

6.1 Introduction

In this second part of the book, we will consider some *modal* extensions of the base relevant logics. For this, we will extend the language with modal operators. A *relevant modal logic* adds additional axioms and rules governing the new modal operators to the base relevant logic.[1] In this chapter, we will cover the axiomatics of relevant modal logics and some matrix techniques for them. In the next chapter (Chapter 7), we will look at their frame theory.

In a sense, relevant modal logics have been around since the start of the study of relevant logics in the Anderson–Belnap tradition. The logic E was billed as the logic of entailment, combining *relevance* and *necessity*.[2] One neat feature of E is that it provides a theory of logical necessity, where $\Box A$ is defined as $(A \to A) \to A$, or, equivalently, as $\mathsf{t} \to A$.[3] The defined modal operator secures the typical axioms of S4, namely $\Box A \to \Box\Box A$, $\Box A \to A$, and $\Box(A \to B) \to (\Box A \to \Box B)$, along with the rule of necessitation, $A \Rightarrow \Box A$.

As another way to investigate the idea of combining relevance and necessity, Meyer proposed a logic, $\mathsf{R}^{\Box}$, that added a primitive S4-ish necessity operator to R.[4] The hope was that the implication of E, $\to_{\mathsf{E}}$, would be definable using necessity and the implication of R, $\to_{\mathsf{R}}$, using the definition $A \to_{\mathsf{E}} B =_{Df} \Box(A \to_{\mathsf{R}} B)$. As noted by Anderson and Belnap (1975, 351), the fit was not exact, and the translation of E into $\mathsf{R}^{\Box}$ was too weak, omitting theorems of $\mathsf{R}^{\Box}$. The divergence indicated that there was some mismatch between the motivating intuitions of E and the logic itself. Anderson and Belnap (1975, 351) say, "we predict that if in fact it is found that $\mathsf{R}^{\Box}$ and E diverge, then we shall, with many a bitter tear, abandon E." In the paragraph immediately following

[1] In the literature, one finds the term "modal relevant logic" as well. As far as we can tell, there is no important difference between the use of the two terms.

[2] Consider the title of Anderson and Belnap (1975), for example.

[3] The second definition is, in the terminology of Humberstone (2021), linear, whereas the first definition is not.

[4] Meyer called this logic NR. In the nomenclature adopted below, the logic is R.MKT4Nec. Why do we say S4-*ish*, rather than just saying that it is an S4 extension of R? While NR contains the distinctive axioms of S4, namely (K), (4), and (T) along with the rule (Nec), these axioms are formulated in the vocabulary of relevant modal logics. The main connective of those axioms is a relevant implication, rather than the classical material conditional. As we will see, NR does not contain every theorem of S4, even when all the classical material conditionals are replaced with disjunction and negation, which is to say replacing the classical $A \to B$ with $\neg A \vee B$.

that, Anderson and Belnap note that there has been a confirmation that the two logics do diverge, to which point we will return later. While not all working in the tradition did abandon E, many did.[5] For this chapter, we will follow the spirit of Meyer, and the work of many others in the area, in adding primitive modal operators to the language, leaving implication as conceptually separate.

Before getting to the technical details, it is worth pausing to consider why one might want to add modal operators to relevant logics. There are formal reasons and there are philosophical reasons. On the formal side, modal logics are technically very interesting and have a wide range of applications.[6] The combination of intensional modal operators with the intensional implications of relevant logics opens the door to many possibilities for interactions, as ones sees also with intuitionstic modal logics.[7] On the philosophical side, modal concepts are used extensively in discussions in metaphysics, epistemology, and elsewhere.[8] If you want a relevant logic to be your general purpose logic (or at least one of them), then you will need some account of how those modal concepts function in a relevant logical setting. Given that relevant logics draw many distinctions that TV does not, and that they tolerate inconsistency in a natural way, one would expect there to be many new options opened up by considering modal concepts against a base relevant logic.

We will provide a brief overview of the chapter before diving into the technical details. In Section 6.2, we will present the basic axioms and rules for relevant modal logics and some of their standard extensions. There are some distinctive features of theories for relevant modal logics that are important to note, and we do so in Section 6.3. This will be followed by a presentation in Section 6.4 of how to extend the matrices of Chapter 5 to work in a modal context. This is followed in Section 6.5 with a short coda on the logic E, commenting on its contemporary status. Finally, Section 6.6 closes the chapter with a comparison between the relevant modal logics and TV-based modal logics, focusing on the distinctively modal aspects of the comparison.

6.2 Axiomatics

For this chapter and the next, we will extend the language $\mathfrak{L}$ with a singulary necessity connective, $\Box$. We will call this extension $\mathfrak{L}_{\Box}$. We will also consider the extension with another singulary connective, $\blacklozenge$. The extension of the language with this connective will be $\mathfrak{L}_{\Box\blacklozenge}$.

All the relevant modal logics considered in this chapter and next will be CONJUNCTIVELY REGULAR in the sense of Fuhrmann (1990), namely they will have the following two principles.

(RM) $A \to B \Rightarrow \Box A \to \Box B$.
($\Box\wedge$) $(\Box A \wedge \Box B) \to \Box(A \wedge B)$.

5 See Meyer (1985) or the critical discussion scattered throughout Routley et al. (1982).
6 See Blackburn et al. (2002) for more.
7 See Božić and Došen (1984) for an example of intuitionistic modal logic.
8 See Humberstone (2016) for a good overview.

The extension of a base relevant logic L with (RM) and ($\Box\wedge$) is obtained by adding (RM) and ($\Box\wedge$) to the basic definition of proof.[9] We will need some nomenclature. Following the convention of Fuhrmann (1990), given a base relevant logic L, the extension of L with (RM) and ($\Box\wedge$) will be L.M.[10] A formula A is a theorem of L.M iff there is a finite sequence of formulas ending in A, each of which is either an axiom of L.M or is the conclusion of one of the rules of L.M, where the premises of the rule all occur earlier in the sequence.

When the possibility operator is in the language, we also add the following rule and axiom.

(RM$\blacklozenge$) $A \to B \Rightarrow \blacklozenge A \to \blacklozenge B$.
($\blacklozenge\vee$) $\blacklozenge(A \vee B) \to (\blacklozenge A \vee \blacklozenge B)$.

When possibility is primitive, L.M will include the preceding axiom and rule as well as the axiom and rule for the necessity operator.

Later we will consider extensions of L.M with additional modal axioms and rules. The result of adding the additional rules and axioms $\mathsf{M}_1, \ldots, \mathsf{M}_n$ to L.M will be called L.MM$_1, \ldots,$ M$_n$. When we want to talk about extensions of L.M without specifying the extending principles, we will write L.MZ, where Z can be any selection of principles from later in this section. We will use "L" to refer to relevant modal logics, and when there is a need to distinguish the non-modal base logic from the extension, we will use the more informative nomenclature, L.M.

Given these rules and axioms, L.M enjoys substitution of equivalents.

THEOREM 6.1 *Let $C(\cdot)$ be a formula context. If $\vdash_{\mathsf{L.M}} A \leftrightarrow B$, then $\vdash_{\mathsf{L.M}} C(A) \leftrightarrow C(B)$.*

Proof The proof proceeds by induction on the construction of the formula context $C(\cdot)$ as in Chapter 2. There are two new cases, for the modals. They are handled using the rules (RM) and (RM$\blacklozenge$) with the assumption. □

We can define operators using negation and the modals. The defined possibility operator is $\Diamond A =_{Df} \neg\Box\neg A$. The defined necessity operator is $\blacksquare A =_{Df} \neg\blacklozenge\neg A$. Using these definitions, the defined operators obey the required rule and axiom of their primitive counterparts.

LEMMA 6.2 *The following are theorems of* L.M.

- $(\blacksquare A \wedge \blacksquare B) \to \blacksquare(A \wedge B)$.
- $\Diamond(A \vee B) \to (\Diamond A \vee \Diamond B)$.

The logic L.M *is closed under the following rules.*

- $A \to B \Rightarrow \blacksquare A \to \blacksquare B$.
- $A \to B \Rightarrow \Diamond A \to \Diamond B$.

Proof For the first axiom, we have $(\Box\neg A \wedge \Box\neg B) \to \Box(\neg A \wedge \neg B)$ by ($\Box\wedge$). Using (RM) and basic facts about L, we have $(\neg\neg\Box\neg A \wedge \neg\neg\Box\neg B) \to \Box\neg(A \vee B)$. Using (R3) and De Morgan equivalences, we have $\Diamond(A \vee B) \to (\Diamond A \vee \Diamond B)$, as desired.

[9] The axiom ($\Box\wedge$) is also called (Agg) in the literature.
[10] Fuhrmann uses the label "L.C."

Table 6.1. Some common modal axioms

(T)	$\Box A \to A$
(D)	$\Box\neg A \to \neg\Box A$
(4)	$\Box A \to \Box\Box A$
(5)	$\Diamond A \to \Box\Diamond A$
(B)	$A \to \Box\Diamond A$
(K)	$\Box(A \to B) \to (\Box A \to \Box B)$
(Nec)	$A \Rightarrow \Box A$
(MC)	$\Box(A \vee B) \to (\Diamond A \vee \Box B)$

For the second axiom, we have $\blacklozenge(\neg A \vee \neg B) \to (\blacklozenge\neg A \vee \blacklozenge\neg B)$ by ($\blacklozenge\vee$). Using (RM$\blacklozenge$) and (R3), we have $\neg(\blacklozenge\neg A \vee \blacklozenge\neg B) \to \neg\blacklozenge(\neg A \vee \neg B)$. Using De Morgan equivalences and the definitions, we have $(\blacksquare A \wedge \blacksquare B) \to \blacksquare(A \wedge B)$.

For the rules, we use (R3), (RM$\blacklozenge$), or (RM), as appropriate, and (R3) again. □

As with TV-BASED MODAL LOGICS, namely modal logics whose base logic is TV, one can add extra axioms and rules governing the modalities. Some common axioms and rules are presented in Table 6.1. When both necessity and possibility are primitive, there will be a choice of which operator will be used for each. We will present them uniformly as using $\Box$ and $\Diamond$, but the other modalities can be used as appropriate. In later chapters we will introduce additional modal axioms.

Most of the axioms and rules in Table 6.1 are standard in TV-based modal logics. Nonetheless, they deserve a few comments. First, the basic modal extension of a relevant logic L, L.M, does not contain (K), and it is not closed under (Nec), which contrasts with the normal TV-based modal logics. We will say a bit more about this difference in Chapter 7. Second, the final axiom, (MC) is not usually considered separately in TV-based modal logics. The axiom, whose name is short for MODAL CONFINEMENT, is a standard theorem of TV-based modal logics, but it need not be a theorem of relevant modal logics.[11] It is added to R.KT4Nec by Mares and Meyer to obtain further connections to the TV-based modal logic S4. It also has important consequences for the frame theory, which we will discuss in Chapter 7.

Let us turn from axioms to some issues concerning theories in relevant modal logics.

6.3 Theories

The results on L-theories from the earlier chapters extend to theories for modal logics. The main results here concern the *modal cores* of theories.

[11] Ferenz (2021) uses a different convention for naming modal relevant logics, where logics without (MC) have a dot in their names, while those with (MC) do not have a dot. For example, B.MTMC, in our nomenclature, would be BMT under the conventions of Ferenz. We are not adopting this convention here, but we note that it is an alternative, systematic convention one finds in the literature.

DEFINITION 6.3 (Modal core, possibilization) *The MODAL CORE of an* L.MZ*-theory X is defined as follows.*

- $X^{-\Box} = \{B \in \mathfrak{L}_{\Box\blacklozenge} : \Box B \in X\}$.

The POSSIBILIZATION of an L.MZ*-theory X is defined as*

- $X^{\blacklozenge} = \{\blacklozenge B \in \mathfrak{L}_{\Box\blacklozenge} : B \in X\}$.

LEMMA 6.4 *Let X be an* L.MZ*-theory. Then $X^{-\Box}$ is a* L*-theory.*

Proof Let X be an L.MZ-theory.

To show that $X^{-\Box}$ is an L.MZ-theory, suppose $B \in X^{-\Box}$ and $\vdash_{\mathsf{L.MZ}} B \to C$. Then by (RM), $\vdash_{\mathsf{L.MZ}} \Box B \to \Box C$. From the definition of $X^{-\Box}$, $\Box B \in X$. Therefore, $\Box C \in X$, so $C \in X^{-\Box}$.

Next, suppose $B \in X^{-\Box}$ and $C \in X^{-\Box}$. So, $\Box B \in X$ and $\Box C \in X$. As X is a theory, $\Box B \wedge \Box C \in X$. From ($\Box\wedge$), $\Box(B \wedge C) \in X$, so $B \wedge C \in X^{-\Box}$, as desired. □

Next, we will prove a lemma about prime extensions of modal cores that will be useful in the Completeness proof of Chapter 7.

LEMMA 6.5 (Modal extension lemma) *Let X be an* L.MZ*-theory such that $A \notin X^{-\Box}$. Then there is a prime* L.MZ*-theory Y such that $X^{-\Box} \subseteq Y$, $A \notin Y$.*

Proof Suppose X is an L.MZ-theory such that $A \notin X^{-\Box}$. Since $X^{-\Box}$ is a theory, $\langle X^{-\Box}, \{A\}\rangle$ is a disjoint pair. By the Prime extension lemma, we can obtain a prime L.MZ-theory Y such that $X^{-\Box} \subseteq Y$ and $\langle Y, \{A\}\rangle$ is a disjoint pair, which suffices for $A \notin Y$. □

We can view $(\cdot)^{-\Box}$ as an operator from theories to theories. As an operator, it is monotonic in the following sense.

LEMMA 6.6 *Let X and Y be* L.M*-theories such that $X \subseteq Y$. Then $X^{-\Box} \subseteq Y^{-\Box}$.*

Proof Suppose $A \in X^{-\Box}$. Then $\Box A \in X$. By assumption, $X \subseteq Y$, so $\Box A \in Y$, whence $A \in Y^{-\Box}$. □

Viewed as an operator on theories, $(\cdot)^{\blacklozenge}$ is monotonic in a similar manner. In the exercises at the end of the chapter, there is a question exploring further properties of $(\cdot)^{-\Box}$ as an operator.

Let us next turn to matrices appropriate to relevant modal logics.

6.4 Matrices

The matrix techniques from Chapter 5 can be extended to cover relevant modal logics. The way to do this is to enrich the L-groupoids with additional operators.

DEFINITION 6.7 (Modal groupoid) *A MODAL* L.M*-GROUPOID G is a sextuple $\langle K, \cdot, \to, \wedge, \vee, -, \mathbb{1}, \Box\rangle$ where $\langle K, \cdot, \to, \wedge, \vee, -, \mathbb{1}\rangle$ is an* L*-groupoid and $\Box$ is a singulary operation satisfying the following*

Table 6.2. Conditions on modal groupoids

	Axiom		Condition
(T)	$\Box A \to A$	(CT)	$\Box a \leq a$
(D)	$\Box A \to \Diamond A$	(CD)	$\Box a \leq -\Box - a$
(4)	$\Box A \to \Box\Box A$	(C4)	$\Box a \leq \Box\Box a$
(5)	$\Diamond A \to \Box\Diamond A$	(C5)	$-\Box - a \leq \Box - \Box - a$
(B)	$A \to \Box\Diamond A$	(CB)	$a \leq \Box - \Box - a$
(K)	$\Box(A \to B) \to (\Box A \to \Box B)$	(CK)	$\Box(a \to b)\Box a \leq \Box b$
(Nec)	$A \Rightarrow \Box A$	(CNec)	$\mathbb{1} \leq \Box\mathbb{1}$
(MC)	$\Box(A \vee B) \to (\Diamond A \vee \Box B)$	(CMC)	$\Box(a \vee b) \leq -\Box - a \vee \Box b$

(MG1) If $x \leq y$, *then* $\Box x \leq \Box y$.
(MG2) $\Box x \wedge \Box y \leq \Box(x \wedge y)$.

A FULL MODAL L.M-GROUPOID *is a modal groupoid equipped with an additional singulary operation* $\blacklozenge$ *obeying the following.*

(MG3) If $x \leq y$, *then* $\blacklozenge x \leq \blacklozenge y$.
(MG4) $\blacklozenge(x \vee y) \leq \blacklozenge x \vee \blacklozenge y$.

It follows by argument similar to those earlier that $\blacksquare$ obeys analogs of (MG1) and (MG2), while $\Diamond$ obeys analogs of (MG3) and (MG4). We will drop "full modal" and "modal" when it is clear that a modal groupoid is under consideration.

The class of modal L.M-groupoids can be cut back to the class of modal L.MZ-groupoids by requiring additional conditions, corresponding to the additional modal axioms in Z. These are listed in Table 6.2.[12] The definitions of satisfying a formula, counterexample, holding, and validity carry over from L-groupoids to modal L.M-groupoids. We then have another matricial Soundness theorem.

THEOREM 6.8 (Matricial Soundness) *If* $\vdash_{\mathsf{L.MZ}} A$ *then A is valid in the class of (full) modal* L.MZ-*groupoids.*

We can use modal groupoids to obtain several useful results concerning relevant modal logics. Before getting to those results, we will define an important concept.

DEFINITION 6.9 (Modal reduct, L-fitting) *Given a formula A, the* MODAL REDUCT *of A is the result of deleting all occurrences of* $\Box$ *and* $\blacklozenge$ *in A.*

Let L *be a base relevant logic. An axiom A* FITS L *iff the modal reduct of A is a theorem of* L. *Such an axiom is* L-*fitting.*

A rule $A_1, \ldots, A_n \Rightarrow B$ FITS L *iff whenever the modal reducts of* $A_1, \ldots, A_n$ *fit* L, *so does the modal reduct of B.*

Some examples are in order. The modal reduct of $\Box A \to \Box\Box A$ is $A \to A$, and the modal reduct of $\Box A \to \Diamond A$ is $A \to \neg\neg A$, as $\Diamond A$ is $\neg\Box\neg A$. The axiom $\Box A \to \Box\Box A$ fits B. The rule $A \Rightarrow \Box B \to A$ does not fit B. The axiom $\Box A \to (\Diamond(A \to B) \to \Box B)$ fits R but does not fit T.

12 If $\blacklozenge$ is in the language, then analogous conditions with $\blacklozenge$ can be considered.

As will be clear from inspecting the axioms and rules, (RM), ($\Box\wedge$), (RM$\blacklozenge$), and ($\blacklozenge\vee$) all fit every base relevant logic. This leads to our first lemma.

LEMMA 6.10 (Modal reduct values) *Let G be an* L.M-*groupoid such that* $\Box \boldsymbol{x} = \blacklozenge \boldsymbol{x} = \boldsymbol{x}$. *Let B be the modal reduct of A. Then for all valuations v on G, $v(B) = v(A)$.*

Proof The proof is by induction on the construction of A. The base cases are immediate, and all the induction steps are immediate from the inductive hypothesis, apart from the cases where A is $\Box C$ or $\blacklozenge C$. In those cases, the stipulation that the modal operations are identity operators together with the inductive hypothesis suffice for the desired conclusion. □

LEMMA 6.11 *Let G be an* L-*groupoid. G can be extended to a full modal* L.M *groupoid by setting $\Box$ and $\blacklozenge$ to be the identity operators, that is,* $\Box \boldsymbol{x} = \blacklozenge \boldsymbol{x} = \boldsymbol{x}$.

Proof The identity operators satisfy conditions (MG1)–(MG4). The other operators still satisfy the conditions for being an L-groupoid. □

LEMMA 6.12 *Suppose all the principles in* Z=$\mathsf{M}_1, \ldots, \mathsf{M}_n$ *fit* L. *Let G be an* L-*groupoid. Let H be G equipped with the operators $\Box$ and $\blacklozenge$, and set* $\Box \boldsymbol{x} = \blacklozenge \boldsymbol{x} = \boldsymbol{x}$. *Then H is a full modal* L.MZ-*groupoid.*

Proof By the preceding lemma, H is an L.M-groupoid. Suppose that some axiom C from Z has a counterexample on H, so $\mathbb{1} \not\leq v(C)$, for some valuation v. Let D be the modal reduct of C. Since the modal axioms fit L, D is a theorem of L. From the algebraic Soundness theorem, $\mathbb{1} \leq u(D)$, for all valuations u on G. By lemma 6.10, $w(C) = w(D)$, for all valuations w on H. Therefore, $\mathbb{1} \not\leq v(D)$, but v is also valuation on G, so v is a counterexample to D on G, which is a contradiction. Therefore, C has no counterexamples on H, so H is a full modal L.MZ-groupoid. □

These lemmas suffice for our initial results, which will require a definition.

DEFINITION 6.13 (Conservative extension) *Suppose for languages $\mathfrak{L}$ and $\mathfrak{M}$, $\mathfrak{L} \subseteq \mathfrak{M}$. Suppose that* $\mathsf{L}_1 \subseteq \mathfrak{L}$ *and* $\mathsf{L}_2 \subseteq \mathfrak{M}$. *Then* L_2 *is a conservative extension of* L_1 *iff whenever* $\vdash_{\mathsf{L}_2} A$ *and* $A \in \mathfrak{L}$, $\vdash_{\mathsf{L}_1} A$.

The preceding lemmas yield conservative extension results.

THEOREM 6.14 (Conservative extension) *Suppose* L $\subseteq$ R *is a base relevant logic and* L.MZ *is a modal extension of* L *such that the principles in* Z *fit* L. *Then* L.MZ *is a conservative extension of* L.

Proof Suppose A has a counterexample v on some L-groupoid G. By lemma 6.12, G can be extended to an L.MZ groupoid, and A will have the same counterexample. □

The lemmas also suffice for a variable-sharing theorem.

THEOREM 6.15 (Modal variable sharing) *Suppose* L *is a base relevant logic and* L.MZ *is a modal extension of* L *such that the principles in* Z *fit* L. *Then* L.MZ *has the variable-sharing property.*

Proof Suppose that $A \rightarrow B$ is a theorem L.MZ such that A and B share no propositional atoms and neither contains t. Using the matrix CL from Chapter 5, set $v(p) = \boldsymbol{c}$ for all atoms p in A and $v(q) = \boldsymbol{d}$ for all atoms q in B, and set $\Box \boldsymbol{x} = \boldsymbol{x}$. By induction on the construction of A and B, $v(A) = \boldsymbol{c}$ and $v(B) = \boldsymbol{d}$. It follows that $v(A \rightarrow B) = \boldsymbol{f}$, so v is a counterexample to $A \rightarrow B$, which is a contradiction. Therefore, A and B share a propositional atom and so L.MZ has the variable-sharing property. □

The earlier simple lemmas suffice for proving two important results about relevant modal logics. When the modal additions fit the underlying base logic, conservative extension is immediate and the extension enjoys the variable-sharing property as well. We can say a bit more about some of the logics that enjoy the variable-sharing property. Let (TRIV) be the axiom $A \leftrightarrow \Box A$. In other words, (TRIV) is (T) and its converse. In L.TRIV, the interpretation of □ is forced to be identity, so L.TRIV satisfies the conditions for the preceding theorem.

COROLLARY 6.16 *Let* M *be sublogic of* L.TRIV, *where* L *is a relevant logic. Then* M *enjoys the variable-sharing property.*

Many standard modal logics considered in philosophical contexts are sublogics of TRIV. All of the axioms listed in Section 6.2 are contained in the TRIV extension of a relevant logic.

Not all modal extensions that have been considered by logicians, however, fit the underlying logic. We will note two here, both of which we will return to in Chapter 7. The first example is from Mares (2000b), who presented a logic he calls RGL, which in our nomenclature is R.MK4Nec with the addition of (L) $\Box(\Box A \rightarrow A) \rightarrow A$. The axiom (L) does not fit R. The second example is due to Goble (2000), who studied a logic he calls G, which is R.MKTNec plus the axiom (G) $\Box A \rightarrow (B \rightarrow B)$. (G) does not fit any base relevant logic, and as Goble notes, the addition of (5) from Table 6.1 results in the dreaded $A \rightarrow (B \rightarrow A)$ becoming provable.

6.5 With Many a Bitter Tear: Coda on E

In the opening of this chapter, it was noted that there was a divergence between E and an S4-ish extension of R, $\mathsf{R}^{\Box}$, on the translation τ that sets $\tau(A \rightarrow_{\mathsf{E}} B) = \Box(\tau(A) \rightarrow_{\mathsf{R}} \tau(B))$, where $\rightarrow_{\mathsf{L}}$ is the implication of L. Before we get to that divergence, let us pause on $\mathsf{R}^{\Box}$.

The logic $\mathsf{R}^{\Box}$ is, in our nomenclature, R.MKT4Nec. It is an S4-ish extension of R that was first proposed by Meyer (1966), and it was given a frame semantics by Routley and Meyer (1972a). More recently, the extension of $\mathsf{R}^{\Box}$ with (MC) has been studied by Mares and Meyer (1993) and Meyer and Mares (1993). The extension has important connections, developed by Meyer and Mares, to the TV-based modal logic S4. $\mathsf{R}^{\Box}$ is, in many ways, an interesting and elegant logic.

It was hoped, in the early days of relevant logic research, that $\mathsf{R}^{\Box}$ and E would align, under the translation that took the implication of E to be the strict conditional of $\mathsf{R}^{\Box}$. It was, however, not to be. The Mints–Maksimova formula

$$\big((A \to (B \to C)) \land (B \to (A \lor C))\big) \to (B \to C),$$

or rather its translation, is a theorem of $\mathsf{R}^{\Box}$.[13] Maksimova (1973) showed that the formula is *not* a theorem of E by providing a countermodel.

There is a divergence between the combination of relevance and necessity found in the implication of E and that found in the strict conditional of NR, which separates the two conceptual components into distinct connectives. Interest in E did not disappear entirely, although it is fair to say that the focus moved away from it. In the remainder of this section, we will highlight some of the developments in the study of E. Belnap (1992, §62) presented a display calculus for E. The system is remarkable because many display systems depend on the presence of Boolean negation for their success, but that route is not available. Boolean negation, $\sim$, is a connective whose verification condition is

- $a \Vdash {\sim}A$ iff $a \nVdash A$.

We will talk more about Boolean negation in Chapter 10. At the time of the publication of *Entailment* volume II, it was unknown whether E was conservatively extended by the addition of Boolean negation, as many other relevant logics are known to be. It turned out that E was distinctive in not being conservatively extended by Boolean negation, as shown by Mares (2000a). The formula shown by Mares to become valid with the addition of Boolean negation is $\Box(p \supset q) \supset (\Box p \supset \Box q)$, which is a material conditional version of (K), where the $\Box$ in question is the defined necessity of E.

More recently, Bimbó (2007a) presented a sequent system for the $\{\to, \mathsf{t}\}$-fragment of E. Standefer (2018) presented a tree-style natural deduction system for E that used the ideas of Prawitz (1965, ch. 6) on natural deduction for $\mathsf{S4}$ necessity. Imposing an analog of Prawitz's side conditions for the necessity introduction rule on the implication introduction rule results in a system for E. Standefer and Brady (2019) present two Fitch-style natural deduction systems for E that do not rely on a restricted reiteration rule, like the system of Anderson and Belnap, in order to fit it into the general format of Brady's earlier work (1984b), which lacks a system for E.

The necessity of E is an $\mathsf{S4}$-ish necessity. Mares and Standefer (2017) address the question of how one weakens the built-in necessity. Different logics are presented, both axiomatically and in Fitch-style natural deduction, with $\mathsf{E.K}$ being the weakest. Frames for which the logics are sound and complete are also presented. More recently, Mares (2024) explored entailment logics in the setting of a frame theory closer to that of Fine frames, and showed how E arises naturally in that setting, as do many of its relatives that can plausibly be said to capture kinds of entailment.

[13] See Read (1988, 76–77) for a derivation of an instance of the formula, in sequent form, there called the "Minc sequent."

6.6 Comparison

In Chapter 2, we compared relevant logics and TV, and we will close this chapter by similarly comparing relevant modal logics and TV-based modal logics. For ease of comparison, we will focus on R as the base relevant logic initially, varying only the modal principles. As with the earlier section, we will suppose that the classical language contains both $\rightarrow$ and $\circ$, with $A \rightarrow B$ equivalent to $\neg A \vee B$ and $A \circ B$ equivalent to $A \wedge B$.

Let us start with the minimal modal logic over R, R.M. The following are theorems of the TV-based modal logic K that are *not* theorems of R.M.[14]

(1) $\Box(A \rightarrow B) \rightarrow (\Box A \rightarrow \Box B)$.
(2) $\Box(A \vee B) \rightarrow (\Diamond A \vee \Box B)$.
(3) $\Box(A \rightarrow A)$.
(4) $\Box\neg(A \rightarrow A) \vee \Diamond(B \rightarrow B)$.

The first two are distinguished in Section 6.2 as axioms that can be added. The third witnesses the failure of the rule (Nec) for R.M.[15] The fourth, due to van Benthem and Humberstone (1983), is a source of Halldén-incompleteness for K, a topic to which we will return in Chapter 7.

While the (K) axiom is not a theorem of R.M, there is a relative that is, namely $(\Box(A \rightarrow B) \wedge \Box A) \rightarrow \Box B$, which follows from (RM), ($\Box\wedge$), and (I1). This becomes invalid if one moves to a non-contractive base logic, such as RW. In fact, in RW.MKNec, this remains invalid. With weaker base logics, there will be more divergences between the relevant and TV-based modal logics.

The modal logics R.M and K are, respectively, the minimal modal logics over the modal frames for R and the Kripke frames for TV. They diverge substantially, although these divergences can, to a degree, be mitigated by adding axioms and rules to R.M to bring it closer to K. In particular, the four validities discussed earlier can be obtained by extending R.M with (K), (MC), and (Nec). The modal logics R.MKNecMC and K are closer, although there will still be divergences. Examples of remaining divergences include the following, which are theorems of K but not of R.MKNecMC.

(1) $A \rightarrow \Box(B \rightarrow B)$.
(2) $\Box(B \rightarrow B) \rightarrow (A \rightarrow A)$.
(3) $\Box(A \rightarrow A) \rightarrow \Box(B \rightarrow B)$.
(4) $\Box A \rightarrow (\Diamond\neg A \rightarrow \Diamond B)$.
(5) $\Box(A \circ B) \rightarrow \Box A$.
(6) $(A \wedge \Box B) \rightarrow \Box(A \rightarrow B)$.
(7) $(\Box A \wedge \Box B) \rightarrow \Box(A \rightarrow B)$.

[14] In our nomenclature, K would be TV.MKNecMC, although this is slightly redundant.

[15] In Chapter 7, we will see how (K) and (Nec) can fail, once we have introduced models. For now we will say that both (K) and (Nec) require more coordination among the parts of the frames than are guaranteed by the basic definitions.

While these are all logical truths of K, none are logical truths of R.MKNecMC. They must fail for R.MKNecMC, since they lead, more or less directly, to violations of variable sharing.

Another fruitful point of comparison is between R.M and M, the minimal monotonic TV-modal logic.[16] The logic M extends TV with the rule (RM). Importantly, ($\Box\wedge$) is not valid in M, while it is adopted as an axiom in R.M. There are other divergences between the logics. For example, the following are all logical truths of M that fail in R.M.

(1) $\Box(A \rightarrow A) \rightarrow \Box(B \rightarrow B)$.
(2) $\Box A \rightarrow \Box(\neg A \rightarrow B)$.
(3) $\Box(A \circ B) \rightarrow \Box A$.

The first of these violates variable sharing. Whether the latter two violate variable sharing will depend on the other principles of the logic. Nonetheless, they are close to principles known to lead to violations.

6.7 Exercises

(1) Show that in B.M4T, (B), and (5) are intederivable.
(2) Show that the axioms listed in Table 6.2 are sound for the class of modal groupoids satisfying the corresponding conditions.
(3) Using some of the matrices from Chapter 5, define modal groupoids to show that the following are not theorems of B.M.
    (1) $\Box(A \rightarrow B) \rightarrow (\Box A \rightarrow \Box B)$
    (2) $\Box(A \vee B) \rightarrow (\Diamond A \vee \Box B)$
    (3) $\Box(A \rightarrow A)$
    (4) $\Box\neg(A \rightarrow A) \vee \Diamond(B \rightarrow B)$
(4) Suppose one adds to the definition of L.MK the following: Whenever A is an axiom of the logic, so is $\Box A$. Show that the resulting logic is identical to L.MKNEc.
(5) Say that a positive modality is a strong of operator $O_1, \ldots, O_n$, $n \geq 1$ such that for each i, $O_i \in \{\Box, \Diamond\}$. Show that in B.MK4T5Nec, there are at most two positive modalities, $\Box$ and $\Diamond$, in the sense that for any positive modality O, either $\vdash_L OA \leftrightarrow \Box A$ or $\vdash_L OA \leftrightarrow \Box A$, where L = B.MK4T5Nec.
(6) Show that in B.MK4TNec, all positive modalities are equivalent to one of the following: $\Diamond, \Box, \Box\Diamond, \Diamond\Box, \Box\Diamond\Box, \Diamond\Box\Diamond$.
(7) Identify a logic L and a set of modal axioms, that is, axioms containing $\Box$, such that the addition of those axioms to L.M provably results in a violation of variable sharing.

[16] In the present nomenclature, M would be TV.MMC. See Pacuit (2017) for more on neighborhood models and the minimal monotonic logic.

(8) Identify a logic L and a set of modal axioms, that is, axioms containing $\Box$, such that the addition of those axioms to L.M results in a non-conservative extension of L that has the variable-sharing property.

(9) Show that in L.M4, $X^{-\Box-\Box} \subseteq X^{-\Box}$.

(10) Show that in L.MT, $X^{-\Box} \subseteq X$. Show that if the logic L.MZ does not have (T) among its theorems, then there is an L.MZ-theory X such that it is not the case that $X^{-\Box} \subseteq X$.

(11) What modal axioms need to be in a logic in order for $(\cdot)^{-\Box}$ to be a closure operator?

7 Modal Frames

7.1 Introduction

As noted in Chapter 6, relevant logics have, in a sense, been concerned with modal considerations from the start. Anderson and Belnap studied the logic E of entailment, which was supposed to capture necessity and relevance. They were concerned with rejecting both fallacies of relevance, such as weakening, $p \to (q \to p)$, and fallacies of modality.[1] To explain the fallacies of modality, we need to introduce some terminology from Anderson and Belnap. A formula A is a NECESSITIVE iff it is equivalent to a formula of the form $\Box B$. In E , all implications are necessitives, and logical truths, such as $p \vee \neg p$, are also necessitives. A fallacy of modality is an improper implication between a non-necessitive and a necessitive, such as $p \to ((p \to p) \to p)$.[2] The logic E was supposed to reject all the fallacies of modality, unlike R.

Meyer, in his PhD thesis (1966), defined and studied proof systems for the logic NR, adding an S4-ish necessity operator to R. Despite this, the investigation of frame theory for relevant modal logics is a more recent development. The area of frames for relevant modal logics really gets going with the work by Fuhrmann (1990), Meyer and Mares (1993), Mares and Meyer (1993), and Mares (1994). Most of the work discussed in the sections below builds on these foundational papers. The basic frames for relevant modal logics add to the ternary relational frames a binary modal accessibility relation, S.[3] This binary relation can, and does, interact with the other parts of the ternary relational frames, as we will see.

It is worth noting that the area has similarities to work on frames for intuitionistic modal logic, as studied, for example, by Božić and Došen (1984) and Ewald (1986). In particular, in both intuitionistic modal logic and relevant modal logic, frames have two sorts of relations. There are relations that interpret the implication of the base logic and there are additional relations that are used to interpret the modal connectives. There can be, and often are, important interactions between these two kinds of relations. In both areas, necessity and possibility need not be interpreted on the same modal relation, and in both cases there are some residual difficulties with possibility

[1] Weiss (2025) investigates fallacies of modality, as well as many related issues.
[2] Sylvan and Plumwood (1969) critically discuss Anderson and Belnap on the fallacies of modality, with further discussion by Humbestone (1982).
[3] In TV-based modal logics, one typically uses "R" for this binary relation, but, in the present context, that label is taken.

operators. This results in a necessity-centric approach for some work, an approach which we will, for the most part, adopt.

In this chapter, we will present the basic frames for relevant modal logics and some of their metatheory. We begin, in Section 7.2, by presenting the basic modal extensions of ternary relational frames, focusing on the language with a primitive necessity operator. In Section 7.3, we will look at the additions that need to be made to interpret a primitive possibility operator. The verification conditions for the primitive and defined possibility operators will differ, in general, but they can be brought into alignment, which will be presented in Section 7.4. Then, in Section 7.5, we will investigate modal frames extended with a set of propositions, in the manner of general frames. In Section 7.6, we will present several examples of modal frames satisfying various frame conditions. We will then turn to issues of metatheory, proving some Soundness theorems (Section 7.7) and some Completeness theorems (Section 7.8), and we will prove Completeness for logics with the modal confinement axiom (MC) (Section 7.9). Following this, we will turn to issues of Incompleteness (Section 7.10) and Halldén-completeness (Section 7.11). We will close the chapter, in Section 7.12, by briefly presenting the basics of neighborhood frames for relevant modal logics.

7.2 Basic Frames

We will begin with the basic modal frames for the minimal logic over B, which will be called B.M.

DEFINITION 7.1 (Modal frames) *A* MODAL FRAME *for* B *is a quintuple* $\langle K, N, R, ^*, S\rangle$ *where the first four components make up a ternary relational frame and* $S \subseteq K \times K$, *and which obeys the following condition, for all* $a, b, c \in K$.

(i) *If* $a \leq b$ *and* Sbc, *then* Sac.

One can add frame conditions to strengthen the base logic or one can add frame conditions to strengthen the modal extension, leaving the base alone. We will focus on the modal strengthenings below.

Following the conventions of Chapter 3, modal principle (A) will have a corresponding frame condition (FA). The principles and conditions are presented in Table 7.1, with the convention that the conditions are all universally quantified. As with ternary relational frames, we will call the modal frames that obey all of the conditions for a logic L the L-modal frames. As an example, a DW.MT-modal frame would be a modal frame that obeys conditions (FN1) and (FT).

We can add a valuation to an L-modal frame in order to obtain an L-modal model.

DEFINITION 7.2 (Valuation, models) *A* VALUATION V *for a frame* F *is a function from* At $\times K$ *to* $\{0, 1\}$.

A MODAL TERNARY RELATIONAL MODEL *for* L *is a pair* $\langle F, V\rangle$ *of an* L*-frame* F *and a valuation* V *where if* $a \leq b$ *and* $V(p, a) = 1$ *then* $V(p, b) = 1$.

A ternary relational model $\langle F, V\rangle$ *is said to be built on the frame* F.

Table 7.1. Common modal axioms and their frame conditions

Name	Axiom	Frame condition
(FT)	$\Box A \to A$	Saa
(FD)	$\Box\neg A \to \neg\Box A$	$\exists b \in K(Sab$ and $Sa^*b^*)$
(F4)	$\Box A \to \Box\Box A$	if Sab and Sbc, then Sac
(F5)	$\Diamond A \to \Box\Diamond A$	If Sa^*c and Sab, then Sb^*c
(FB)	$A \to \Box\Diamond A$	If Sab, then Sb^*a^*
(FK)	$\Box(A \to B) \to (\Box A \to \Box B)$	If $\exists f \in K(Rabf$ and $Sfc)$, then $\exists d, e \in K(Rdec$ and Sad and $Sbe)$
(FNec)	$A \Rightarrow \Box A$	If $a \in N$ and Sab, then $b \in N$
(FMC)	$\Box(A \vee B) \to (\Box A \vee \Diamond A)$	If Sab, then $\exists c \in K(c \leq b$ and Sac and $Sa^*c^*)$

Table 7.2. Verification conditions

$a \Vdash p$	iff	$V(p, a) = 1$
$a \Vdash \mathbf{t}$	iff	$a \in N$
$a \Vdash \neg B$	iff	$a^* \nVdash B$
$a \Vdash B \wedge C$	iff	$a \Vdash B$ and $a \Vdash C$
$a \Vdash B \vee C$	iff	$a \Vdash B$ or $a \Vdash C$
$a \Vdash B \to C$	iff	$\forall b, c \in K($if $Rabc$ and $b \Vdash B$, then $c \Vdash C)$
$a \Vdash B \circ C$	iff	$\exists b, c \in K(Rbca$ and $b \Vdash B$ and $c \Vdash C)$
$a \Vdash \Box B$	iff	$\forall b \in K($if Sab, then $b \Vdash B)$

We can extend the valuation to a verification relation, $\Vdash$, on the whole language as listed in Table 7.2. The Heredity lemma extends to the present setting.

LEMMA 7.3 (Heredity lemma) *Let M be an* L*-modal model. For all $a \in K$, if $a \Vdash A$ and $a \leq b$, then $b \Vdash A$.*

Proof We will do the new case, $\Box$. Suppose that $a \leq b$, $a \Vdash \Box B$, but $b \nVdash \Box B$. Then there is a $c \in K$ such that Sbc and $c \nVdash B$. As $a \leq b$, by the condition on modal frames, Sac, which implies $a \nVdash \Box B$, which is a contradiction. □

The Verification lemma still holds for these models, as it is a consequence of the Heredity lemma.

As noted by Božić and Došen (1984), albeit in a different setting, condition (i) on modal frames is slightly stronger than is needed to secure Heredity. The following weaker condition will suffice.

(Alt i) $\forall a, b, c \in K$, if $a \leq b$ and Sbc, then there is a $d \in K$ such that $d \leq c$ and Sad.

Later we will see that the canonical frame obeys (i), so we do not need to appeal to (Alt i), although it presents some interesting possibilities, such as in the mostly Meyer modal frames of Mares (1994).

The definitions of counterexamples, holding, and validity all carry over to the present setting without any change. With the definition of counterexample in hand, we are in a position to see why the basic relevant modal logic differs from the basic TV-based modal logic. In particular, we can see why (K) and (Nec) fail. The former fails because there is no need for R and S to be coordinated in the way required to validate (K). The rule (Nec) fails because validity is evaluated only in terms of N, and, without further conditions, S can take one outside of the set N of regular points.[4]

7.3 Possibilities

Our focus is on the monomodal language $\mathfrak{L}_{\Box}$ with a primitive necessity operator and a defined possibility operator, $\Diamond$. In Chapter 6, however, we noted that one could introduce a primitive possibility operator, $\blacklozenge$, for a multimodal language. We can obtain frames for the language $\mathfrak{L}_{\Box\blacklozenge}$ by adding a second binary accessibility relation, $S_{\blacklozenge}$, to a modal frame. The resulting frames must obey the condition

- For all $a,b,c \in K$, if $S_{\blacklozenge}ac$ and $a \leq b$, then $S_{\blacklozenge}bc$.

For this primitive possibility operator, we use the following verification condition.

- $a \Vdash \blacklozenge B$ iff $\exists b \in K(S_{\blacklozenge}ab$ and $b \Vdash B)$.

As sometimes happens with non-classical logics, a primitive possibility operator and a possibility operator defined in terms of necessity and negation may not coincide. In general, $\blacklozenge$ will not be equivalent to $\Diamond$. We can see this by considering the derived verification condition for the latter.

- $a \Vdash \Diamond B$ iff $\exists b \in K(Sa^*b$ and $b^* \Vdash B)$.

The latter condition involves star points and the S relation, whereas the former condition involves $S_{\blacklozenge}$ and no stars. Given those interactions, it would be surprising if the two possibility operators were equivalent without further conditions being imposed.

With a primitive possibility operator, one can define a related necessity operator, $\blacksquare$, as $\neg\blacklozenge\neg$.

- $a \Vdash \blacksquare B$ iff $\forall b \in K($if $S_{\blacklozenge}a^*b$, then $b^* \Vdash B)$.

As one might expect based on the verification conditions, $\Box$ and $\blacksquare$ are not generally equivalent. Seki (2003a; 2003b; 2009; 2011a; 2012a; 2013; 2015) considers relevant logics whose modal axioms involve a wide variety of combinations of operators from among $\Box, \Diamond, \blacklozenge$, and $\blacksquare$. Since we are not developing the general techniques used by Seki, we are not covering the same wide variety of combinations.

While in the basic frames, the two accessibility relations, S and $S_{\blacklozenge}$, are independent, one can obtain frames where the relations are not entirely independent by imposing frame conditions. A natural requirement is to take $S_{\blacklozenge}$ to be the converse of S:

(FCA) For all $a,b \in K$, Sab iff $S_{\blacklozenge}ba$.

[4] This observation was made by Fuhrmann (1990).

We will call this condition (FCA) for "converse accessibility." The result is a kind of Priorean tense, or temporal, logic where one has the rules

- $A \rightarrow \Box B \Rightarrow \blacklozenge A \rightarrow B$, and
- $\blacklozenge A \rightarrow B \Rightarrow A \rightarrow \Box B$

valid.[5]

LEMMA 7.4 *In the class of modal frames for which* (FCA) *holds, the displayed rules are valid.*

Proof Let $\mathcal{C}$ be the class of modal frames for which (FCA) holds. Let F be a frame in $\mathcal{C}$, and let M be a model on F.

Suppose $\models_{\mathcal{C}} A \rightarrow \Box B$. Suppose that $a \Vdash \blacklozenge A$. Then there is a b such that $S_{\blacklozenge}ab$ and $b \Vdash A$. By (FCA), Sba, so, as $b \Vdash \Box B$, it follows that $a \Vdash B$, as desired.

Suppose $\models_{\mathcal{C}} \blacklozenge A \rightarrow B$. Suppose that $a \Vdash A$. Let $b \in K$ be such that Sab. From the assumption and (FCA), $b \Vdash \blacklozenge A$, so, again by the assumption, $b \Vdash B$. It then follows that $a \Vdash \Box B$, which was to be proved. □

Apart from the temporal logic motivation, a natural motivation for the inclusion of a backward-looking possibility operator arises from display logic, introduced by Belnap (1982).[6] Display logic is a general sequent-based proof system for a wide range of logics that exploits the sort of residuation connections found in the two earlier, rules, and in the fusion and arrow rules from Chapter 2. Restall (2000) develops a sequent-based proof system for substructural logics with necessity and a backward-looking possibility operator based on this idea.[7]

A backward-looking possibility operator will have substantial connections to the forward-looking necessity. If $A \rightarrow B$ is a theorem, then $\blacklozenge A \rightarrow \blacklozenge B$ will be, by (RM$\blacklozenge$). By the above rules, $A \rightarrow \Box\blacklozenge B$ will be a theorem, and similar reasoning shows $\blacklozenge\Box A \rightarrow B$ will be a theorem. As will be noted below, there are some issues that arise when (MC) is considered, and those issues will be multiplied when there are different options for the possibility displayed operator in that axiom.

7.4 Generic Verification

As one might expect from grinding through the verification condition for $\Diamond$, $a \Vdash \Diamond B$ need not always imply that there is a b such that Sab and $b \Vdash B$. This is despite the familiar verification condition for $\Box$. Define Tab as Sab and Sa^*b^*. Following Mares and Meyer (1993), the verification condition,

- $a \Vdash \Diamond B$ iff $\exists b \in K(Tab$ and $b \Vdash B)$,

[5] See Humberstone (2016) for more on tense logics.

[6] An expanded version of that paper was printed in Anderson et al. (1992, §62). There are important connections between display logic and the gaggle theory of Bimbó and Dunn (2008), some of which are explored by Restall (1995).

[7] Wansing (1998) shows how flexible the display logic framework is for the proof theory of modal logic.

will be called the GENERIC condition, and it differs from the DEFINED condition,

- $a \Vdash \Diamond B$ iff $\exists b \in K(Sa^*b$ and $b^* \Vdash B)$.

The corresponding generic verification condition for the $\Box$ is

- $a \Vdash \Box B$ iff $\forall b \in K$(if Tab, then $b \Vdash B$).

One upshot of the frame condition (FMC) is that the defined verification condition is equivalent to the generic verification condition.

LEMMA 7.5 *Let M be a model built on a modal frame that obeys* (FMC). *Then the following are equivalent.*

1. $\exists b \in K(Sa^*b$ *and* $b^* \Vdash B)$.
2. $\exists b \in K(Tab$ *and* $b \Vdash B)$.

Proof For (1) to (2), suppose $\exists b \in K(Sa^*b$ and $b^* \Vdash B)$. By (FMC), there is a $c \in K$ such that $c \leq b$, Sa^*c, and $Sa^{**}c^*$. Since $b^* \Vdash B$, $b \nVdash \neg B$, so $c \nVdash \neg B$. It follows that $c^* \Vdash B$, so $\exists d \in K(Tad$ and $d \Vdash B)$.

For (2) to (1), suppose $\exists b \in K(Tab$ and $b \Vdash B)$. Unpacking the definition yields $\exists b \in K(Sab$ and Sa^*b^* and $b \Vdash B)$. But then b^* is a c such that Sa^*c and $c^* \Vdash B$, which is what we wanted to prove. $\Box$

While we will not present the details here, the generic verification condition for $\Box$ is equivalent to the condition given earlier in this chapter in the presence of (FMC).[8]

We can obtain the equivalence with the generic conditions if we use the stronger condition

(FMM) For all $a, b \in K$, if Sab, then Sa^*b^*,

provided that the condition on modal frames is weakened to (Alt i). This is, essentially, what Mares (1994) does in defining mostly Meyer modal models, although those models include several more frame conditions corresponding to R axioms and other modal principles. Those features are not required for the argument that we will present here.

LEMMA 7.6 *Let M be a model built on a modal frame that obeys (Alt i), rather than (i), and that obeys* (FMM). *Then the following are equivalent.*

1. $\exists b \in K(Sa^*b$ *and* $b^* \Vdash B)$.
2. $\exists b \in K(Tab$ *and* $b \Vdash B)$.

Proof (FMM) clearly implies (FMC), which suffices for the result. $\Box$

The reason that (Alt i) needs to be adopted rather than (i) in the presence of (FMM) is that one can prove that if $a \leq b$ and $b \Vdash \Box B$, then $a \Vdash \Box B$. This would be disastrous, so (i) is weakened enough to avoid this result while still permitting a proof of the Heredity lemma.

[8] See Mares and Meyer (1993).

7.5 General Modal Frames

We can expand the range of operations on propositions that yield new propositions to take account of the modal operators.

LEMMA 7.7 *Let F be a modal* L.M*-frame and let X be a proposition. Then the following are propositions.*

(i) $\Box X = \{a \in K : \forall b \in K(\textit{if } Sab, \textit{ then } b \in X)\}$.
(ii) $\blacklozenge X = \{a \in K : \exists b \in K(S_{\blacklozenge}ab \textit{ and } b \in X\}$.

Proof For the first, suppose $a \leq b$ and $a \in \Box X$. Then for all $c \in K$, if Sac, then $c \in X$. Suppose $b \notin \Box X$. Then for some $d \in K$, Sbd, but $d \notin X$. From the condition on modal frames and $a \leq b$, Sad. It follows from the assumption that $d \in X$, which is a contradiction.

For the second, suppose that $a \leq b$ and $a \in \blacklozenge X$, so that there is a $c \in K$ such that $S_{\blacklozenge}ac$. From the condition on the possibility accessibility relation, it follows that $S_{\blacklozenge}bc$, which suffices for $b \in \blacklozenge X$. □

As with basic L-frames, one can define algebras based on modal L.M-frames that can interpret the modal connectives. The arguments are largely similar to those of the earlier chapter, so we will not recapitulate them here.

In Chapter 3, we defined propositional frames, as a pair $\langle F, \mathsf{Prop}\rangle$ of an L-frame F and a model-like set Prop of propositions on F. This can be extended to a propositional modal frame, also called a general modal frame.

DEFINITION 7.8 (Propositional modal frame) *Let* $F = \langle K, N, R, ^*, S\rangle$ *be a modal* L*-frame and let* Prop *be a model-like set of propositions on F.*

The set Prop *is* MODAL *iff for all* $X \in \mathsf{Prop}$, $\Box X \in \mathsf{Prop}$. *If* $\blacklozenge$ *is in the language, then for all* $X \in \mathsf{Prop}$, $\blacklozenge X \in \mathsf{Prop}$ *as well.*

Then $\langle F, \mathsf{Prop}\rangle$ *is a propositional modal* L*-frame iff F is a modal* L*-frame, and* Prop *is both model-like and modal.*

A model M on a propositional modal L*-frame* $\langle F, \mathsf{Prop}\rangle$ *is a pair of* $\langle\langle F, \mathsf{Prop}\rangle, V\rangle$ *where V is a valuation admissible with respect to* Prop.

As before, the definitions of counterexample, holding, etc. carry over in a straightforward way.

One reason that one would want to define propositional modal frames, or general frames, is to obtain results on duality theory relating modal frames and algebras, as done by Seki (2003a). Seki's frames, called ENLARGED FRAMES, include an extra element, e, which is the empty state, which appears in several conditions. It is included to obtain completeness results for some substructural logics that are incomplete over the class of frames without it.[9]

[9] See Seki (2003b) for more details. The example Seki provides is the logic obtained by adding $(A \wedge \neg A) \to B$ as an axiom, which is not a relevant logic. Routley et al. (1982, 380ff.) also discuss enlarged frames.

Another reason to define general frames is to obtain a general Sahlqvist-style theorem that yields completeness of a range of modal logics with respect to their frame conditions.[10] Seki (2003b) syntactically defines classes of modal axioms and rules and shows that any logics defined using certain selections from those classes are complete with respect to their propositional modal frames. The details of the proofs are beyond the scope of the book, but they demonstrate how logically well-behaved many relevant modal logics are.

7.6 Examples

It will be useful to have some example modal frames. In this section, we will look at a few examples with differing modal extensions. The examples will extend the frames from Chapter 3 with a binary modal accessibility relation. The accessibility relations on different frames will obey different conditions. Following on the convention for R in the table, an entry of multiple points under S for a point x means that Sxy, for each y in the entry. For example, in Table 7.3, the entry of bc in the row for a under S means that Sab and Sac hold.

For the example of a B.M-frame, let $K = \{a,b,c,d\}$, $N = \{b,c\}$, and define the star, R, and S as in Table 7.3.

The B.M-frame, along with the next several frames, are distinctive in different ways. For present purposes, one of the most distinctive features is that the four points in the frames are linearly ordered, $d \leq a \leq b \leq c$. Given the condition on modal frames that requires S to be downward closed in the first position, this restricts the options for modal accessibility relations for the underlying ternary relational frames. The modal frames of this chapter do not require the condition that if $b \leq c$ and Sab then Sac, as this would further constrain the modal extensions available for the underlying ternary relational frames.[11] For the example of a DW.MT-frame, let $K = \{a,b,c,d\}$, $N = \{a,b,c\}$, and define the star, R, and S as in Table 7.4.

For the example of a DJ.MD-frame, let $K = \{a,b,c,d\}$, $N = \{b,c\}$, with the star, R, and S defined in Table 7.5.

Table 7.3. B.M-frame

R	a	b	c	d	$*$	S
a	$abcd$	$abcd$	bc	$abcd$	b	bc
b	abc	bc	c	$abcd$	a	b
c	c	c	c	$abcd$	d	$\emptyset$
d	$abcd$	$abcd$	$abcd$	$abcd$	c	bc

[10] See Blackburn et al. (2002, 157ff.) for more on Sahlqvist theorems.

[11] Adding this condition would not change the logics generated by most of the classes of frames obeying the various frame conditions of this chapter. The reason is that the canonical frame for the logics obeys that condition. The exception is the frames obeying (FMC) but not (FD), as the canonical frames for those logics require an alternative definition for the canonical accessibility relation. This point will be discussed in Section 7.9.

Table 7.4. DW.MT-frame

R	a	b	c	d	$*$	S
a	abc	bc	c	$abcd$	b	abc
b	bc	c	c	$abcd$	a	bc
c	c	c	c	$abcd$	d	c
d	$abcd$	$abcd$	$abcd$	$abcd$	c	$abcd$

Table 7.5. DJ.MD-frame

R	a	b	c	d	$*$	S
a	$abcd$	$abcd$	abc	$abcd$	b	abd
b	abc	bc	c	$abcd$	a	ad
c	c	c	c	$abcd$	d	d
d	$abcd$	$abcd$	$abcd$	$abcd$	c	$abcd$

Table 7.6. TW.MB-frame

R	a	b	c	d	$*$	S
a	abc	bc	c	$abcd$	b	abc
b	abc	bc	c	$abcd$	a	abc
c	c	c	c	$abcd$	d	$\emptyset$
d	$abcd$	$abcd$	$abcd$	$abcd$	c	abc

Table 7.7. RW.M4Nec-frame

R	a	b	c	d	$*$	S
a	abc	bc	c	$abcd$	b	ab
b	bc	c	c	$abcd$	a	ab
c	c	c	c	$abcd$	d	ab
d	$abcd$	$abcd$	$abcd$	$abcd$	c	ab

For the example of a TW.MB-frame, let $K = \{a,b,c,d\}$, $N = \{b,c\}$, and define the star, R, and S as in Table 7.6. As the reader may have noticed, the relation S is not symmetric, as would be required in frames for TV-based modal logics containing (B). Wherever a point x relates to c, we have d relating to x^*. The result is a striking failure of symmetry, since three points bear S to c, which itself bears S to no points.

For the example of a RW.M4Nec-frame, let $K = \{a,b,c,d\}$, $N = \{a,b,c\}$, and define the star, R, and S as given in Table 7.7. The transitivity condition for (4) is familiar and there are no surprises here. The condition (FNec) means that for no accessibility relation S on this frame can have a, b, or c relate to d. We can extend this frame to be a frame for Meyer's RW.MKT4Nec by using the relations in Table 7.8.

For the example of a T.M5-frame, let $K = \{a,b,c,d\}$, $N = \{b,c\}$, and define the star, R, and S as in Table 7.9. While the example of a frame for (B) didn't obey symmetry, this frame does obey the Euclidean condition corresponding to (5) in frames for TV-based modal logics. The Euclidean condition says that if Sab and Sac, then Sbc. As should be clear from the frame condition (F5), not all frames for (5) will obey the Euclidean condition. If they did, then all the frames for R.MT45 would have

Table 7.8. RW.MKT4Nec-frame

R	a	b	c	d	$*$	S
a	abc	bc	c	$abcd$	b	abc
b	bc	c	c	$abcd$	a	abc
c	c	c	c	$abcd$	d	abc
d	$abcd$	$abcd$	$abcd$	$abcd$	c	$abcd$

Table 7.9. T.M5-frame

R	a	b	c	d	$*$	S
a	$abcd$	$abcd$	$abcd$	$abcd$	b	ab
b	abc	bc	c	$abcd$	a	ab
c	c	c	c	$abcd$	d	a
d	$abcd$	$abcd$	$abcd$	$abcd$	c	ab

Table 7.10. R.MMC-frame

R	a	b	c	d	$*$	S
a	abc	abc	c	$abcd$	b	ab
b	abc	bc	c	$abcd$	a	ab
c	c	c	c	$abcd$	d	b
d	$abcd$	$abcd$	$abcd$	$abcd$	c	ab

Table 7.11. R4-frame

R	a	b	c	d	$*$	S
a	abc	abc	c	$abcd$	b	abc
b	abc	bc	c	$abcd$	a	bc
c	c	c	c	$abcd$	d	bc
d	$abcd$	$abcd$	$abcd$	$abcd$	c	$abcd$

Table 7.12. E.MK-frame

R	a	b	c	d	$*$	S
a	$abcd$	$abcd$	$abcd$	$abcd$	a	d
b	abc	bc	c	$abcd$	b	d
c	c	c	c	$abcd$	d	d
d	$abcd$	$abcd$	$abcd$	$abcd$	c	d

accessibility relations that were equivalence relations, but this is not the case, as is clear from the work of Standefer (2023b), to be discussed in Section 8.2.2.

For an example of an R.MMC-frame, let $K = \{a,b,c,d\}$, $N = \{b,c\}$, and define the star, R, and S as in Table 7.10. We can extend this frame to a frame for Mares and Meyer's logic R4 by using the relations in Table 7.11. This uses essentially the same trick to satisfy the (FK) condition as the frames in Tables 7.8 and 7.12.

For the example of an E.MK-frame, let $K = \{a,b,c,d\}$, $N = \{b,c\}$, and define the star and R as in Table 7.12. This frame takes advantage of the fact that when the point d occupies either of the first two positions in the ternary relation, the final position is

Table 7.13. C.MT45NecMC-frame

R	a	b	c	d	$*$	S
a	ac	bc	c	$abcd$	b	$abcd$
b	$abcd$	$abcd$	$abcd$	$abcd$	a	$abcd$
c	c	c	c	$abcd$	d	$abcd$
d	$abcd$	$abcd$	$abcd$	$abcd$	c	$abcd$

universal. The existential condition in the consequent of (FK) will be satisfied so long as every point bears S to d. For this reason, we can extend S freely and the result will still satisfy (FK).

For a final example of a modal frame built from a ternary relational frame from Chapter 3, we will use a somewhat simple example. For a C.MT45MC-frame, let $K = \{a,b,c,d\}$, $N = \{a,c\}$, and define the star and R as in Table 7.13. The binary relation on this frame is the universal relation. N is not closed under S, as that would be disastrous, from a relevance point of view. In Section 8.2.2, we will discuss frames with a universal modal accessibility relation. While in frames for TV-based modal logics, the universal relation provides models for the logic S5, in the present context, the universal modal accessibility relation cannot deliver all the usual S5 principles.

7.7 Soundness

In this section we will prove the Soundness of L.MZ with respect to L.MZ-frames. Although we will prove Soundness for the basic modal logic with possibility axioms and rules, for the extensions we will focus on the $\Box$ axioms, with the possibility operators defined in terms of negation and $\Box$.

LEMMA 7.9 (Soundness for L.M) *For all formulas A, if $\vdash_{\mathsf{L.M}} A$ then $\models_{\mathsf{L.M}} A$.*

Proof The proof is by induction on the proof of A. The axioms for L are sound from the results of Chapter 4. We will focus on the additional axioms and rules.

Let F be L.MZ-frame and let M be a model on that frame. There are then four cases.

For the ($\Box\wedge$) case, suppose $a \Vdash \Box B \wedge \Box C$. Then for all b such that Sab, $b \Vdash B$ and $b \Vdash C$. Therefore, $b \Vdash B \wedge C$, so $a \Vdash \Box(B \wedge C)$.

For the ($\blacklozenge\vee$) case, suppose $a \Vdash \blacklozenge(B \vee C)$. Then there is a b such that $S_{\blacklozenge}ab$ and $b \Vdash B \vee C$. It follows that $b \Vdash B$ or $b \Vdash C$, so $a \Vdash \blacklozenge B \vee \blacklozenge C$.

For the (RM) case, suppose $\models_{\mathsf{L.M}} B \rightarrow C$, and suppose $a \Vdash \Box B$. Then for all b such that Sab, $b \Vdash B$. Therefore, $b \Vdash C$, so $a \Vdash \Box C$.

For the (RM$\blacklozenge$) case, suppose $\models_{\mathsf{L.M}} B \rightarrow C$, and suppose $a \Vdash \blacklozenge B$. Then for some b such that $S_{\blacklozenge}ab$, $b \Vdash B$. Therefore, $b \Vdash C$, so $a \Vdash \blacklozenge C$. □

The Soundness proof for modal extensions of L.M has additional cases for the extra axioms and rules.

LEMMA 7.10 (Soundness for modal extensions) *Let Z be a set of additional modal axioms added to* L.M. *For all axioms A from* Z, *if $\vdash_{\mathsf{L.MZ}} A$, then $\models_{\mathsf{L.MZ}} A$.*

Proof The cases for (T) and (4) are straightforward.

For (D), suppose F is a frame satisfying (FD), and suppose $a \Vdash \Box\neg A$. By (FD), there is a b such that Sab and a^*b^*. Suppose $a \nVdash \neg\Box A$. Then $a^* \Vdash \Box A$, which implies $b^* \Vdash A$. As Sab, $b \Vdash \neg A$, whence $b^* \nVdash A$, which is a contradiction.

For (5), suppose F is a frame satisfying (F5), and suppose $a \Vdash \neg\Box\neg A$. Suppose that $a \nVdash \Box\neg\Box\neg A$. Then there is a b such that Sab and $b \nVdash \neg\Box\neg A$. So, $b^* \Vdash \Box\neg A$. From the initial assumption, $a^* \nVdash \Box\neg A$, so there is c such that Sa^*c and $c \nVdash \neg A$. From (F5), it follows that Sb^*c, so $c \Vdash \neg A$, which is a contradiction.

The cases for (B), (K), (Nec), and (MC) are left as an exercise. □

The previous two lemmas suffice for our general Soundness result.

THEOREM 7.11 (Soundness for L.MZ) *Let* Z *be a subset of the additional modal principles from Section 7.2 added to* L.M. *For all A, if* $\vdash_{\mathsf{L.MZ}} A$, *then* $\models_{\mathsf{L.MZ}} A$.

The Soundness theorem is to be expected, and, as expected, it is less involved than Completeness, to which we will now turn.

7.8 Completeness

The proof of the Completeness theorem will use the canonical frame method of Chapter 4, following on the work of Furhmann (1990) and Restall (2000, ch. 11). We will reuse many of the definitions, such as L-theory and primeness, from the earlier chapter.

We will begin with the definition of the canonical model, where much of it is carried over from the earlier chapter. As before, we will omit the subscript on the frame and its components to reduce clutter.

DEFINITION 7.12 (Canonical frame) *Define a modal* L.MZ *frame* $\langle K, N, R, ^*, S\rangle$ *as follows.*

- *K is the set of all prime* L.MZ*-theories.*
- $a \in N$ *iff* $\mathsf{t} \in a$.
- *$Rabc$ iff for all formulas $B, C \in \mathfrak{L}$, if $B \to C \in a$ and $B \in b$, then $C \in c$.*
- $a^* = \{A \in \mathfrak{L} : \neg A \notin a\}$.
- *Sab iff* $a^{-\Box} \subseteq b$.

As before, we can show that $a \leq b$ iff $a \subseteq b$.

We have to show that the canonical frame obeys the conditions on modal frames.

LEMMA 7.13 *The canonical frame obeys condition (i): If $a \leq b$ and Sbc, then Sac.*

Proof Suppose that $a \leq b$ and Sbc. The former is equivalent to $a \subseteq b$. The latter is equivalent to $b^{-\Box} \subseteq c$. By lemma 6.6, $a^{-\Box} \subseteq b^{-\Box}$, so $a^{-\Box} \subseteq c$, which is equivalent to Sac. □

The canonical valuation is defined as usual, via containment in theories.

- $V(p,a) = 1$ iff $p \in a$.

To get Completeness for the logic extended with the modal axiom or rule (M), the canonical model needs to satisfy the frame condition (FM).

LEMMA 7.14 *Suppose that* L.MZ *has the formula* (C) *as a theorem. Then the canonical frame for* L.MZ *obeys the frame condition* (FC).

Proof There are several cases. The cases for (FT) and (F4) are straightforward, so they will be omitted.

For (FD), let a be an L.MZ-theory. The pair $\langle a^{*-\Box}, \{A : \Box\neg A \in a\}\rangle$ is a disjoint pair. Suppose not, then $\vdash_{\mathsf{L.MZ}} A \rightarrow B_1 \vee \cdots \vee B_n$, where A is a conjunction of formulas from $a^{*-\Box}$ and $\Box\neg B_1, \ldots, \Box\neg B_n \in a$. So, $\Box(\neg B_1 \wedge \cdots \wedge \neg\Box B_n) \in a$, so $\Box\neg(B_1 \vee \cdots \vee B_n) \in a$. Since a is a theory and (D) is a theorem, $\neg\Box(B_1 \vee \cdots \vee B_n) \in a$. It follows that $\Box(B_1 \vee \cdots \vee B_n) \notin a^*$, so $B_1 \vee \cdots \vee B_n \notin a^{*-\Box}$. This, however, is a contradiction, since $A \in a^{*-\Box}$ because $a^{*-\Box}$ is a theory.

We can then use the Pair extension lemma to obtain a prime L.MZ-theory b such that $a^{*-\Box} \subseteq b$ and $b \cap \{A : \Box\neg A \in a\} = \emptyset$.

Finally, suppose that $\Box A \in a$, so $\Box\neg\neg A \in a$. So, $\neg A \notin b$, by construction of $\Box$, so $A \in b^*$, as required.

For (F5), suppose Sa^*c and Sab. Then $a^{*-\Box} \subseteq c$ and $a^{\Box} \subseteq b$. Let $A \in b^{*-\Box}$, so $\Box A \in b^*$. Since b^* is an L.MZ-theory, $\Box\neg\neg A \in b^*$. So, $\neg\Box\neg\neg A \notin b$, so $\neg\Box\neg\neg A \notin a^{-\Box}$. Therefore, $\Box\neg\Box\neg\neg A \notin a$. Since $\vdash_{\mathsf{L.MZ}} \neg\Box\neg\neg A \rightarrow \Box\neg\Box\neg\neg A$ and a is an L.MZ-theory, $\neg\Box\neg\neg A \notin a$, and so $\neg\Box A \notin a$. It follows that $\Box A \in a^*$, so $A \in a^{*-\Box}$, and so $A \in c$, as desired.

For (FB), suppose $a^{-\Box} \subseteq b$. Suppose $A \in b^{*-\Box}$ but $A \notin a^*$. Then $\neg A \in a$, so $\Box\neg\Box\neg\neg A \in a$, as (B) is a theorem. It follows that $\Box\neg\Box A \in a$, so $\neg\Box A \in b$. Therefore, $\Box A \notin b^*$, so $A \notin b^{*-\Box}$, which contradicts an assumption.

For (FK), suppose there is an f such that $Rabf$ and Sfc. Suppose that there are formulas A and B such that $A \rightarrow B \in a^{-\Box}$, $A \in b^{-\Box}$, but $B \notin c$. By definition, $\Box(A \rightarrow B) \in a$ and $\Box A \in b$. Since (K) is provable, $\Box A \rightarrow \Box B \in a$, so $\Box B \in f$, whence $B \in c$, which is a contradiction. Therefore, there are no such formulas A and B. Using lemma 2.39, we can obtain prime L.MZ theories $d \supseteq a^{-\Box}$ and $e \supseteq b^{-\Box}$ such that $Rdec$, as desired.

For (FNec), suppose $a \in N$ and Sab. From the former, it follows that $\mathsf{t} \in a$, so by (Nec), $\Box\mathsf{t} \in a$. From the latter, $\mathsf{t} \in b$, so $b \in N$. □

Completeness with (MC) appears to require a slightly different definition of accessibility, so we will leave that till later.

The next step is the Truth lemma.

LEMMA 7.15 (Truth lemma) *For all formulas A, $A \in a$ iff $a \Vdash A$.*

Proof The modification to previous proofs is the addition of the $\Box$ step.

Suppose $\Box A \in a$. Let $b \in K$ be an arbitrary prime L.MZ-theory such that Sab. Then $A \in a^{-\Box}$, so $A \in b$. By the inductive hypothesis, $b \Vdash A$. Therefore, $a \Vdash \Box A$, as desired.

Suppose that $\Box A \notin a$. Then $A \notin a^{-\Box}$. By lemma 6.5, there is a prime L.MZ-theory b such that $a^{-\Box} \subseteq b$ and $A \notin b$. Therefore, by the inductive hypothesis, $b \not\Vdash A$, so $a \not\Vdash \Box A$, as desired. □

The results of the section so far were obtained without ♦ in the language. For the remainder of the section, we will consider how to adapt the results to accommodate the addition. The first step is to add a canonical accessibility relation.

- $S_{\blacklozenge}ab$ iff $b^{\blacklozenge} \subseteq a$.

Then we verify that the canonical frame obeys the appropriate condition.

LEMMA 7.16 *If ♦ is in the language, the canonical frame obeys the condition that if $a \leq b$ and $S_{\blacklozenge}ac$, then $S_{\blacklozenge}bc$.*

Proof Suppose that $a \leq b$ and $S_{\blacklozenge}ac$. These are equivalent to $a \subseteq b$ and $c^{\blacklozenge} \subseteq a$, respectively. The transitivity of $\subseteq$ yields $c^{\blacklozenge} \subseteq b$, which is $S_{\blacklozenge}bc$, as desired. □

If one has additional frame conditions for $S_{\blacklozenge}$ then one can verify that the canonical frame satisfies those conditions.

7.9 Completeness with Confinement

Completeness for logics containing (MC) presents some complications in general. These complications can be avoided if the logic has (D) among its theorems, in which case we can use the standard definition for accessibility in the canonical frame.[12] For the first part of this section, we will work in a logic that has (D). After that, we will consider the more general case without (D).

For this section, define $X^{-\Diamond} = \{A : \Diamond A \in X\}$, where X is an L.MZ-theory, and Tab iff Sab and Sa^*b^*. These two definitions will be crucial in the proofs below, and we start by proving a lemma about their relationship.

LEMMA 7.17 *Tab iff $a^{-\Box} \subseteq b \subseteq a^{-\Diamond}$.*

Proof Suppose Tab, so Sab and Sa^*b^*. From the first part, we have $a^{-\Box} \subseteq b$. Suppose that $A \in b$ but $A \notin a^{-\Diamond}$. It follows that $\neg\Box\neg A \notin a$, so $\Box\neg A \in a^*$. This implies $\neg A \in a^{*-\Box}$, so, as Sa^*b^*, $\neg A \in b^*$. This, however, implies $A \notin b$, which is a contradiction.

Suppose $a^{-\Box} \subseteq b \subseteq a^{-\Diamond}$. The first subset relation suffices for Sab. For the second, suppose $A \in a^{*-\Box}$ but $A \notin b$. Then $\neg A \in b$, so by assumption, $\neg A \in a^{-\Diamond}$. This implies $\neg\Box\neg\neg A \in a$, so, as a is a theory, $\neg\Box A \in a$. From the assumption $A \in a^{*-\Box}$, $\Box A \in a^*$, so $\neg\Box A \notin a$, which is a contradiction. □

Note that if (D) is a theorem, then $a^{-\Box} \subseteq a^{-\Diamond}$, but if it is not a theorem, then the indicated containment relation can fail. This observation is needed for the crucial next lemma.

[12] See Mares (1993).

LEMMA 7.18 *If the logic* L.MZ *has* (D) *and* (MC) *as theorems, then the canonical frame satisfies* (FMC).

Proof For (FMC), suppose Sab, so $a^{-\Box} \subseteq b$. To obtain a prime theory c such that $a^{-\Box} \subseteq c \subseteq a^{-\Diamond}$, we will show that

$$\langle a^{-\Box}, (\mathfrak{L}_{\Box} \setminus b) \cup (\mathfrak{L}_{\Box} \setminus a^{-\Diamond})\rangle$$

is a disjoint pair. Note that this argument requires $a^{-\Box} \subseteq a^{-\Diamond}$. If that fails, then there is a formula $A \in a^{-\Box}$ such that $A \notin a^{-\Diamond}$, so $A \in (\mathfrak{L}_{\Box} \setminus a^{-\Diamond})$, which ensures that the displayed pair is *not* a disjoint pair.

Suppose that it is isn't. Then there is a conjunction A of formulas from $a^{-\Box}$ and a disjunction of formulas $B_1, \ldots, B_n \in (\mathfrak{L} - b)$, $C_1, \ldots, C_m \in (\mathfrak{L}_{\Box} \setminus a^{-\Diamond})$ such that

$$\vdash_{\mathsf{L.MZ}} A \rightarrow (B_1 \vee \cdots \vee B_n \vee C_1 \vee \cdots \vee C_m).$$

From (RM), it follows that

$$\vdash_{\mathsf{L.MZ}} \Box A \rightarrow \Box(B_1 \vee \cdots \vee B_n \vee C_1 \vee \cdots \vee C_m).$$

By using (MC) and distribution laws, we have

$$\vdash_{\mathsf{L.MZ}} \Box A \rightarrow (\Box(B_1 \vee \cdots \vee B_n) \vee \Diamond C_1 \vee \cdots \vee \Diamond C_m).$$

Since $A \in a^{-\Box}$, it follows that $\Box A \in a$, so $\Box(B_1 \vee \cdots \vee B_n) \vee \Diamond C_1 \vee \cdots \vee \Diamond C_m \in a$, as a is a theory. Since a is prime, either $\Box(B_1 \vee \cdots \vee B_n) \in a$ or $\Diamond C_i \in a$, for $1 \leq i \leq m$. Suppose $\Box(B_1 \vee \cdots \vee B_n) \in a$. Then, as $a^{-\Box} \subseteq b$, $B_1 \vee \cdots \vee B_n \in b$. Since b is prime, $B_i \in b$, for some i such that $1 \leq i \leq n$. This is, however, a contradiction, as $B_i \in (\mathfrak{L}_{\Box} \setminus b)$. Suppose, instead, that $\Diamond C_i \in a$. But then $C_i \in a^{-\Diamond}$, which is impossible, as $C_i \in (\mathfrak{L}_{\Box} \setminus a^{-\Diamond})$. Therefore the initial pair is a disjoint pair.

It follows that there is prime L.MZ-theory $c \supseteq a^{-\Box}$ such that $\langle c, (\mathfrak{L}_{\Box} \setminus b) \cup (\mathfrak{L}_{\Box} \setminus a^{-\Diamond})\rangle$ is a disjoint pair. To see that $c \subseteq b$, suppose not, so there is $A \in c$ such that $A \notin b$. This means that $A \in (\mathfrak{L}_{\Box} \setminus b)$, which contradicts the fact that $\langle c, (\mathfrak{L}_{\Box} \setminus b) \cup (\mathfrak{L}_{\Box} \setminus a^{-\Diamond})\rangle$ is a disjoint pair.

To see that $c \subseteq a^{-\Diamond}$, suppose not. Then there is $A \in c$ such that $A \notin a^{-\Diamond}$, which implies $A \in (\mathfrak{L}_{\Box} \setminus a^{-\Diamond})$. As before, this contradicts the fact that $\langle c, (\mathfrak{L}_{\Box} \setminus b) \cup (\mathfrak{L}_{\Box} \setminus a^{-\Diamond})\rangle$ is a disjoint pair.

Since $a^{-\Box} \subseteq c \subseteq a^{-\Diamond}$ is equivalent to Tac, c is a prime theory such that $c \leq b$ and Sac and Sa^*c^*, as desired. □

The foregoing proof works when (D) is a theorem of the logic. It is needed to ensure that $a^{-\Box} \subseteq a^{-\Diamond}$ so that we can find an intervening prime theory c. Completeness for logics with (MC) can be had without (D) around, though. As Mares (1992b) showed, what is needed is a redefinition of the canonical accessibility relation. Rather than Sab defined as $a^{-\Box} \subseteq b$, we instead use

- Sab iff both $a^{-\Box} \subseteq a^{-\Diamond}$ and $a^{-\Box} \subseteq b$.

Clearly, when (D) is available, the two definitions coincide. In weaker logics that lack (D), however, differences emerge. The extra conjunction is needed to ensure the containment relation obtained from (D) works in the canonical frame.

Restall (2000, 265ff.) observes some additional difficulties that arise with (MC) if the possibility operator involved has an accessibility relation that is the converse of the necessity operator's relation. In frames where Sab iff $S_{\blacklozenge}ba$, both (MC) and $(\Box A \wedge \blacklozenge B) \to \blacklozenge(A \wedge B)$ are valid.[13] This means that, in general, the straightforward definition of the canonical accessibility relation as Sab iff $a^{-\Box} \subseteq b$ will not work. This point could lead to some difficulties or limitations in future developments of certain relevant temporal logics. As we will see, (MC) leads to further surprising problems.

7.10 Incompleteness

So much for Completeness. In this section we will turn to *failures* of Completeness. In TV-based modal logics, it is well known that there are many logics that are not complete with respect to any class of Kripke frames.[14] One example of this incompleteness is the extension of the TV-based modal logic K with the axiom (H), $\Box(\Box A \leftrightarrow A) \to \Box A$, a relative of the axiom (L), $\Box(\Box A \to A) \to \Box A$, discussed below.[15] In the present setting of modal ternary relational frames, there are some additional potential sources of incompleteness. Indeed, there have been some striking results obtained on incompleteness in relevant modal logics, and we will cover three of the important incompleteness results in this section.

The first result due to Mares (2000b). The logic RGL, adds to R.MK4Nec the axiom

(L) $\Box(\Box A \to A) \to \Box A$

to obtain a relevant analog of provability logic.[16] The main result is that all modal frames for R.MK4Nec that validate (L) validate the formula $\Box\neg\Box\neg A \to \Box B$. Thus, the logic defined model-theoretically is not a relevant logic, as it violates the variable-sharing property. Further, that formula is not a theorem of RGL, as shown via a matrix argument. So, RGL is not complete with respect to the natural class of modal frames.

The second result is due to Goble (2000). Goble extends the logic R.MKTNec with the axiom

(G) $\Box A \to (B \to B)$,

which is a surprising addition to a relevant modal logic. Since (G) violates variable sharing, the resulting logic will as well. If one defines $\Box A$ as $\mathsf{t} \wedge A$, then (G) falls out

[13] These two axioms are called the Dunn conditions following on the work of Dunn (1995). Dunn showed that versions of these axioms are important for axiomatizing positive modal logics, modal logics without negation. Humberstone (1990) explored positive modal logics, as well as more generally expressively impoverished modal logics.

[14] See Blackburn et al. (2002, ch. 4) for discussion and more examples.

[15] For the incompleteness of the logic KH, see Boolos (1993, 149ff.).

[16] See Boolos (1993) for more on provability logic.

as a theorem and the violation of variable sharing is merely apparent. Goble considers what happens if one considers $\Box$ as a primitive operator with the usual verification condition. Goble provides a class of modal frames for which the logic is sound and complete, crucially using a non-standard verification condition,

- $a \Vdash \Box A$ iff $a \in N$ and $\forall b \in K$(if Sab, then $b \Vdash A$).

He goes on to show that using the standard verification condition, namely the condition presented at the start of this chapter, there is no class of modal frames for which the axiomatic logic is sound and complete. The reason is that every class of modal R.MKTNec-frames on which (G) is valid will validate $A \to (B \to B)$. The latter formula, however, is not a theorem of the axiomatic system. In fact, the logic valid on the class of all modal R.MKTNecG-frames is the TV-based modal logic KT. An analogous collapse result is obtained for S4.

The third result is a result of Furhmann (1990). Fuhrmann shows that the logic R.MKT4Nec is not complete with respect to reduced modal R.MKT4Nec-frames, that is, those modal frames that obey the reduction condition.[17] He shows that the reduced frames interact in a surprising way with the rule (RM). The result is a formula, $(\Box A \to \Box(A \to A)) \vee (\Box A \to \Box\neg(A \to A))$, valid over all the reduced frames for the logic that is not a theorem of the logic.

Furhmann's result was extended and strengthened by Standefer (2021b). Standefer showed that for a wide range of base logics, namely any logic between B and TV that has the scheme $(A \to B) \vee \neg(A \to B)$ among its theorems, one will have incompleteness with respect to reduced frames for any modal extension that uses axioms and rules that are contained in S5 upon a straightforward, homophonic translation. This, in fact, includes reduced models for S5.

The main idea behind the proof is that, by theorem 3.21, if the logic of a class of reduced frames is closed under a rule $A \Rightarrow B$, then the logic is also closed under its disjunctive form, $A \vee C \Rightarrow B \vee C$. This means that the rule (RM) gives us the disjunctive rule, $(A \to B) \vee C \Rightarrow (\Box A \to \Box B) \vee C$. If $(p \to q) \vee \neg(p \to q)$ is a theorem of the logic, then $(\Box p \to \Box q) \vee \neg(p \to q)$ will be as well. The latter formula, however, is not, upon translation, a theorem of S5, as a simple Kripke model with two worlds can demonstrate. Let $W = \{w, x\}$, $|p| = W$, and $|q| = \{w\}$, using the universal accessibility relation. In this model, $(\Box p \to \Box q) \vee \neg(p \to q)$ will fail at w. This incompleteness result holds even for modal logics whose base logics are complete with respect to their reduced frames, such as C, R, and TV.

7.11 Halldén-Completeness

Next, we will briefly look at the work that has been done on Halldén-completeness for relevant modal logics. For this section, we will consider the language $\mathfrak{L}_\Box$ without t.

[17] Fuhrmann's reduction condition is slightly different than the one used in this book, which was adopted for uniformity. The differences are not relevant for the results.

DEFINITION 7.19 (Halldén-completeness) *A logic* L *is* HALLDÉN-COMPLETE *iff whenever* $\vdash_L A \vee B$ *and* A *and* B *do not share any atoms, then either* $\vdash_L A$ *or* $\vdash_L B$.

As Mares (2003) remarks, Halldén-completeness is similar in spirit to the variable-sharing property. There is nothing about Halldén-completeness that restricts us to considering only modal logics. Indeed, Routley and Meyer (1973, 241) establish that R is Halldén-complete.

The topic of Halldén-completeness has been much studied for TV-based modal logics, but less so for relevant modal logics. In TV-based modal logics, the minimal normal modal logic K turns out not to be Halldén-complete, as $\Diamond(p \vee \neg p) \vee \Box q$ is a theorem, while neither disjunct is. Van Benthem and Humberstone (1983) provide two sufficient conditions on classes of Kripke frames determining Halldén-complete modal logics. One condition is closure of the class under an operation of a certain kind of "fusion" operation on frames, p-morphic fusion. The other condition applies to classes of serial frames, where a frame is *serial* iff for all w, there is a u such that Swu. In Kripke frames, seriality is the frame condition corresponding to (D). The sufficient condition is that if a class of serial frames is closed under direct products, then the logic of the class is Halldén-complete. The upshot is that many extensions of TV-based KD are Halldén-complete.

Mares (2003) extends the results of van Benthem and Humberstone to relevant modal logics extending R. Mares shows that R.M with the addition of any of (T), (D), (4), (5), (B), (K), or (Nec) is Halldén-complete. Seki (2015) extends these results to a wide range of logics, including those with weaker base logics and extensions with disjunctive rules, as well as logics formulated in the language $\mathfrak{L}_{\Box\blacklozenge}$ with t.

To present things in a little more detail, let us define an rp-morphism.[18]

DEFINITION 7.20 (Rp-morphism) *Given two modal ternary relational frames* $F = \langle K, N, R, ^*, S\rangle$ *and* $F' = \langle K', N', R', ^{*\prime}, S'\rangle$, *an* RP-MORPHISM f *is a function* $K \mapsto K'$ *such that*

- $f^{-1}(N') = N$,
- *if* $Rabc$, *then* $R'f(a)f(b)f(c)$,
- *if* $R'f(a)bc$, *then there are* $d, e \in K$ *such that* $Rade$, $b \leq' f(d)$, *and* $c \leq' f(e)$,
- *if* Sab, *then* $S'f(a)f(b)$, *and*
- *if* $S'f(a)b$, *then there is* $c \in K$ *such that* Sac *and* $b \leq' f(c)$.
- $f(a^*) = a^{*\prime}$.

Using this, we can define rp-morphic fusions of frames.

DEFINITION 7.21 (Rp-morphic fusions) *Let* $F = \langle K, N, R, ^*, S\rangle$, $F' = \langle K', N', R', ^{*\prime}, S'\rangle$, *and* $F'' = \langle K', N', R'', ^{*\prime\prime}, S''\rangle$ *be modal frames. Suppose* $a \in K$, $b \in K'$ *and* $c \in K''$. *Then* (F, a) *is an* RP-MORPHIC FUSION *of* (F', b) *and* (F'', c) *iff there are rp-morphisms* $f_1\colon K \mapsto K'$ *and* $f_2\colon K \mapsto K''$ *such that* $f_1(a) = b$ *and* $f_2(a) = c$.

A class of frames is closed under rp-morphic fusions iff for any two frames F' *and* F'' *in the class, there is a frame* F *that is an rp-morphic fusion of them.*

[18] The definition used is that of Seki (2015), which is slightly different from that of Mares.

If a class of frames is closed under rp-morphic fusions, then an analog of the van Benthem–Humberstone argument for Halldén-completeness can be given. The key idea of this argument is the following. Suppose that there are two invalid formulas A and B that do not share any propositional atoms. Then there are two frames F' and F'' that have models on them that serve as counterexamples to A and B, respectively. Taking an rp-morphic fusion of these two frames yields a new frame that can then be used to obtain a counterexample to $A \vee B$.

More information is available about classes of frames that are closed under rp-morphic fusions. To state those, we need some definitions.

DEFINITION 7.22 (Completely serial) *A modal frame is* COMPLETELY SERIAL *if both*

- *for all* $a \in K$, *there are* $b, c \in K$ *such that* $Rabc$ *and*
- *for all* $a \in K$, *there is* $b \in K$ *such that* Sab.

As Mares notes, the seriality condition on S falls short of the condition (FD), and, perhaps surprisingly, it does not validate any additional theorems over R.M. The reason for this is that the canonical frame for R.M is completely serial. Indeed, the canonical frames for most of the relevant logics discussed in this book are completely serial. The reason is that there is a trivial theory, ω, in each frame such that ω contains every formula.

DEFINITION 7.23 (Frame products) *Let* $F_1 = \langle K_1, N_1, R_1, {}^{*1}, S_1\rangle$ *and* $F_2 = \langle K_2, N_2, R_2, {}^{*2}, S_2\rangle$ *be modal frames.* $F = \langle K, N, R, {}^*, S\rangle$ *is the product of* F_1 *and* F_2 *iff*

- $K = K_1 \times K_2$,
- $\langle a_1, a_2\rangle \in N$ *iff* $a_1 \in N_1$ *and* $a_2 \in N_2$,
- $R\langle a_1, a_2\rangle\langle b_1, b_2\rangle\langle c_1, c_2\rangle$ *iff* $R_1a_1b_1c_1$ *and* $R_2a_2b_2c_2$,
- $S\langle a_1, a_2\rangle\langle b_1, b_2\rangle$ *iff* $S_1a_1b_1$ *and* $S_2a_2b_2$, *and*
- $\langle a_1, a_2\rangle^* = \langle a_1^{*1}, a_2^{*2}\rangle$.

Mares and Seki both show the following.

THEOREM 7.24 *If a class of completely serial modal frames is closed under products, then it is closed under rp-morphic fusions.*

Proof Sketch: Let $F_1 = \langle K_1, N_1, R_1, {}^{*1}, S_1\rangle$ and $F_2 = \langle K_2, N_2, R_2, {}^{*2}, S_2\rangle$ be modal frames, and let $F = \langle K, N, R, {}^*, S\rangle$ be their product. Define the functions f_1 and f_2 as projections, as follows.

- $f_1(\langle a_1, a_2\rangle) = a_1$, for all $\langle a_1, a_2\rangle \in K$, and
- $f_2(\langle a_1, a_2\rangle) = a_2$, for all $\langle a_1, a_2\rangle \in K$.

By stepping through the definitions, one can show that F is an rp-morphic fusion of F_1 and F_2. □

Since the condition of complete seriality can be imposed on classes of frames for many logics without altering the logic obtained from the class, one can use the earlier proof methods to demonstrate Halldén-completeness for many relevant modal logics that do not contain the (D) axiom.

Mares (2003) shows that, notably, the Halldén-completeness results do not extend to all logics with (MC). In fact, R.MKNec with the addition of (MC) is not Halldén-complete, as it has as a theorem the formula $(\neg\Box p \vee \Diamond p) \vee \Box q$, and neither disjunct is a theorem. On the frame side, the canonical frame for this logic is not completely serial. Seriality, and Halldén-completeness, can be regained with the addition of (D).

It is worth noting that, much like weak variable sharing, Halldén-completeness is not automatically preserved to sublogics. The reason for this is that Halldén-completeness involves a disjunction of provability claims, which need not be satisfied as one moves to a sublogic.

Additionally, it is worth noting how much of the argumentation for Halldén-completeness for TV-based modal logics carries over to the present setting: It is a significant amount. This might prompt the reader to wonder about some related concepts, such as bisimulations, which are generalizations of p-morphisms.[19] Relevant bisimulations have been studied by Restall (2000, ch. 13.3) and by Badia (2016a; 2016b; 2017a; 2017b). Bisimulations are relations between models that capture a sense of similarity that is weaker than isomorphism. As in the case of TV-based modal logic, one can obtain preservation results between bisimilar models showing that bisimilar points satisfy the same formulas. Compared to TV-based modal logics, relevant bisimulations have been relatively underexplored.

7.12 Neighborhoods

We will close this chapter by considering an alternative type of frame for modal logics, neighborhood frames.[20] Neighborhood frames provide ways to model modal logics that lack one or both of (RM) and ($\wedge\Box$). In TV-based modal logics, neighborhoods are sets of propositions, that is, subsets of $\wp(W)$. In the setting of ternary relational frames, propositions are hereditarily closed subsets of K, rather than arbitrary subsets, but the basic idea still works.

DEFINITION 7.25 (Neighborhood frames) *A neighborhood* L*-frame* F *is a sextuple* $\langle K, N, R, ^*, \mathcal{N}, \mathsf{Prop}\rangle$ *where*

- *the first four components make up an* L*-frame,*
- Prop *is a set of hereditarily closed subsets of* K*,*
- $\mathcal{N}\colon K \mapsto \wp(\mathsf{Prop})$*, and*
- *if* $a \leq b$*, then* $\mathcal{N}(a) \subseteq \mathcal{N}(b)$*.*

19 See Blackburn et al. (2002, ch. 2.2) or van Benthem (2010, ch. 3) for more on bisimulations.
20 See Pacuit (2017) for more on neighborhood frames.

A model M on a neighborhood L*-frame* $\langle F, \mathsf{Prop}\rangle$ *is a pair of* $\langle\langle F, \mathsf{Prop}\rangle, V\rangle$ *where* V *is a valuation admissible with respect to* Prop.

The verification condition for $\Box$ in neighborhood models is

- $a \Vdash \Box A$ iff $|A| \in \mathcal{N}(a)$.

The neighborhood frames validate the modal rule

- (RE) $A \leftrightarrow B \Rightarrow \Box A \leftrightarrow \Box B$.

Let us call the logic obtained by adding the rule (RE) to L the logic L.E.

Further modal principles can be validated with the adoption of frame conditions. Some of these further modal principles require the presence of Prop for Completeness, using the method below. The basic logic L.E can be shown complete without appeal to Prop, as none of the conditions in the basic neighborhood frames quantify over propositions.

Many of the optional conditions are shared by classical neighborhood models. To simplify the presentation, we will adopt the definition

- $\Box X = \{a \in K : X \in \mathcal{N}(a)\}$.[21]

for use with notation from Chapter 3. We will present some of the conditions here.

- (FRM): for $X, Y \in \mathsf{Prop}$, if $X \subseteq Y$, then $\Box X \subseteq \Box Y$.
- (F$\Box\bigwedge$): for $X, Y \in \mathsf{Prop}$, $\Box X \cap \Box Y \subseteq \Box(X \cap Y)$.
- (F4): for $X \in \mathsf{Prop}$, $\Box X \subseteq \Box\Box X$.
- (FT) for $X \in \mathsf{Prop}$, $\Box X \subseteq X$.
- (FB) for $X \in \mathsf{Prop}$, $X \subseteq \Box - \Box - X$.
- (F5) for $X \in \mathsf{Prop}$, $-\Box X \subseteq \Box - \Box X$.
- (FK): for $X, Y \in \mathsf{Prop}$, $\Box(X \Rightarrow Y) \subseteq (\Box X \Rightarrow \Box Y)$.[22]
- $\Box$t: $N \subseteq \Box N$.
- (FNec): for $X \in \mathsf{Prop}$, if $N \subseteq X$, $N \subseteq \Box X$.

Soundness for L.E, and extensions, with respect to the appropriate class of frames, is proved by a straightforward induction. The canonical model Completeness proof likewise carries over without too many surprises. For the construction, we have to define the canonical neighborhood function and the canonical set Prop, which we do as follows.

- $\mathsf{Prop} = \{\|A\| : A \in \mathfrak{L}_\Box\}$.
- $N(a) = \{X \in \mathsf{Prop} : \exists A \in \mathfrak{L}_\Box, X = \|A\| \text{ and } \Box A \in a\}$.

[21] We take this from Ferenz and Tedder (2022).

[22] We can write this out in an alternative notation, due to Fuhrmann. In that notation, the condition is: If $Rabc$, then $\hat{R}\mathcal{N}(a)\mathcal{N}(b)\mathcal{N}(c)$, where for $X, Y, Z \in \mathsf{Prop}$, $\hat{R}XYZ$ iff for all $U, V \in \mathsf{Prop}$, if both
(i) $\{x \in K : \forall u \forall v(\text{if } Rxuv \text{ and } u \in U, \text{ then } v \in V)\} \in X$, and
(ii) $U \in Y$,
then $V \in Z$.

One can then verify that these definitions work. The resulting canonical frame is a neighborhood frame, and the Truth lemma extends to the canonical model, which means it is a model.

LEMMA 7.26 (Truth lemma) *Let M be the canonical model for* L.E. *Then for all $A \in \mathfrak{L}_{\Box}$ and for all $a \in K$, $A \in a$ iff $a \Vdash A$.*

This suffices for Completeness.

THEOREM 7.27 *Let C be the class of* L.E *models. If $\models_C A$, then $\vdash_{\mathsf{L.E}} A$.*

We will note that neighborhood frames for relevant logics have been used by many logicians. Not all of them have been used in the context of modal logics. Routley and Meyer (1975; 1976), Lavers (1985), Goble (2003), and Tedder (2021b) all use neighborhoods in a modified form of the the ternary relational frames to give models for the basic relevant logical language. This provides a way to model the weak logic BB, from the appendix of Chapter 2. Furhmann (1988), Standefer (2019), Ferenz (2021), Tedder and Ferenz (2022), and Ferenz and Tedder (2022) have used neighborhoods to model relevant modal logics.

7.13 Exercises

(1) Show that in modal frames obeying (FMC), the generic verification condition for $\Box$, given in Section 7.4, is equivalent to the standard verification condition.
(2) Complete the proof of lemma 7.10 by proving Soundness for (B), (K), (Nec), and (MC) with respect to classes of frames obeying the appropriate conditions.
(3) Does symmetry for S, if Sab, then Sba, imply (FB)? Does imposing symmetry as a frame condition validate any new formulas or does it not affect validities, like making S serial?
(4) Using a frame from Section 7.6, come up with a model that shows that $\blacklozenge A \leftrightarrow \Diamond A$ is invalid.
(5) Using some of the frames in Section 7.6, come up with counterexamples to the principles (K), (Nec), and (MC).
(6) Prove the Soundness of $\mathsf{B.EC}_1,\ldots,\mathsf{C}_n$ with respect to the class of neighborhood frames obeying the frame conditions $(\mathsf{FC}_1),\ldots,(\mathsf{FC}_n)$.
(7) The next three problems explore a special case of the construction of Seki (2011b). Given a modal frame $F = \langle K, N, R, {}^*, S\rangle$ for the logic T.MT, define the NORMALIZATION $\langle K', N', R', {}^{*'}, S'\rangle$ of F for a point $o \in N$ as follows, where $0 \notin K$.
   - $K' = K \cup \{0\}$.
   - For $a, b \in K$: $R'000$;
     - $R'00a$ iff $R'ooa$;
     - $R'0a0$ iff $R'oao^*$;
     - $R0'ab$ iff $R'oab$;
     - $R'a0b$ iff $R'aob$;
     - $R'ab0$ iff $R'abo^*$.

- For $a, b, c \in K$, $R'abc$ iff $Rabc$.
- For $a \in K$: $S'00$; $S'0a$ iff Soa; $S'a0$ iff $S'ao^*$.
- For $a, b \in K$, $S'ab$ iff Sab.
- $0^{*'} = 0$ and for $a \in K$ $a^{*'} = a^*$.

Show that $\langle K', N', R', ^{*'}, S' \rangle$ is a modal frame for T.MT.

(8) Show that for the frame $\langle K', N', R', ^{*'}, S' \rangle$, defined in the previous problem, the following are true, where $x \leq' y$ is defined as $\exists z \in N' R'zxy$.

- For all $a, b \in K$, $a \leq' b$ iff $a \leq b$.
- For $a \in K$, $0 \leq' a$ iff $o \leq a$.
- For all $a \in K$, $a \leq' 0$ iff $a \leq o^*$.

Conclude that $o^* \leq' 0 \leq' o$.

(9) Say that a class of frames ADMITS NORMALIZATION iff whenever a frame is in the class, so is its normalization with respect to any element in its set N. Show that if a class of frames admits normalization, then it is closed under (γ). Conclude that T.MT is closed under (γ).

8 Applications of Modal Logics

8.1 Introduction

TV-based modal logics have been used to investigate issues in epistemology, metaphysics, and normative systems for some time, but there has been comparatively little development of the relevant modal logics in these areas. This is a pity, since part of the interest in the relevant logic project is to develop philosophical and mathematical theories whose underlying logics are relevant logics. In order for the relevant-logic-inclined philosopher, or the philosophically inclined relevant logician, to engage with debates in different areas of philosophy, they need some formal development of their desired relevant logical tools.

In the previous two chapters, we looked at the basic axiomatics and frame theory for relevant modal logics. In this chapter, we will consider some of the applications that have been developed. There are three topics that we will discuss: alethic modal logics, epistemic logics, and deontic logics.[1]

We will begin by discussing alethic modal logics in Section 8.2. We will present work, primarily by Mares and Meyer, on S4-ish logics (Section 8.2.1). Then we will present some of the varieties of S5-ish relevant logics (Section 8.2.2), and this section will close by discussing the addition of an actuality operator to the language of relevant logics (Section 8.2.3). The next application we look at is epistemic relevant logics (Section 8.3). We begin by discussing Wansing's response to the Fitch–Church knowability paradox (Section 8.3.1). Then we will look at a distinctive account of knowledge available to the relevant logician (Section 8.3.2). In the context of relevant logics, there are different ways of developing the formal framework for justification logics, and we will present the extant work in that area (Section 8.3.3). We will close the epistemic logic section by briefly looking at some of the other epistemic relevant logics that have been developed (Section 8.3.4). Following epistemic logics, we will look at some applications in deontic relevant logics (Section 8.4). There are two main divisions in approaches to logics of obligation – those that take obligation as a primitive operator (Section 8.4.1) and those that opt for the so-called Andersonian reduction that defines

[1] The reader may be surprised that we do not discuss temporal logics here. The reason is that there has, at the time of writing, been little sustained work on relevant temporal logics. Cheng (1998; 2004; 2009) considers some axiom systems for relevant temporal logics. Sylvan (1989; 1992) briefly discusses frames for relevant temporal logics, and some of this material is reproduced by Brady (2003, 23–24). Kamide (2018) investigates temporal logics over RW.

obligation in terms of implication and a special propositional constant (Section 8.4.2). Finally, we close the section and the chapter by looking at Goble's work on conditional obligation in the context of relevant logics (Section 8.4.3).

8.2 Alethic Modal Logics

Alethic modal logics are modal logics where the necessity concerns truth. In this section, we will look at three different developments in the area of alethic modal logic. The first concerns S4, which is perhaps the most studied relevant modal logic. Then we will turn to look at different conceptions of necessity and their connections to S5. Finally, we will explore the addition of actuality operators to the language, which is an important step toward engagement with two-dimensional semantics.

8.2.1 R and S4

In many ways, S4-ish logics are surprisingly well behaved. As noted in Section 6.5, the logical necessity definable in E is S4-ish. The modal logic of NR is also S4-ish. After a period of relatively little investigation concerning NR, there was a flurry of activity. In a series of papers, Meyer and Mares investigated many feature of S4-ish modal logics over R.[2]

Meyer's logic NR is, in our nomenclature, R.MCKT4Nec. The logic R4 extends this with the addition of (MC), $\Box(A \vee B) \to (\Diamond A \vee \Box B)$. The reason for the interest in R4 is that NR does not contain a direct translation of TV-based S4. The translation τ commutes with the connectives in the set $\{\neg, \wedge, \vee, \Box\}$, but the translation of a classical conditional is a disjunction in the relevant language, that is, $\tau(A \to B) = \neg\tau(A) \vee \tau(B)$. Among the theorems of S4 that fail, upon translation, is $\neg\Box(\neg A \vee B) \vee (\neg\Diamond A \vee \Diamond B)$. The addition of (MC) remedies this problem.

Mares and Meyer (1993) examine an extension of R4 with Boolean negation, which they call CR4. While the addition of Boolean negation to NR is not a conservative extension, as $\neg\Box(\neg A \vee B) \vee (\neg\Diamond A \vee \Diamond B)$ becomes valid, the addition of Boolean negation to R4 results in a conservative extension of R4. They provide models for CR4 and show that they can be analyzed into combinations of models for R with Boolean negation and Kripke models for TV-based S4.

Mares and Meyer (1992) use metavaluational techniques to show that (γ) is admissible in R4. They use this to show that R4 is a conservative extension of S4. Mares and Meyer (1993) provide models for which R4 is sound and complete. They strengthen their earlier results relating R4 and S4. Using the metavaluational techniques, one can show that every prime regular R4-theory contains a prime regular R4-theory that is also negation-consistent, in the sense that for no formula A are both A and $\neg A$ members of the theory. These theories are called *normal* theories. In terms of models, normal theories can be used as members of N obeying the condition $x = x^*$. Models of R4

[2] Mares and Meyer (1992; 1993) and Meyer and Mares (1993).

where all members of N obey the condition $x = x^*$ are called *normal models* of R4. In normal models of R4, the members of N act like possible worlds in the sense of Kripke models for TV-based modal logics, in the sense that exactly one of A and $\neg A$ holds. They show that every Kripke model for S4 can be embedded into a normal model of R4, where the accessibility relation of the S4 model is reproduced over the members of N in the R4-model. Further, every normal model of R4 determines a Kripke model of S4 by considering a reflexive, binary relation on N. The logic R4 then conservatively extends both R and S4, combining relevant implication with an S4-ish necessity in a very natural way.

8.2.2 Conceptions of Necessity and S5

It is well known that there are three different, but equivalent, presentations of TV-based S5. The first, building on David Lewis's idea of necessity being truth at every possible world, uses Kripke frames $\langle W, S \rangle$ in which the binary relation S is a universal relation, relating every world to every world. The second generalizes the first, permitting S to be an equivalence relation.[3] A formula $\Box A$ is true at a world, on that conception, when it is true at all accessible worlds, which may not exhaust W. The third presents the logic axiomatically, adding the axioms (T), (4), (5), and (B) to the basic (TV-based) modal logic K. These three presentations have different conceptions of necessity: UNIVERSAL, EQUIVALENCE RELATION, and AXIOMATIC. Over the basic modal language, $\mathfrak{L}_{\Box}$, they coincide, so the differing conceptions do not affect the formal details.[4]

Standefer (2023b) investigates these three conceptions of necessity over the base logic R and shows that they come apart. Following that work, let us use the symbols

- $\mathbb{U}$ for the necessity of the universal conception,
- $\mathbb{E}$ for the necessity of the equivalence relation conception, and
- $\Box$ for the necessity of the axiomatic conception.

We will consider different languages, $\mathfrak{L}_{\mathbb{U}}$, $\mathfrak{L}_{\mathbb{E}}$, and $\mathfrak{L}_{\Box}$, but we can talk about the relation between the logics by using translations between the languages that only change the necessity operator. For example, the translation from $\mathfrak{L}_{\mathbb{U}}$ to $\mathfrak{L}_{\mathbb{E}}$ would change all occurrences of $\mathbb{U}$ to $\mathbb{E}$ but not otherwise alter the connectives or atoms.

To model $\mathfrak{L}_{\mathbb{U}}$ and $\mathfrak{L}_{\mathbb{E}}$, we use modal frames where S is, respectively, a universal relation or an equivalence relation. The respective verification conditions are

- $a \Vdash \mathbb{U}A$ iff for all b, $b \Vdash A$, and
- $a \Vdash \mathbb{E}A$ iff for all b, if Sab, then $b \Vdash A$.

In the first condition, we are suppressing Sab, since S is, in that case, the universal relation.

The axiomatic logic will add to R.M the following: (T), (4), (5), (B), (K), and (Nec). Let us say call this logic RS5 for this section. The logic of universal conception will

[3] S is an equivalence relation iff S is reflexive, symmetric, and transtiive.

[4] If one extends the language with other operators, such as adding an actuality operator, then they can come apart. See Crossley and Humberstone (1977) for more.

Table 8.1. S5 principles in the logics U, Eq, and RS5

Name	Modal principle	U	Eq	RS5
$(\Box\wedge)$	$(\Box A \wedge \Box B) \rightarrow \Box(A \wedge B)$	✓	✓	✓
(K)	$\Box(A \rightarrow B) \rightarrow (\Box A \rightarrow \Box B)$	✓		✓
(T)	$\Box A \rightarrow A$	✓	✓	✓
(4)	$\Box A \rightarrow \Box\Box A$	✓	✓	✓
(B)	$A \rightarrow \Box\neg\Box\neg A$	✓		✓
(5)	$\neg\Box A \rightarrow \Box\neg\Box A$	✓		✓
(MC)	$\Box(A \vee B) \rightarrow (\Box A \vee \neg\Box\neg B)$	✓		
(OM)	$\Box(\Box A \vee B) \rightarrow (\Box A \vee \Box B)$	✓	✓	
(Nec)	$A \Rightarrow \Box A$			✓
(RM)	$A \rightarrow B \Rightarrow \Box A \rightarrow \Box B$	✓	✓	✓

be U, the set of formulas valid in the class of universal modal R-frames. The logic of equivalence relation conception will be Eq, the set of formulas valid in the class of modal R-frames where S is an equivalence relation. Let us call these two classes of frames universal and equivalence frames, respectively.

How do these logics relate? Standefer (2023b) shows the following, with Table 8.1 recording results concerning some of the important modal principles.

- Eq $\subsetneq$ U.
- RS5 and Eq are incomparable.
- RS5 and U are incomparable.

In fact, U does not have the variable-sharing property, as $(\mathbb{U}A \wedge \neg\mathbb{U}A) \rightarrow B$ is valid in universal frames. Both Eq and RS5 enjoy the variable-sharing property.[5]

The failure of the variable-sharing property for U illustrates a feature of interest to the relevant logician.

DEFINITION 8.1 (Ubiquitous truth, ubiquitous falsity) *A formula A is* UBIQUITOUSLY TRUE *in a model iff for all* $a \in K$, $a \Vdash A$.

A formula A is UBIQUITOUSLY FALSE *in a model iff for all* $a \in K$, $a \nVdash A$.

A formula is UBIQUITOUS *in a model iff it is either ubiquitously false or ubiquitously true.*

The phenomenon of ubiquitous formulas was discussed by Standefer (2022c), building on an idea of Fuhrmann (1990).[6] If a formula B is ubiquitous in all models, then both $B \rightarrow B$ and $B \wedge \neg B$ will be ubiquitous in all models. These lead straightaway to violations of variable sharing, as there will be no point in any models that falsify $C \rightarrow (B \rightarrow B)$ or $(B \wedge \neg B) \rightarrow C$. No connective that leads to ubiquitous truths can be accepted by the relevant logician, a point to which we will return in chapter 10. By design, $\mathbb{U}A$ is ubiquitous in all models, and it leads to violations of variable sharing.

[5] See Standefer (2023b; 2025a) for details.
[6] Fuhrmann attributes the term "ubiquitous" to Lloyd Humberstone.

What are the logics U and Eq like? In both, (Nec) fails. In Eq, (B), (5), (K), and (MC) are invalid, while (T), (4), (RM), and ($\mathbb{E}\wedge$) hold. Standefer and Mares (2025) show that adding the axiom (OM) $\Box(\Box A \vee B) \to (\Box A \vee \Box B)$ to the preceding results in a complete axiomatization of Eq.[7] In U, (B), (5), (K), (MC), (T), (4), (RM), ($\mathbb{U}\wedge$), and $\mathbb{U}(\mathbb{U} \to A)$ hold. A complete axiomatization for U is, as yet, unknown. A summary of which logics satisfy which principles can be found in Table 8.1.

One interesting feature of TV-based S5 is its connection to classical monadic first-order logic. One can define a translation τ from the modal language to a first-order language as follows, where different atoms are translated to different unary predicates.

- $\tau(p) = Px$, for atoms p.
- $\tau(B\#C) = \tau(B)\#\tau(C)$, for $\# \in \{\wedge, \vee, \to\}$.
- $\tau(\neg B) = \neg\tau(B)$.
- $\tau(\Box B) = \forall x \tau(B)$.

Intuitively, atoms are translated to predicates on worlds, and the modal operators are translated to quantifiers over worlds. As shown by Mints (1992), a formula A is valid in S5 iff $\tau(A)$ is valid in classical FOL. This sort of connection can be had in non-classical logics as well, as shown by Caicedo et al. (2019; 2022). Ferenz (2024) establishes a similar result in the context of relevant logics. The modal logic he works with adds the modal axioms and rules stated earlier for RS5, that is, (K), ($\Box\wedge$), (T), (4), (B), (5), (Nec), (RM), and (MC), to any relevant logic L containing permutation, which is called L5. Ferenz shows that L5 has the desired connection with the monadic relevant first-order logic.

There is more to say about these logics, but we will end this section with two issues raised by Standefer. The first concerns the logic Eq. Equivalence relation frames, or rather their Kripke model counterparts, are used extensively in TV-based epistemic logic. Since we have a complete axiomatization of the logic, a natural next step is to compare how Eq handles different epistemic logic applications and to enrich the logic with the addition of more epistemic operators, such as common knowledge or distributed knowledge.

The second issue concerns the universal conception of necessity. As Parks and Byrd (1989) noted, there was a suspicion among relevant logicians that there was something amiss with that conception, although nothing specific was provided. As demonstrated, that conception of necessity leads to violations of the variable-sharing property, and, in fact, those violations happen regardless of the base logic. The relevant logician has good reason to reject the universal conception of necessity. Relevant logicians cannot accept the Lewisian view of necessary truth as truth in all possible worlds.

[7] The axiom (OM) was investigated by Ono (1977) in the context of intuitionistic modal logics, in particular it is axiom A_3. The name (OM) is a homage to Ohnishi and Matsumoto (1957), and it is connected to their sequent system for S5.

We should also add that Standefer and Mares demonstrate completeness for the analog of Eq over B-frames, rather than R-frames, but completeness carries over with the additional frame conditions.

Where does this leave the relevant logician? While the simple, Lewisian view of necessity is off the table, a logic that offers a lot of the attractive features of TV-based S5 is still available. The axiomatic conception, whether in the form of RS5 or one of its close relatives, presents a modal logic that offers many of the same features as the logic of TV-based S5. It has modal reduction principles that offer a simple way of reasoning with necessity and possibility. It enjoys Belnap's variable-sharing property, and there is a class of frames for which it is sound and complete. While the frames for RS5 are not all universal frames, that need not be a problem if the primary interest is in the logic, rather than the frames themselves. The axiomatic conception captures most of the good features of TV-based S5 that the relevant logician needs.

8.2.3 Actuality

In some presentations of the ternary relational models, there is a distinguished point, which is supposed to represent the real world. In reduced models, this is often how the point 0 is described. For example, in an early paper, Routley and Meyer describe it in this way.

> The *real world [0]* plays a distinguished role in our semantical postulates. (Routley and Meyer, 1973, 201)

Routley and Meyer use similar descriptions in other places, as well.

> A final word concerns the role of [g] . . . [g] could also be thought of as a normal set-up – *a candidate, so to speak, to be the 'real' world,* in that each sentence or its negation could be made to turn out true in [g] without ever making both true; logical truth then turns out to be truth in all 'real' worlds. (Routley and Meyer, 1972b, 197)

Routley et al. (1982, 294) uses the term "factual world," but that seems close enough to "real world" or "actual world" to include here. Slaney and Giambrone use similarly suggestive language in talking about reduced models for contraction-free relevant logics.

> It is splendid that the main contractionless relevant systems allow the possibility of models in which there are no contingent truths, for logic is in a way impure when it claims there are truths about matters which are not its business; but it is also desirable that there be models looking like reality, in which there are many truths beyond what is given by Pure Reason. (Slaney, 1987a, 406)
>
> Slaney has provided reduced models (ones in which there is but *one "real" world*). (Giambrone, 1992, 442)
>
> In particular, *the real worlds* of our "real" models are large in that appropriate if somewhat obscure sense of [Slaney (1987a)]: there really are "models looking like reality, in which there are many truths beyond what is given by Pure reason." (Giambrone, 1992, 449)

Finally, Fuhrmann echoes the earlier papers by talking about the distinguished world of reduced models as being a representative for the real world.

> It is sometimes thought to be intuitively more satisfactory, if the set N of "distinguished worlds" could be reduced to a singleton set containing just *the "real world", [g].* In such a

reduced model *a sentence is true (in a model) just in case it is true in the real world (of that model)*. (Fuhrmann, 1990, 501)

It is a natural thought to add to the language an operator that is supposed to represent truth at the actual, or real, world. Indeed, this thought has been explored in two-dimensional logics, following on the work of Crossley and Humberstone (1977).

We can add to the language $\mathfrak{L}$ a singulary operator, $\mathbb{A}$, to form $\mathfrak{L}_{\mathbb{A}}$. Many of the issues with actuality arise already without any additional modal operators in the language, so we omit them for now. The question, then, is how the new connective should be interpreted in the models. The simplest idea is suggested by the earlier quotations: Equip the frames with a distinguished point, $g \in N$, which can, and plausibly should, be 0 if one is using reduced frames. The distinguished point is then used to interpret $\mathbb{A}$.

- $a \Vdash \mathbb{A}B$ iff $g \Vdash B$.

The issue with this approach is that if $\mathbb{A}B$ holds anywhere in a model, it holds everywhere in that model. Thus, with this verification condition, $\mathbb{A}$ leads to ubiquitous truths in all models, and so violations of variable sharing. The situation is the same if a designated proposition is used rather than a designated point. Thus, the simple idea does not work for relevant logics.

Standefer (2020) proposes, instead, permitting each point to have different options considered to be candidates for actuality.[8] In terms of frames, this involves adding a binary accessibility relation T for $\mathbb{A}$, subject to the same condition as the relation used for $\Box$, namely that if $a \leq b$ and Tbc, then Tac.[9] Let us call such frames actuality frames.[10] In this setting, the verification condition is

- $a \Vdash \mathbb{A}B$ iff for all b, if Tab, then $b \Vdash B$.

The extended language obeys the Heredity lemma, and the basic logic obeys the analogs for $\mathbb{A}$ of the principles (RM) and ($\Box\wedge$).

The basic logic seems to be too weak for a plausible logic of actuality. The natural move is to add some frame conditions to validate more formulas to strengthen the logic. Table 8.2 lists some axioms that are distinctive for actuality and their corresponding frame conditions. Adding these, or any of the analogous modal principles from Chapter 6, to the basic modal logic for actuality over L will result in a logic that is Sound and Complete with respect to the class of actuality frames obeying the corresponding frame conditions for the modal principles and for the axioms of L. One upshot of this approach to actuality is that all the logics of actuality obtained in this way have the variable-sharing property.

Nothing prohibits the addition of an alethic necessity operator, $\Box$, to $\mathfrak{L}_{\mathbb{A}}$ to form $\mathfrak{L}_{\mathbb{A}\Box}$. The frames for that language combine the modal frames of Chapter 7 with the actuality frames of this chapter by adding a binary modal accessibility relation

8 This leaves open the option of there being multiple candidates for actuality, which Sylvan (1988) argues for.

9 NB: This is not the defined relation T of Chapter 7.

10 Standefer (2020) calls these RMS-frames, using S for the binary accessibility relation and T for an additional modal relation, as the focus is on the actuality operator.

Table 8.2. Distinctive actuality axioms and frame conditions

Name	Axiom	Frame condition
(Dc)	$\neg\mathbb{A}A \rightarrow \mathbb{A}\neg A$	$\forall x\forall y\forall z$(if $(Tx^*y$ and $Txz)$, then $y^* \leq z$)
(U)	$\mathbb{A}(\mathbb{A}A \rightarrow A)$	$\forall x \in N\forall y$(if Txy, then $\forall z\forall u$(if $Ryzu$, then Tzu))
(Uc)	$\mathbb{A}(A \rightarrow \mathbb{A}A)$	$\forall x \in N\forall y$(if Txy, then $\forall u\forall v$(if $Ryuv$, then $\forall z$(if Tvz, then $u \leq z$))

S subject to the familiar condition. Call these frames MODAL ACTUALITY FRAMES. Unlike the case with the TV-based two-dimensional modal logics, none of the usual interaction principles are forced upon us. Standefer (2020) considers two interaction principles from TV-based two-dimensional logics, $\Box A \rightarrow \mathbb{A}A$ and $\mathbb{A}A \rightarrow \Box\mathbb{A}A$, and notes that one can have the former without the latter in a natural logic of actuality. Beyond this, investigation of interaction principles is almost entirely open.[11]

One further line of investigation is suggested by Standefer (2020), and this line involves further connections to work on two-dimensional semantics. In double-indexing models for TV-based logics of actuality, one often evaluates formulas with respect to pairs, w, u, where w is the world of evaluation and u is the world considered as actual. In such models, $\mathbb{A}$ and $\Box$ have the following verification conditions.

- $w, u \Vdash \mathbb{A}A$ iff $u, u \Vdash A$.
- $w, u \Vdash \Box A$ iff for all $v \in W$, $v, u \Vdash A$.

It is a natural thought to add another singulary connective, fixedly ($\mathbb{F}$), to the language to vary the world considered as actual.[12]

- $w, u \Vdash \mathbb{F}A$ iff for all $v \in W$, $w, v \Vdash A$.

The "fixedly" connective has played an important role in two-dimensional semantics.[13] In modal actuality models, the verification condition for $\mathbb{A}$ uses a binary relation, rather than a distinguished point, but incorporating an operator like "fixedly" into the relevant logics would be good for making connections with two-dimensional semantics. How double-indexing should work in the context of modal actuality models is an interesting, albeit unexplored, question.

8.3 Epistemic Logics

Epistemic logics are logics that add operators for various epistemic concepts.[14] While *knowledge* is perhaps the most studied concept in epistemic logic, there are other concepts, such as justification and evidence, that get formalized.[15]

[11] Weiss (2025) investigates interaction principles in many different modal extensions of relevant logics.
[12] The "fixedly" connective is suggested by Crossley and Humberstone (1977).
[13] For a discussion, see Restall and Standefer (2023b, ch. 8).
[14] Hendricks and Symons (2015).
[15] See Artemov and Fitting (2016) and Artemov and Fitting (2019) for overviews of justification logics.

Ternary relational frames have natural connections to epistemic logics in virtue of a standard interpretation of them in terms of bodies of information. On this interpretation, the points are bodies of information. The ternary relation connects points based on the informational connections between the bodies of information.[16] This gives the models for the base logics an epistemic flavor that naturally extends to the modal case. We think that relevant logics fit particularly neatly with epistemic extensions for this reason.

8.3.1 The Fitch–Church Knowability Paradox

The first work on epistemic logic that we will discuss is by Wansing (2002). The Fitch–Church knowability paradox is an argument that shows that from the principle that every truth is *knowable*, $A \rightarrow \Diamond KA$, one can derive the conclusion that every truth is *known*, $A \rightarrow KA$. A version of the argument for TV, reconstructed by Wansing (2002, 593), is the following, where $\supset$ is the classical material conditional.

(1) $\neg\Diamond(KA \wedge \neg KA)$, plausible axiom.
(2) $A \wedge \neg KA$, assumption.
(3) $\Diamond K(A \wedge \neg KA)$, from (2) and $A \supset KA$.
(4) $\Diamond(KA \wedge K\neg KA)$, from (3) and (RM) for K and $\Diamond$.
(5) $\Diamond(KA \wedge \neg KA)$, from (4) and (T) and (RM) for $\Diamond$.
(6) $\neg(A \wedge \neg KA)$, negation introduction from the contradiction between (1) and (5).

Wansing remarks that (1) is required as an additional axiom if the underlying epistemic logic does not provide it, but it is an intuitively plausible axiom. There is another version of the argument that Wansing considers, as well as other versions in the literature, although this will be enough or the current discussion.

Wansing's proposed solution is an epistemic logic whose underlying logic combines R with Nelson's strong negation.[17] This uses a modification of the ternary relational frames to handle the strong negation. The alethic modal operators are interpreted using a binary accessibility relation, and the K operator is interpreted using a neighborhood function, resulting in a fairly weak logic for K. To accommodate constructive negation, different verification clauses are used, along with falsification clauses, all of which will be omitted here. Wansing provides an axiomatization that is sound and complete with respect to the provided frames.

The upshot of Wansing's proposal is that the version of the earlier paradoxical argument does not go through, even though (1) is adopted as an axiom. The reason is that the negation introduction rule is invalid, so the final step is blocked. The second version of the argument Wansing considers is, likewise, blocked. That argument depends on a contraposition principle that is invalid. Wansing provides a model where $p \rightarrow \Diamond Kp$ is verified but $p \rightarrow Kp$ is not, thereby showing that his logic is safe from

[16] For more on channel theory and relevant logics, see Mares (1996, 2021), Restall (2005), Mares et al. (2010), and Tedder (2021b).
[17] See Kamide and Wansing (2015) and the citations therein for more on constructive negation.

Fitch's paradox. The result is a strong, well-motivated epistemic alethic modal logic in which knowability does not imply being known.

8.3.2 Existential Knowledge

The second work on epistemic logic we will present is by Bílková et al. (2010; 2016). In these papers, Bílková et al. present a new conception of knowledge. Rather than a universally quantified conception of knowledge, according to which an agent knows something if that holds in *all* epistemic alternatives, they develop an existential conception of knowledge, according to which an agent knows something if it is verified by *some* information source possessed by the agent.

They use models based on ternary relational frames with a modal accessibility relation S and a compatibility relation C, rather than the Routley star. The points in the frames are interpreted as information states, representations of the data available to an agent. A subset of the points act as *sources of information* that can provide the verification needed for knowledge. Bílková et al. (2016) present the basic conditions on accessibility for epistemic frames, using the convention that s and s' are used for sources.

(1) Ssa only if $s \leq a$.
(2) Ssa only if Cas.
(3) If $a \leq b$, $c \leq d$, and Sbc, then Sad.

In these models, the knowledge operator has the verification condition

- $a \Vdash KB$ iff $\exists s \in K(Ssa \text{ and } s \Vdash B)$.

From the preceding conditions and the other frame conditions, one can derive some properties of sources.

(1) Ssa only if Css.
(2) Ssa only if there is no A such that $s \Vdash A$ and $s \Vdash \neg A$.
(3) Ssa and $Ss'a$ implies Css'.
(4) If $a \Vdash KA$, then for any s such that Ssa, $s \nVdash \neg A$.

What these properties reveal is that sources exhibit some strong consistency features not required by other points. The first says that sources are compatible with themselves, so they cannot verify contradictions, as the second says. The third says that any two sources for a given point will be compatible. The fourth says that if it is verified at a point a that the agent knows A, then there is no source s accessible to a that verified $\neg A$.

These models are, then, appropriate for modeling the sort of scenarios where an agent will only be getting consistent information from data sources. the primary example provided by Bílková et al. is that of a scientist working in a lab. The scientist does some experiments, which act as the reliable sources, and knowledge claims are baed on those. The experiments deliver consistent information, although they do not have to deliver total information, which is good since lab experiments will fall short of settling *everything*.

The properties of sources provide distinctive features to the epistemic logic. The basic axioms and rules for knowledge are the following.

- $KA \rightarrow A$.
- $(\neg A \wedge KA) \rightarrow B$.
- $K(A \vee B) \rightarrow (KA \vee KB)$.
- $A \rightarrow B \Rightarrow KA \rightarrow KB$.

Some comments are in order. The first axiom is factivity of knowledge, which is standard in epistemic logics of knowledge. The second is a strong consistency condition that leads to violations of the variable-sharing principle. The third axiom says that if an agent knows a disjunction, then the agent knows one disjunct or the other. This is to say that if a source provides verification for a disjunction, it must provide verification for a disjunct. This is a surprising principle in a logic of knowledge. One usually has the agglomeration of knowledge over conjunction, the analog of ($\Box\wedge$), although that need not hold in the present setting. The validity of this final axiom is tied to the existential verification condition for knowledge.

Other axioms that are sometimes found in TV-based epistemic logics are invalid here. In particular, (K), (4), (5), and (Nec) all fail. They can be added with additional frame conditions.[18] Bílková et al. show that the base logic can be weak, and it too can be strengthened with standard frame conditions.

Although our focus is on relevant logics, Bílková et al. (2016, 1829) make some interesting remarks concerning extensions to intuitoinistic and classical logic. They define frames for intuitionistic logic using an underlying poset $(K, \leq)$ on which C and R are defined, and S can be any binary relation monotonic on $\leq$. The extension to TV is obtained by adding conditions to the ternary relational frames. The conditions together imply that if Sab then $a = b$. So, if a point has a source, then it is its own source and there are no others. Not all points, however, have sources. The result is a much stronger epistemic logic, for example, validating (4), although it does not validate (Nec) and (5).

8.3.3 Justification Logics

Justification logics are modal logics that add explicit modalities that represent a formula being justified by a particular bit of evidence.[19] For example, $[\![x]\!]A$ is meant to express that x is evidence for A. There are, additionally, operators on the justification terms that provide ways to combine them to obtain new justification terms. This provides the machinery to analyze many fine-grained epistemic distinctions.

18 Bílková et al. (2016, 1823).

19 Artemov and Fitting (2016, 2019) and Kuznets and Studer (2019) provide overviews of justification logics.

For the language of justification logics, we use a countably infinite set of justification variables $x_1, \ldots, x_n, \ldots$, and one can use a set of justification constants $c_1, \ldots, c_n, \ldots$ and define the set of Terms as follows.

$$s,t := x_i \mid c_i \mid s \cdot t \mid s + t.$$

One can add additional operators on justification terms, but we will not pursue that here. A note on the intuitive interpretation: The dot operator acts much like theory fusion – whenever an implication is justified by s and the antecedent if justified by t, then the consequent is justified by $s \cdot t$. The plus operator is more like set union, preserving to $s+t$ what is justified by s and by t individually. For each term $t \in$ Terms, there is a justification modal, $[\![t]\!]$, and $[\![t]\!]A$ is a formula.[20]

In TV-based justification logics, a standard kind of model for the logics is a Fitting model, $\langle W, S, \mathcal{E}, V\rangle$, where $\langle W, S\rangle$ makes up a Kripke frame, $\mathcal{E}$ is an evidence function from pairs in Terms $\times$ W to sets of formulas, and V is a valuation function. These models are subject to the condition that if $A \to B \in \mathcal{E}(s,w)$ and $A \in \mathcal{E}(t,w)$, then $B \in \mathcal{E}(s \cdot t, w)$. Additional conditions are required for the plus operator as well as any other term-forming operators, although these will be omitted here. The verification condition for the justification modals is

- $w \Vdash [\![t]\!]A$ iff $A \in \mathcal{E}(t,w)$ and for all $u \in W$, if Swu, then $u \Vdash A$.

The basic justification logic adds to TV the axiom

- $[\![s]\!](A \to B) \to ([\![t]\!]A \to [\![s \cdot t]\!]B)$,

a relative of the (K) axiom. Unlike normal modal logics, the rule (Nec) is not generally valid.

To obtain models for relevant justification logics, a natural thought is that on the frame side, one simply combines the modal frames of earlier chapters with evidence functions, and on the logic side, one simply adds the analog of the (K) axiom stated earlier to a base relevant logic. The frames are then sextuples $\langle K, N, R, {}^*, S, \mathcal{E}\rangle$ and they require some additional conditions governing the interaction of the evidence functions, the heredity relation, and the ternary relation. The key frame condition is the condition for the (K) axiom,

(FK) If $\exists f(Rabf$ and $Sfc)$, then $\exists d, e(Rdec$ and Sad and $Sbe)$.

Standefer (2019; 2023a) shows that the canonical frame generated by the basic justification logic fails to satisfy the condition (FK), providing a serious stumbling block to the usual method of proving Completeness. The problem, in essence, is the following. Let a^- be the set $\{A : \exists t \in \text{Terms}, [\![t]\!]A \in a\}$. The canonical accessibility relation, S, is defined as

- Sab iff $a^- \subseteq b$,

[20] In the justification logic literature, the notation "$t : A$" is more standard.

but the set a^- need not be a theory, as it need not be closed under provable implications or under adjunction.

There are three alternatives to the straightforward frames. The first alternative, called RMJ-frames by Standefer (2019), drops the evidence functions and instead adds a modal accessibility relation S_t for each $t \in \mathsf{Terms}$, $\langle K, N, R, ^*, \{S_t \subseteq K^2 : t \in \mathsf{Terms}\}\rangle$. The accessibility relation S_t is used to interpret $[\![t]\!]B$.

- $a \Vdash [\![t]\!]B$ iff for all $b \in K$, if $S_t ab$, then $b \Vdash B$.

To axiomatize the logic, one needs to add, for each $t \in \mathsf{Terms}$, analogs of ($\Box\wedge$) and (RM),

- $([\![t]\!]A \wedge [\![t]\!]B) \rightarrow [\![t]\!](A \wedge B)$,
- $A \rightarrow B \Rightarrow [\![t]\!]A \rightarrow [\![t]\!]B$

in addition to the (K) axiom given earlier. The resulting logics are sound and complete with respect to appropriate classes of RMJ-frames.

The second alternative, explored by Standefer (2023a), sticks with the basic frames, $\langle K, N, R, ^*, S, \mathcal{E}\rangle$, and expands the language with an operator for implicit commitment, $\Box$. The verification conditions for the modals are

- $a \Vdash [\![t]\!]A$ iff $A \in \mathcal{E}(t, a)$ and for all $b \in K$, if Sab, then $b \Vdash A$, and
- $a \Vdash \Box A$ iff for all $b \in K$, if Sab, then $b \Vdash A$.

On the logic side, one adds ($\Box\wedge$), (RM), and (K) for $\Box$, the justification logical (K) axiom, and the axiom $[\![t]\!]A \rightarrow \Box A$, for all $t \in \mathsf{Terms}$. In the canonical frame construction, $\Box$ is used to define the canonical accessibility relation, and it satisfies the required conditions. Thus, the problem identified by Standefer can be solved by extending the language. A Completeness result can be obtained for the logics in the language with the implicit commitment operator.

The final alternative, explored by Savić and Studer (2019) and by Standefer (2023c), is to eschew the modal accessibility relation entirely. Instead, the justification modals are evaluated using only the evidence function.

- $a \Vdash [\![t]\!]A$ iff $A \in \mathcal{E}(t, a)$.[21]

With some conditions on the frames to ensure that R, $\leq$, and $\mathcal{E}$ interact appropriately, Soundness and Completeness results can be had for the justification logic obtained by adding the justification logical (K) axiom to the base logic.

Both Savić and Studer (2019) and Standefer (2023c) add term-forming operators to the language, although the latter adds more.

$$s, t := x_i \mid c_i \mid (t \cdot s) \mid (t + s) \mid \mathfrak{a}(t, s) \mid \mathfrak{c}t \mid \mathfrak{p}t \mid \mathfrak{s}t \mid \mathfrak{i}t \mid \mathfrak{e}t \mid \mathfrak{t}s.$$

Each of the additional operators corresponds to a rule of the axiom system for B. For each singulary operator $\mathfrak{o}$, there is an axiom, $[\![t]\!]A \rightarrow [\![\mathfrak{o}t]\!]B$, where $A \Rightarrow B$ is the

[21] Both Savić and Studer (2019) and Standefer (2023c) use the notation ♠ for the evidence function, although we will stick with the notation "$\mathcal{E}$" we have used throughout this subsection.

rule to which the operator corresopnds. For each binary operator $\mathfrak{o}$, there is an axiom, $([\![t]\!]A \wedge [\![s]\!]B) \rightarrow [\![\mathfrak{o}(t,s)]\!]C$, where $A, B \Rightarrow C$ is the corresponding rule. This allows one to prove an internalization theorem for the logic provided a constant specification. Without getting into all the details, the constant specification provides a limited form of the (Nec) rule, permitting one to conclude $[\![c_1]\!] \cdots [\![c_n]\!]A$, for $n \geq 1$, where A is an axiom. Say that the basic justification logic over L is L.J and the logic with the addition of a constant specification is L.JC. Then the following internalization theorem is provable via induction on proofs.

THEOREM 8.2 *If* $\vdash_{L.JC} A$, *then for some* $t \in$ Terms, $\vdash_{L.JC} [\![t]\!]A$.

The other approaches to relevant justification logics do not permit this sort of internalization theorem to be proven. The internalization theorem says that the justification logic can represent its owns proofs via its justification terms. Further work on relevant justification logic remains to be done. In the program of justification logics, realization theorems are important. Realization theorems provide a precise sense in which a particular justification logic is connected with a familiar modal logic by showing

- if A is derivable in the justification logic, then A' is derivable in the modal logic, where A' is obtained from A by changing all justification modals to $\Box$, and
- if A is derivable in the modal logic, then A' is derivable in the justification logic, where A' is a formula that replaces each $\Box$ with a justification modal in an algorithmic way.

Realization theorems for relevant justification logics are yet to be proven. Nonetheless, they are promising logics for tracking reasons and evidence.

8.3.4 Further Epistemic Logics

Relevant epistemic logic is a very active area, and there are several more papers that we have not yet discussed. In this section, we will briefly highlight some of the additional work, to give the reader a sense of how the area is developing.

Sedlár (2015; 2016) presents logics for epistemic extensions of substructural logics. Sedlár uses ternary relational frames with compatibility relations. A distinctive feature of the frames is that some points are *worlds*. They satisfy five conditions, the upshot of which is that the verification conditions for $\neg$, $\rightarrow$, and $\circ$ are all essentially classical at worlds. Further, the set of worlds is a subset of N. An epistemic frame G over a ternary relational frame F is a quadruple $\langle F, W, S, \sigma(\cdot)\rangle$, where W is a non-empty set of worlds, S is a binary accessibility relation obeying both the usual condition and if Sab and $a \in W$, then $b \in W$, and $\sigma(\cdot)$ is a monotone function on K, that is, $a \leq b$ only if $\sigma(a) \leq \sigma(b)$. Holding in a model on an epistemic frame is defined with respect to the set W of worlds, rather than just normal points.

In models over epistemic frames, there are two new singulary connectives that are interpreted using the additional machinery: $\mathbb{I}$ (implicit belief) and $\mathbb{S}$ (support). As intuitive glosses, $\mathbb{I}A$ means "the agent implicitly believes A" and $\mathbb{S}A$ means "the body of evidence available to the agent supports that A."[22] These connectives have the following verification conditions.

- $a \Vdash \mathbb{I}A$ iff $\forall b \in K$(if Sab, then $a \Vdash A$).
- $a \Vdash \mathbb{S}A$ iff $\sigma(a) \Vdash A$.

These connectives permit the definition of another connective, $\mathbb{B}A$, as $\mathbb{I}A \wedge \mathbb{S}A$, meaning "the agent justifiably believes that A."[23]

Sedlár (2015) provides an axiomatization of the logic that is sound and complete with respect to the basic frames. This result holds under extensions of the underlying logic as well as the epistemic components.

Information pooling is a way of modeling the combination of information had by different agents. In TV-based modal logics, this is sometimes done by representing an agent's information as a set of worlds, and then the operation of information pooling is simply the intersection of all such propositions. The result is that if two agents have contradictory beliefs, then pooling their information will be the empty set of worlds, which will entail everything. Punčochář and Sedlár (2017) develop a substructural logic of information pooling that avoids this problem and problems analogous to the paradoxes of implication. The account uses information frames, similar to Urquhart's semilattice frames and Humberstone's (1988) modified semilattice frames, with some additional features, including a compatibility relation for negation.[24] The information belonging to each group G of agents (at a point a) is represented using a singulary modality, $\Box_G$, which is interpreted in the models via a function $f_G(a)$ subject to three frame conditions. The verification condition is

- $a \Vdash \Box_G A$ iff $f_G(a) \Vdash A$.

There has not yet been an adaptation of this work to the ternary relational frames we have used in this book.

Punčochář (2019; 2020) builds on the information frames to extend relevant logics, and other substructural logics, with the logical machinery of inquisitive semantics.[25] In inquisitive semantics, there are two disjunctions, the inquisitive disjunction, $\mathbb{V}$, and the regular disjunction, $\vee$. Inquisitive disjunction is used to form questions, whereas regular disjunction is used to form new assertoric formulas.

The models follow Humberstone (1988) in using two operations on points, one, $\cdot$, to interpret implication and one, $+$, to interpret (regular) disjunction.[26] Without getting further into the details of the models, inquisitive disjunction has the verification condition

- $a \Vdash B \mathbb{V} C$ iff $a \Vdash B$ or $a \Vdash C$,

22 NB: Sedlár uses "$\mathbb{A}$" where we use "$\mathbb{S}$," which we adopt to avoid a clash with the notation used in Section 8.2.3.

23 Sedlár (2016, 1792).

24 See Punčochář (2017) for more on these frames.

25 See Ciardelli et al. (2019) for more on inquisitive semantics, albeit in a more classical setting.

26 NB: In this context, "$\cdot$" does not denote the same operation as in the groupoids of Chapter 5.

while the regular disjunction has the verification condition

- $a \Vdash B \vee C$ iff there are b, c such that $a \leq b + c$, $b \Vdash B$ and $c \Vdash C$.[27]

Punčochář provides sound and complete axiomatizations for a range of relevant logics equipped with inquisitive disjunction. He also proves a duality result, between his informational frames and a class of ternary relational frames, specifying the conditions under which the two frameworks coincide. Nonetheless, the task of developing inquisitive semantics on ternary relational frames is still open.

Belief revision is an important area of epistemic logic.[28] In belief revision, the goal is to formalize how an agent should update their beliefs when given new information, which can, potentially, conflict with the beliefs they already have. The standard approach to belief revision is the AGM theory.[29] This approach relies, in important ways, on the consistency of belief sets for handling revision, but the reliance on consistency is somewhat inappropriate in the context of relevant logics.[30]

Belief revision has been explored in the context of relevant logics by Mares and by Sedlár, albeit in different ways. Mares (2002) takes a consequence-theoretic approach, defining operators of belief revision in terms of consequence operators and operations on belief sets. Sedlár (2016), by contrast, takes a model-theoretic approach. Using the epistemic frames mentioned earlier, Sedlár defines revision operations with models. The question of a sound and complete axiomatization of the belief revision logic is left open.

Development of relevant epistemic logics is ongoing. As a recent example, Punčochář et al. (2023) showed how to combine public announcements and common knowledge in models for R. A great deal of work has been done on TV-based epistemic logics, much of which has not been revisited from the perspective of relevant logic.

8.4 Deontic Logics

Deontic logics are logics that contain operators for expressing normative concepts, such as obligation ($\mathbb{O}$) and permission ($\mathbb{P}$), among others.[31] The combination of deontic logic with relevant logics is a natural idea. Goble (1999) motivates the combination by pointing to several issues that are neatly resolved in the relevant logical setting. In TV-based deontic logics, conflicting obligations result in triviality. Relevant logics can handle contradictions without triviality, so problems with conflicting obligations will not arise. Similarly, in TV-based deontic logics, there are contrary-to-duty obligations, which say that if it ought to be the case that A, then it ought to be the

27 See Humberstone (2019) for a general discussion of this sort of verification condition, with connections to his work on operational models as well as work on inquisitive semantics.

28 Hansson (2017).

29 The AGM theory was first presented by Alchourrón et al. (1985).

30 The reliance on consistency is inappropriate in paraconsistent logics generally, a point made by Restall and Slaney (1995) and by Priest (2001).

31 McNamara and Van De Putte (2021).

case that if $\neg A$, then B, in symbols $\mathbb{O}A \rightarrow \mathbb{O}(\neg A \rightarrow B)$, which arise from the TV-theorem $A \rightarrow (\neg A \rightarrow B)$. Additionally, in TV-based deontic logics, one often gets that tautologies are obligatory. While this is not as much of a problem as the previous two issues, it is counterintuitive and not regarded as a virtue. Luckily, relevant deontic logics avoid that issue easily as well.

There are two approaches to deontic logic. According to one, the concept of obligation is added as a primitive singulary operator, $\mathbb{O}$. According to the second, following Anderson (1958a; 1958b; 1967a; 1967b; 1968) and others, one analyzes obligation in terms of implication and a special constant, Ω, so that a claim A is obligatory if $\neg A$ implies the bad state: $\neg A \rightarrow \Omega$. This is known as the Andersonian reduction. The implication in question may be a strict implication, in which case it is the *modalized Andersonian reduction* and the definition would be $\Box(\neg A \rightarrow \Omega)$, for the appropriate notion of alethic necessity. We can define the good state as $\mho =_{Df} \neg\Omega$, so we can think of a claim being obligatory if the good state implies that claim, $\mho \rightarrow A$. Variations on these definitions are found in the literature.

To fix some terminology, the language with the primitive $\mathbb{O}$ operator will be $\mathfrak{L}_{\mathbb{O}}$, and the language with the constants Ω or $\mho$ will be $\mathfrak{L}_{\Omega}$ or $\mathfrak{L}_{\mho}$. Since the constants are interdefinable, we will not worry about which is in the language if either are. Where there is an additional alethic modality around, a subscripted $\Box$ will be added. Let us begin with the primitive $\mathbb{O}$ approach.

8.4.1 Deontic Logics with Primitive Obligation

Routley and Plumwood (1989) discuss deontic logics and motivate the addition of deontic operators to relevant logics, covering much the same ground as Goble (1999), presented in the introductory subsection. A major focus, however, is moral dilemmas and unrealizable obligations. One of the problems with TV-based deontic logic that they return to repeatedly is the problems caused by conflicting obligations: situations in which one has an obligation that A and an obligation that $\neg A$. In particular, in the face of (D), ($\Box\wedge$), and (RM) for $\mathbb{O}$, conflicting obligations result in triviality. While there are various technical maneuvers that the classical logician can deploy to avoid triviality, they do not find these particularly appealing.

Routley and Plumwood (1989, 674ff.) think that one of the major philosophical issues with thinking about moral dilemmas is that philosophers and deontic logicians take it for granted that one can always reason one's way out of a moral dilemma. Upon full and proper consideration, one or the other horn of a dilemma will yield. They go on to pin the blame for this idea on adherence to the ought-implies-can principle, $\mathbb{O}A \rightarrow \Diamond A$. They recommend rejecting it, along with some supporting philosophical theses, and holding onto the idea that some moral dilemmas may actually be intractable. They do not go on to develop, axiomatically or frame-theoretically, a preferred logic in which the ought-implies-can principle fails. In the next subsection, we will return to the ought-implies-can principle, and we will see that, at least for some Andersonian deontic logics, it falls out naturally and unproblematically.

Goble (1999) explores the addition of deontic principles to R.[32] His methodology is to incorporate the basic principles of standard (TV-based) deontic logic to the base logic R. The basic axioms he considers are the analogs of ($\Box\wedge$), (K), and (RM), (D), and (Nec). He briefly considers some further extensions as well. He provides frame theories for which the different combinations of axioms are sound and complete. The frames use a binary accessibility relation S, where Sab is interpreted to mean that b is one of the ideal or best points from the point of view of a, which is a fairly standard gloss on the accessibility relation in deontic logics.

The details of the frame theory match those of Chapter 7, with no real surprises. Thus, Goble concludes, the intuitions of basic deontic logic can be merged with those of relevant logics without issue. He goes on to define a binary connective of conditional obligation $\mathbb{O}(B/A)$, for which he provides a frame semantics. We will wait till after the discussion of Andersonian deontic logic to return to these frames.

While Goble's focus is on R, there is nothing in that work that relies on the proprietary features of R. His basic approach, both axiomatically and frame-theoretically, could be carried out with weaker logics. As we will see, this is in contrast to the Andersonian deontic logics to follow. Let us turn now to the Andersonian deontic logics.

8.4.2 Andersonian Deontic Logics

As mentioned earlier, Anderson (1958a; 1958b; 1967a; 1967b; 1968) provided a reduction of the deontic modality of obligation to combination of logical connectives and a new constant, Ω, via the definition of $\mathbb{O}A$ as $\neg A \to \Omega$. McArthur (1981) criticized different versions of Anderson's proposal on logical grounds. He argued that the logical features of R meant that the definition of $\mathbb{O}A$ as $\neg A \to \Omega$ has excessively counterintuitive consequences. Among the provable theorems are $\mathbb{O}\mathbb{O}A \to A$, $\mathbb{O}(A \to B) \leftrightarrow (A \to \mathbb{O}B)$, and $(A \to \mathbb{O}B) \to (\neg B \to \mathbb{O}\neg A)$. In light of the theorem $(A \to B) \to (\mathbb{O}A \to \mathbb{O}B)$, he concludes that R cannot be used to express contingent, conditional obligations. He considers using E as the base logic for Andersonian deontic logic, but he concludes that it has too many theorems that are implausible as deontic claims. It is worth noting that the problems McArthur cites for R are due to a combination of permutation, contraction, and suffixing.[33] If one is persuaded by McArthur's criticisms, this suggests that a weaker logic that lacks these principles, such as DJ, TW, or even RW, may be the best place for exploring Andersonian deontic logic.

Mares (1992a) develops the Andersonian idea in $\mathfrak{L}_{\Box\Omega}$ against the backdrop of R.MKTNecMC with the addition of the axiom $\neg\Box\Omega$. The logic is called $\mathsf{RT}_{\Omega\mathsf{mo}}$.[34] The definitions of prohibition, obligation, and permission adopted by Mares are the following.

[32] Goble (2006) explores deontic extensions of close relatives of R that end up not being relevant logics, although they are paraconsistent.

[33] The arguments also use contraposition, but we will leave that one alone for now.

[34] Mares (1992a) calls this logic $\mathsf{RT}_{V\mathsf{mo}}$, but we have adjusted the name for uniformity with the notation.

- $\mathbb{F}A =_{Df} \Box(A \rightarrow \Omega)$.
- $\mathbb{O}A =_{Df} \mathbb{F}\neg A$, that is, $\Box(\neg A \rightarrow \Omega)$.
- $\mathbb{P}A =_{Df} \neg\mathbb{F}A$, that is, $\neg\Box(A \rightarrow \Omega)$.

Using a six-valued matrix, Mares (1992a, 8–11) shows that the logic does not have any deontic analogs of modal fallacies, which is to say formulas of the form $A \rightarrow \mathbb{O}B$, where A does not contain $\Box$ or Ω.[35] Additionally, (Nec$\mathbb{O}$), $A \Rightarrow \mathbb{O}A$, is not valid.

The logic enjoys analogs for $\mathbb{O}$ of (K), (RM), and (D), as can be seen by unpacking the definitions and doing some axiom chopping. The derivation or (RM) for $\mathbb{O}$ is neat, so we will present it here.

Proof Suppose $A \rightarrow B$ is derivable. Then $(\neg\Omega \rightarrow A) \rightarrow (\neg\Omega \rightarrow B)$ is. Using contraposition and double negation axioms, $(\neg A \rightarrow \Omega) \rightarrow (\neg B \rightarrow \Omega)$ is derivable. By (RM), $\Box(\neg A \rightarrow \Omega) \rightarrow \Box(\neg B \rightarrow \Omega)$ is derivable, which is, by definition, $\mathbb{O}A \rightarrow \mathbb{O}B$. □

The derivations for the axioms given earlier and rule appeal to features of the logic R.MKT4Nec, so (MC) can be dropped without losing those features.

The additional axiom, $\neg\Box\Omega$, is used in the derivation of an ought-implies-can principle: $\mathbb{O}A \rightarrow \Diamond A$. It is, perhaps, worth noting that the derivation supplied by Mares appeals to distinctive features of R shared by RW, namely (I6) (permutation) and (N1).[36] Indeed, the ought-implies-can principle is not a theorem if the base logic is T, instead of R, as can be seen through the use of Slaney's MaGIC. The principle also fails to be a theorem if the base logic contains (I6) but not (N1). This suggests that for philosophers that dislike the ought-implies-can principle, such as Routley and Plumwood (1989), the best logic for Andersonian deontic logic would be one no stronger than T. For the philosophers just mentioned, however, this is not much of a concession, as they were independently suspicious of (I4).

Mares supplies a modal frame theory for which the logic is sound and complete. The frames are those of Chapter 7 with the appropriate postulates, with one addition for the special axiom $\neg\Box\Omega$. The additional condition is a model condition, rather than a frame condition:

- If $a \in N$, then $\exists x \in K(Tax$ and $x^* \notin V(\Omega))$.

In these models, Ω is treated as a regular atomic formula, with the proviso that it must fail at certain points.

Goble (2001) explores the non-modal version of the Andersonian reduction in $\mathfrak{L}_{\mho}$ using R, taking $\mho$ as primitive and defining $\mathbb{O}A$ as $\mho \rightarrow A$. $\mathbb{P}A$ can be defined as $\neg\mathbb{O}\neg A$. Let us call this logic R.$\mho$.

The properties of R immediately yield several features for $\mathbb{O}$, including ($\mathbb{O}\wedge$), (K), and (RM). The axiom(D) does not, however, fall out of the basic definitions. To obtain (D), Goble adds an additional axiom,

35 Cf. Weiss (2025) on fallacies of modality in deontic logic.

36 Mares's proof also uses (I4) (suffixing), but that is not required, as its rule form will suffice.

- $(\mho \to A) \to \neg(\mho \to \neg A)$,

which is (D) once the definition of $\mathbb{O}$ is applied.

Whereas Mares showed that formulas of the form $A \to \mathbb{O}B$ are not provable in $\mathsf{R}_{\Omega\mathsf{mo}}$, Goble shows that $\mathbb{O}A \to B$ is not provable in R.$\mho$, provided $\mho$ does not occur in B. Mares's result can be adapted to Goble's system, as long as $\mho$ does not occur in either A or B. In Goble's system, (Nec) for $\mathbb{O}$ also fails.

Goble (2001, 222–227) lists some of the dubious or negative consequences of adopting the Andersonian reduction with R. In particular, the following are provable in R.$\mho$.

(1) $\mathbb{O}(\mathbb{O}A \to A)$.
(2) $\mathbb{O}(A \to \mathbb{P}A)$.
(3) $\mathbb{O}(\mathbb{O}A \to \mathbb{P}A)$.
(4) $(A \to B) \to (\star A \to \star B)$, for $\star \in \{\mathbb{O}, \mathbb{P}\}$.
(5) $\mathbb{O}(A \to B) \leftrightarrow (A \to \mathbb{O}B)$.
(6) $A \to \mathbb{O}\mathbb{P}A$.
(7) $A \to \mathbb{F}\mathbb{F}A$.

The last four are the worst consequences according to Goble. These are, additionally, consequences due to (I4). In a modalized version of the Andersonian reduction, such as $\mathsf{R}_{\Omega\mathsf{mo}}$ adopted by Mares, the strict conditional version of (I4), that is, $\Box(A \to \Box(B \to C)) \to \Box(B \to \Box(A \to C))$, does not hold. Consequently, some of the bad theorems given earlier are avoided with the modalized reduction.

Lokhorst (2006) makes many interesting observations concerning Andersonian reduction over R. He provides an axiomatization of the deontic fragment of R.$\mho$, which is the translation of $\mathfrak{L}_{\mathbb{O}}$ to $\mathfrak{L}_{\mho}$. The translation $\tau(\cdot)$ is defined as follows.

- $\tau(p) = p$.
- $\tau(\neg A) = \neg\tau(A)$.
- $\tau(A\#B) = \tau(A)\#\tau(B)$, for $\# \in \{\to, \wedge, \vee\}$.
- $\tau(\mathbb{O}A) = \mho \to \tau(A)$.

The deontic fragment, so defined, is axiomatized with the $\mathbb{O}$ analogs of ($\Box\wedge$), (K), (RM), as well as $\mathbb{O}(\mathbb{O}A \to A)$, $(A \to B) \to (\mathbb{O}A \to \mathbb{O}B)$, and $A \to \mathbb{O}\mathbb{P}A$. The last three axioms were noted by Goble (2001) as negative consequences of the Andersonian reduction.

Although we have not yet discussed propositional quantification, it will be worth discussing it briefly in order to present some interesting results of Lokhorst (2006). Propositional quantification adds to L axioms and rules for quantifiers over propositions. The axiomatization used by Lokhorst is the following.

(QR1) $\forall p(A \to B) \to (\forall pA \to \forall pB)$.
(QR2) $(\forall pA \wedge \forall pB) \to \forall p(A \wedge B)$.
(QR3) $\forall pA(p) \to A(B)$.
(QR4) $\forall p(A \to B) \to (A \to \forall pB)$, where p is not free in B.
(QR5) $\forall p(A \vee B) \to (A \vee \forall pB)$, where p is not free in A.
(QR6) If A is an axiom, then $\forall pA$ is an axiom.

The logic with these axioms over R is $\mathsf{R}^{\forall}$, and the logic in the language with $\mho$ is $\mathsf{R}^{\forall}_{\mho}$. It is a consequence of these axioms and the definition of $\mathbb{O}$ that $\mho \leftrightarrow \forall p(\mathbb{O}p \rightarrow p)$ is derivable. The constant $\mho$ is then provably equivalent to the claim that if a proposition is obligatory, then it is true. This surprising fact lends weight to the interpretation of $\mho$, supplied by McNamara and Van De Putte (2021), as saying that "all obligations are fulfilled."

It is worth dwelling on what is involved in this proof. Both directions of the proof crucially use (I4). Thus, the interpretation of $\mho$ as the state in which all obligations are fulfilled is not forced in T or a weaker logic.

Perhaps even more surprising is that the derivation works when $\mathbb{O}$ is defined using a strict conditional. It is sufficient for the derivation that the necessity involved obeys (T) and (Nec), as shown by Lokhorst (2006, 391). With this modalized definition, one can still show that $\mho \leftrightarrow \forall p(\mathbb{O}p \rightarrow p)$.

Lokhorst explores a mixed logic of deontic and alethic modality in the language $\mathfrak{L}_{\Box\mathbb{O}}$. He adds to R.MKT4Nec the $\mathbb{O}$ analog of ($\Box\wedge$) and the following axioms.

- $\Box(A \rightarrow B) \rightarrow (\mathbb{O}A \rightarrow \mathbb{O}B)$.
- $\mathbb{O}(\mathbb{O}A \rightarrow A)$.
- $\mathbb{O}A \rightarrow \Box\mathbb{O}A$.

Call this logic R.S4O. The translation of this language to $\mathfrak{L}_{\Box\mho}$ is the same as the earlier translation except for the following clauses.

- $\tau(\Box A) = \Box\tau(A)$.
- $\tau(\mathbb{O}A) = \Box(\mho \rightarrow \tau(A))$.

Using this translation, Lokhorst shows that R.S4O axiomatizes the deontic fragment of R.MKT4Nec with $\mho$. Further, he shows that when both logics are enriched with propositional quantification, they have the same theorems. This equivalence is facilitated by the provability of $\mho \leftrightarrow \forall p(\mathbb{O}p \rightarrow p)$ and $\mathbb{O}A \leftrightarrow \Box(\mho \rightarrow A)$ in the appropriate logic.

Goble (2001) also shows that the Andersonian deontic logics over R are equivalent, under translation, to the systems with a primitive $\mathbb{O}$ operator but without $\mho$, using different methods. The translation used is the same as earlier except for the $\mathbb{O}A$ clause, which is the following

- $\tau(\mathbb{O}A) = \mho \rightarrow \tau(A)$.

Instead of proceeding proof-theoretically, Goble goes via frames. He provides a Sound and Complete frame theory for the former. To obtain a frame for the Andersonian deontic logic, take a modal frame F for the logic with primitive $\mathbb{O}$ and add an additional set E such that $E = \{a \in K : \exists x(x \in N \text{ and } Sax)\}$.[37] Call the resulting frame $F^{\mho}$. We can define a model $M^{\mho}$ on $F^{\mho}$ from a model M on F by setting $a \in V^{\mho}(p)$ iff $a \in V(p)$, for $p \in \mathsf{At}$ apart from $\mho$, and $a \in V^{\mho}(\mho)$ iff $a \in E$. Goble then proves by induction that A is true at a point a in M iff $\tau(A)$ is true at a in $M^{\mho}$.

[37] Goble uses "U" where we use "$\mho$."

8.4.3 Conditional Obligation

The work surveyed so far has focused primarily on obligation, permission, and prohibition, understood as singulary connectives. In this section, we will look at the binary connective of conditional obligation, $\mathbb{O}(\cdot/\cdot)$, considered by Goble (1999).

The language with $\mathbb{O}(\cdot/\cdot)$ will be $\mathfrak{L}_{\mathbb{O}2}$. The frames for this language extend the ternary relational frames with another ternary relation, T, over $K \times \mathfrak{L}_{\mathbb{O}2} \times K$. In addition to the postulates for ternary relational frames and for R-frames, these frames obey the conditions

(G1) if $a \leq b$ and $TbAc$, then $TaAc$, and
(G2) if $\exists x(Rabx$ and $TxAc)$, then $\exists y \exists z(TaAy$ and $TbAz$ and $Ryzc)$.

The verification condition for conditional obligation is then the following.

- $a \Vdash \mathbb{O}(B/A)$ iff $\forall b \in K$(if $TaAb$, then $b \Vdash B$).

The Heredity lemma is provable in this setting.

Goble provides a gloss on the verification condition for $\mathbb{O}(B/A)$, adapting the gloss from the models for $\mathbb{O}$. Goble (1999, 337) says,

> [W]e shall say that $\mathbb{O}(B/A)$ holds (at a point) just in case B holds at all the best A-points (from that same point of view). So, instead of supposing that there is a set of ideal points associated with each point, we now suppose that there is a set of ideal or acceptable points associated with each point *and* with each formula, A.

While there is something to this gloss, it is not clear why the pair of a point and a formula should take a stand on the set of ideal points. The points are often understood as situations or bodies of information. That those things should have some stance on ideal points is plausible, perhaps given some additional background philosophical theory on what situations are ideal relative to other situations. The addition of formulas, however, sits less well philosophically with the picture presented. While the ternary relation T works formally, $TaAb$ holding in a model does not imply that A holds at either a or b. This is, perhaps, the source of some of the issues arising from the detachment principles considered below.

The postulates (G1) and (G2) validate the following axioms and rule, analogs of $(\Box\wedge)$, (K), and (RM).

- $(\mathbb{O}(B/A) \wedge \mathbb{O}(C/A)) \rightarrow \mathbb{O}(B \wedge C/A)$.
- $\mathbb{O}(B \rightarrow C/A) \rightarrow (\mathbb{O}(B/A) \rightarrow \mathbb{O}(C/A))$.
- $B \rightarrow C \Rightarrow \mathbb{O}(B/A) \rightarrow \mathbb{O}(C/A)$.

Analogs of (D) and (Nec)

- $\mathbb{O}(B/A) \rightarrow \neg\mathbb{O}(\neg B/A)$, and
- $B \Rightarrow \mathbb{O}(B/A)$

can be validated by adding the frame conditions

(G3) $\forall a \in K \exists b \in K(TaAb$ and $Ta^*Ab^*)$, and
(G4) for all $a,b \in K$, if $a \in N$ and $TaAb$, then $b \in N$,

respectively.

Some additional deontic principles are validated with the addition of some simple conditions. Notably, some of the axioms and rules require *model* conditions rather than *frame* conditions. The axioms

- $\mathbb{O}(A/A)$,
- $(\mathbb{O}(C/A) \wedge \mathbb{O}(C/B)) \to \mathbb{O}(C/A \vee B)$, and the rule
- $A \leftrightarrow B \Rightarrow \mathbb{O}(C/A) \leftrightarrow \mathbb{O}(C/B)$

are validated, respectively, by frames and models satisfying the following conditions, for all $a, b \in K$ and all formulas A and B,

(G5) if $a \in N$ and $TaAb$, then $b \Vdash A$,
(G6) if $TaAb$ and $TaBb$, then $Ta(A \vee B)b$, and
(G7) if $|A| = |B|$, then $TaAb$ iff $TaBb$.

Goble obtains soundness and completeness results for the basic and extended logics with respect to the class of frames or models satisfying the appropriate conditions.

Let us say that in $\mathbb{O}(C/A)$, the position occupied by A is the antecedent position and the position occupied by C is the consequent position. In the general case, the antecedent position of conditional obligation is non-congruential. Without the imposition of (G7), we need not have $\mathbb{O}(r/p \wedge q) \leftrightarrow \mathbb{O}(r/q \wedge p)$. It clearly does not respect (RM), which is good, since one might have a conditional obligation to go to the store if asked without having a conditional obligation to go to the store if asked and forcibly restrained. In general, the consequent position, however, respects (RM).

There are two further issues that Goble discusses that will interest us. The first is DETACHMENT PRINCIPLES that generate obligation from conditional obligations. For this discussion, we will work in the language $\mathfrak{L}_{\mathbb{O}^2\mathbb{O}}$, which adds to $\mathfrak{L}_{\mathbb{O}^2}$ the singulary $\mathbb{O}$ connective. The are two varieties of detachment principle, factual detachment (FD) and deontic detachment (DD).

(FD) $(\mathbb{O}(B/A) \wedge A) \to \mathbb{O}B$.
(DD) $(\mathbb{O}(B/A) \wedge \mathbb{O}A) \to \mathbb{O}B$.

These principles differ over whether one can detach a conditional obligation $\mathbb{O}(B/A)$ with the *fact* that A or with the *deontic claim* that it ought to be that A.

The frames for the language $\mathfrak{L}_{\mathbb{O}^2\mathbb{O}}$ add to the frames for $\mathfrak{L}_{\mathbb{O}^2}$ a binary accessibility relation S obeying at least the basic principle that if Sbc and $a \leq b$ then Sac. The principles (FD) and (DD) have corresponding model conditions.

(FFD) If Sab and $a \Vdash A$, then $TaAb$.
(FDD) If Sab and $b \Vdash A$, then $TaAb$.

Goble (1999, 339) glosses (FDD) saying,

[I]f b is an ideal point (from the point of view of a) and also an A-point, then it is an ideal A-point (from a's point of view). After all, one might say, ideal points, (from a's point of view) are supposed to be the absolutely best there can be, so naturally if one of them is among the points of A, then it must be among the best A-points.

Despite this intuitive description, it appears that (FD) is not sufficient for Completeness with respect to models satisfying (FDD), as another axiom is needed. By contrast, (FD) is sufficient for Completeness to the models satisfying (FFD).

Before we continue to the additional axiom that appears to be needed for Completeness for the logics with (FD), it would be worth digressing here to consider the axioms on their own merits. The discussion of these axioms is one place where it becomes important that R is under consideration rather than a weaker logic. The reason is that the non-contractive logics, such as RW, will lack the theorem (I1), $((A \rightarrow B) \wedge A) \rightarrow B$, itself a kind of detachment principle. If conditional obligation is supposed to be akin to a conditional, and the conditional of the logic rejects (I1), then it would be odd to aim for one or both of (FD) and (DD). Of course, the conditional aspect of conditional obligation could obey principles not obeyed by the general implication of the logic, but there is an intuitive harmony in keeping the two in step.

In the context of a weaker logic, one lacking (I1), it would make sense to adopt weaker principles. In particular, it would make sense to consider the rule forms of (FD) and (DD).

(RFD) $\mathbb{O}(B/A), A \Rightarrow \mathbb{O}B$.
(RDD) $\mathbb{O}(B/A), \mathbb{O}A \Rightarrow \mathbb{O}B$.

These rules, respectively, are valid on models obeying the following conditions.

(FRFD) If $a \in N$, $N \subseteq |A|$, and Sab, then $TaAb$.
(FRDD) If $a \in N$, $N \subseteq |\mathbb{O}A|$, and Sab, then $TaAb$.

One could trade the model portion of the antecedents here for a claim about provability, since theoremhood is at issue in the rules. That would make the conditions frame conditions, albeit ones defined in terms of proof systems.

That is enough digression. Let us return to the issue at hand, namely, the additional axiom needed to get completeness with (FD). The additional axiom is

(J) $\mathbb{O}(B/A) \rightarrow \mathbb{O}(A \rightarrow B)$,

which supports a model condition, which is stronger than (FDD) in the presence of the other R-frame conditions,

(FJ) if $\exists b \in K(Sab$ and $Rbcd$ and $c \Vdash A)$, then $TaAd$.

Goble notes that (J) with (K) entails (DD). Note, though, that it does so against the backdrop of R, and that entailment can fail in weaker logics.

The combination of (J), (K), and the analog of (Nec) for $\mathbb{O}(\cdot/\cdot)$ is disastrous for the relevant logician. That combination has the consequence that if B is a theorem, then $\mathbb{O}(B/A)$ is, by (Nec). With (J), this yields that $\mathbb{O}(A \rightarrow B)$ is derivable, and from (K), we get that $\mathbb{O}A \rightarrow \mathbb{O}B$ is derivable as well. The variable-sharing property is then lost.

The obvious culprit here is the analog of (Nec), which is not a desired principle even in the setting of TV. If the conditional aspect of conditional obligation is supposed to be a bona fide conditional, in the relevantist sense, then (Nec) is a close cousin of weakening. That is a Bad Thing. It is more obviously bad when (J) is in the mix, since that connections conditional obligation to the implication of the logic, albeit embedded under $\mathbb{O}$.

To end this section, we will note an additional aspect of deontic logic that we have not yet commented upon. The deontic logics we have been looking at have been focused on obligatory states or propositions, rather than obligatory actions and aspects of agency.[38] Sylvan (1992), in an exploratory work, canvasses some possibilities and difficulties for a relevant theory of action and processes. While the work covers much ground, it is fair to say that it does not settle all the questions.[39] To quote Sylvan (1992, 435),

> The conclusion has to be that we have not reached a conclusion. We have simply reached a stopping point, dictated by available time and energy. More is required than variational drill on modal elaboration. Locating satisfactory process logics lies in the future.

We agree with the spirit of what Sylvan says, that there is much more to be done concerning developing relevant logics of normative systems, and for applications of relevant modal logics more generally.

[38] Cf. Belnap (2001) and Horty (2001). One might say we have been focused on ought to be rather than ought to do.

[39] Indeed, Tedder and Bilková (2022) break new ground, building on many of Sylvan's ideas.

9 Metavaluations

9.1 Introduction

Metavaluations, introduced to relevant logics by Meyer (1976b), combine aspects of valuational semantics with provability in a logic, as we will see. One can, sometimes, define a standard, or canonical, metavaluation, and then one shows Metasoundness, that every theorem of the logic gets value 1 relative to the standard metavaluation, and Metacompleteness, that every formula that gets value 1 in the standard metavaluation is a theorem of the logic. Metacompleteness has important consequences, as we will see in Section 9.3.

Metavaluations have been used in the study of basic relevant logics, as well as modal and quantificational extensions, for example, by Mares and Meyer (1992) and Seki (2011a; 2013). Meyer and Restall (1999) use metavaluations to study arithmetic over E with an ω-rule. Brady (2014) uses metavaluations to show that some naive set theories are not only non-trivial, but also consistent. Brady (2017a) provides an excellent overview of the development and use of metavaluations, including connections to related techniques used in the context of intuitionistic logic.[1]

We begin, in Section 9.2, with a presentation of Meyer's basic metavaluations for positive logics and Slaney's metavaluations that work for De Morgan negation, along with basic Metasoundness and Metacompleteness results. In Section 9.3, we explore some of the important consequences of Metacompleteness. Following that, in Section 9.4, we look at a method for using metavaluations to show that the rule (γ) is admissible in relevant logics. Section 9.5 shows some methods of adapting the basic metavaluations to work with relevant modal logics. Meyer formulated a criticism of classical logic using metavaluations, arguing that classical logic is inadequate by its own lights, and we present and discuss this argument in Section 9.6. In Section 9.7, we present a connection between Metacompleteness and reduced frames. Finally, we close the chapter with a short coda on the rule (γ) (Section 9.8). Let us now turn to the basics of metavaluations and Metacompleteness.

[1] Brady (2017a) omits reference to the use of the Kleene slash by Božić and Došen (1984, 221) in the context of intuitionistic modal logic.

9.2 Basic Metacompleteness

The metavaluations of Meyer (1976b) work only for the positive fragments of relevant logics.[2] Slaney (1984) showed how to incorporate De Morgan negation into metavaluations, for some of the weaker logics. We will begin with Meyer's metavaluations.

DEFINITION 9.1 *A POSITIVE METAVALUATION $m(\cdot)$ for* L *is a function from $\mathfrak{L} \mapsto \{0,1\}$ such that*

- $m(p) = 0$, *for* $p \in \mathsf{At}$,
- $m(\mathsf{t}) = 1$,
- $m(A \wedge B) = 1$ *iff* $m(A) = 1$ *and* $m(B) = 1$,
- $m(A \vee B) = 1$ *iff* $m(A) = 1$ *or* $m(B) = 1$, *and*
- $m(A \rightarrow B) = 1$ *iff* $\vdash_{\mathsf{L}} A \rightarrow B$.

Meyer did provide a clause for negation, defined as $A \rightarrow \bot$, but that is not the right clause for De Morgan negation.

Meyer showed that the positive metavaluations for many logics, including many relevant logics and intuitionistic logic, evaluate all theorems as 1 and having value 1 is preserved by the rules. Classical logic does not have this feature, a point to which we will return in Section 9.6.

Slaney (1984) showed how to adjust Meyer's positive valuations to incorporate De Morgan negation. It is Slaney's metavaluations that are used most often in contemporary work. Slaney (1987a, 396–397) distinguishes two classes of logics, M1 and M2 logics, distinguished by their axioms and rules, which included some options that go beyond those found in the relevant logic family. An M1 logic is any logic extending B with zero or more of the principles from Table 9.1. An M2 logic is any logic extending B with the addition of $A \Rightarrow (A \rightarrow B) \rightarrow B$ plus zero or more of the principles from Table 9.2.[3] Of the base logics we have been interested in so far, B, DW, DJ, and TW are M1 logics, and RW is an M2 logic. The difference between these two classes matters for the definition of metavaluation.

DEFINITION 9.2 (Metavaluation) *Let* L *be an* M1 *or* M2 *logic. A METAVALUATION for* L *is a pair of functions $m(\cdot)$ and $m^{\star}(\cdot)$ from $\mathfrak{L} \mapsto \{0,1\}$ such that*

Table 9.1. Principles for M1 logics

(I2)	$((A \rightarrow B) \wedge (B \rightarrow C)) \rightarrow (A \rightarrow C)$	(I9)	$A \rightarrow (A \rightarrow A)$
(N1)	$(A \rightarrow \neg B) \rightarrow (B \rightarrow \neg A)$	(PI1)	$A \rightarrow (B \rightarrow A)$
(I3)	$(A \rightarrow B) \rightarrow ((C \rightarrow A) \rightarrow (B \rightarrow A))$	(RPI1)	$A \Rightarrow B \rightarrow A$
(I4)	$(A \rightarrow B) \rightarrow ((B \rightarrow C) \rightarrow (A \rightarrow C))$		

[2] Meyer provides a negation clause that is appropriate for intuitionistic negation, and we will return to it in Section 9.6.

[3] Slaney also permits the addition of $((A \rightarrow (B \vee C)) \wedge ((A \wedge B) \rightarrow C)) \rightarrow (A \rightarrow C)$ to M2 logics, but this does not correspond to any known frame condition. It is valid in Urquhart's semilattice models.

Table 9.2. Principles for M2 logics, adding $A \Rightarrow (A \rightarrow B) \rightarrow B$

(N1)	$(A \rightarrow \neg B) \rightarrow (B \rightarrow \neg A)$	(PI1)	$A \rightarrow (B \rightarrow A)$
(I3)	$(A \rightarrow B) \rightarrow ((C \rightarrow A) \rightarrow (B \rightarrow A))$	(I7)	$A \rightarrow ((A \rightarrow B) \rightarrow B)$
(I4)	$(A \rightarrow B) \rightarrow ((B \rightarrow C) \rightarrow (A \rightarrow C))$	(I6)	$(A \rightarrow (B \rightarrow C)) \rightarrow (B \rightarrow (A \rightarrow C))$
(I9)	$A \rightarrow (A \rightarrow A)$	(RPI1)	$A \Rightarrow B \rightarrow A$

- $m(p) = 0$, $m^\star(p) = 1$, *for* $p \in$ At.
- $m(\mathrm{t}) = 1$, $m^\star(\mathrm{t}) = 1$.
- $m(\neg A) = 1$ *iff* $m^\star(A) = 0$.
  $m^\star(\neg A) = 1$ *iff* $m(A) = 0$.
- $m(A \wedge B) = 1$ *iff* $m(A) = 1$ *and* $m(B) = 1$.
  $m^\star(A \wedge B) = 1$ *iff* $m^\star(A) = 1$ *and* $m^\star(B) = 1$.
- $m(A \vee B) = 1$ *iff* $m(A) = 1$ *or* $m(B) = 1$.
  $m^\star(A \vee B) = 1$ *iff* $m^\star(A) = 1$ *or* $m^\star(B) = 1$.
- $m(A \rightarrow B) = 1$ *iff*
  - $\vdash_{\mathsf{L}} A \rightarrow B$,
  - *if* $m(A) = 1$ *then* $m(B) = 1$, *and*
  - *if* $m^\star(A) = 1$, *then* $m^\star(B) = 1$.

  For M1 *logics,* $m^\star(A \rightarrow B) = 1$.

  For M2 *logics,* $m^\star(A \rightarrow B) = 1$ *iff* $m(A) = 1$ *only if* $m^\star(B) = 1$.

With those definitions in place, we can now proceed to prove some things about logics using metavaluations. For the two classes of logics, a formula A being provable is equivalent to $m(A) = 1$ in the metavaluation defined earlier.

LEMMA 9.3 (Metacompleteness) *Let* L *be an* M1 *logic or an* M2 *logic. For all* $A \in \mathfrak{L}$, *the following hold.*

(1) If $m(A) = 1$, *then* $\vdash_{\mathsf{L}} A$.
(2) If $m^\star(A) = 0$, *then* $\vdash_{\mathsf{L}} \neg A$.

Proof The proof is by simultaneous induction on the structure of A.

For one base case, suppose A is p. Both cases are immediate from the definition. For the other base case, suppose A is t. This case is also immediate from the definition.

For the inductive cases, suppose that the assumptions hold for B and C less complex than A.

Suppose A is $B \wedge C$. If $m(B \wedge C) = 1$, then $m(B) = 1$ and $m(C) = 1$. By the inductive hypothesis, $\vdash_{\mathsf{L}} B$ and $\vdash_{\mathsf{L}} C$, so $\vdash_{\mathsf{L}} B \wedge C$. Suppopse $m^\star(B \wedge C) = 0$. then $m^\star(B) = 0$ or $m^\star(C) = 0$. By the inductive hypothesis, $\vdash_{\mathsf{L}} \neg B$ or $\vdash_{\mathsf{L}} \neg C$, so $\vdash_{\mathsf{L}} \neg B \vee \neg C$. Therefore, $\vdash_{\mathsf{L}} \neg(B \wedge C)$.

Suppose A is $B \vee C$. This case is similar to the conjunction case.

Suppose A is $\neg B$. Suppose $m(\neg B) = 1$. Then, $m^\star(B) = 0$, so by the inductive hypothesis, $\vdash_{\mathsf{L}} \neg B$. Suppose $m^\star(\neg B) = 0$. Then $m(B) = 1$, so, by the inductive hypothesis, $\vdash_{\mathsf{L}} B$, and so $\vdash_{\mathsf{L}} \neg\neg B$.

Suppose A is $B \rightarrow C$. Suppose $m(B \rightarrow C) = 1$. It follows that $\vdash_{\mathsf{L}} B \rightarrow C$, by definition. Suppose that $m^{\star}(B \rightarrow C) = 0$. If L is an M1 logic, then this cannot happen. Let L be an M2 logic. Then, $m(B) = 1$ and $m^{\star}(C) = 0$. By the inductive hypothesis, $\vdash_{\mathsf{L}} B$ and $\vdash_{\mathsf{L}} \neg C$. From the rule $B \Rightarrow (B \rightarrow C) \rightarrow C$ of M2 logics, we have $\vdash_{\mathsf{L}} (B \rightarrow C) \rightarrow C$. Using contraposition and modus ponens, we have $\vdash_{\mathsf{L}} \neg(B \rightarrow C)$, as desired. □

LEMMA 9.4 (Metasoundness) *Let* L *be an* M1 *logic or an* M2 *logic. For all $A \in \mathfrak{L}$, if $\vdash_{\mathsf{L}} A$, then $m(A) = 1$.*

Proof The proof is by induction on the length of derivation of A in L.

The base case is made of several subcases, one for each axiom. We will only present a few subcases.

Suppose A is (A7). To show that $m(\neg\neg B \rightarrow B) = 1$, we need to establish if $m(\neg\neg B) = 1$, then $m(B) = 1$, and if $m^{\star}(\neg\neg B) = 1$, then $m^{\star}(B) = 1$. Suppose $m(\neg\neg B) = 1$. Then, $m^{\star}(\neg B) = 0$, so $m(B) = 1$, as desired. Suppose that $m^{\star}(\neg\neg B) = 1$. Then, $m(\neg B) = 0$, so $m^{\star}(B) = 1$, as desired.

Suppose A is (I2), $((B \rightarrow C) \wedge (C \rightarrow D)) \rightarrow (B \rightarrow D)$. Suppose that $m((B \rightarrow C) \wedge (C \rightarrow D)) = 1$, so $m(B \rightarrow C) = 1$ and $m(C \rightarrow D) = 1$. From the first subclause for metavaluations for $\rightarrow$, $\vdash_{\mathsf{L}} B \rightarrow C$ and $\vdash_{\mathsf{L}} C \rightarrow D$. It follows that $\vdash_{\mathsf{L}} B \rightarrow D$. Suppose $m(B) = 1$. From the assumption that $m(B \rightarrow C) = 1$, it follows that $m(C) = 1$. From the assumption that $m(C \rightarrow D) = 1$, it follows that $m(D) = 1$. The subsubcase for showing that if $m^{\star}(B) = 1$ then $m^{\star}(D) = 1$ is similar.

Suppose A is (I5), $B \rightarrow ((B \rightarrow C) \rightarrow C)$. Suppose that $m(B) = 1$. By lemma 9.3, $\vdash_{\mathsf{L}} (B \rightarrow C) \rightarrow C$. Suppose that $m(B \rightarrow C) = 1$. It follows that $m(C) = 1$. Suppose that $m^{\star}(B \rightarrow C) = 1$. Then it follows that $m^{\star}(C) = 1$. Thus, $m((B \rightarrow C) \rightarrow C) = 1$, as desired.

For the inductive cases, we need to show that the rules preserve the desired property.

Suppose that (R1) is under consideration. Assume that $m(B) = 1$ and $m(B \rightarrow C) = 1$. From $m(B \rightarrow C) = 1$, it follows that if $m(B) = 1$, then $m(C) = 1$. Therefore, $m(C) = 1$, as desired.

The remaining rules are straightforward and left to the reader. □

Putting together lemmas 9.3 and 9.4 yields the desired equivalence.

THEOREM 9.5 *Let* L *be an* M1 *logic or an* M2 *logic. For all A, $m(A) = 1$ iff $\vdash_{\mathsf{L}} A$.*

Thus, all M1 and M2 logics are Metacomplete. In the next section, we will discuss some of the consequences of Metacompleteness. Before we move on, however, it will be worth remarking on a feature of the proof of lemma 9.4.

The proof of lemma 9.4 covered a case distinctive of M1 logics and a case distinctive of M2 logics. There are some logics that we have discussed that do not fall into either category, such as C. It is worth pausing to consider where the problem lies. In the case of C, the problem lies with (I1), $(B \wedge (B \rightarrow C)) \rightarrow C$. To see the issue, suppose that $m(B \wedge (B \rightarrow C)) = 1$. From the assumption, we have $m(B) = 1$ and $m(B \rightarrow C) = 1$, which together yield $m(C) = 1$. So far so good, but let us suppose

$m^\star(B \wedge (B \to C)) = 1$. This yields $m^\star(B) = 1$ and $m^\star(B \to C) = 1$, but there is no apparent way to obtain $m^\star(C) = 1$. The M1 clause for $m^\star(B \to C) = 1$ provides us with nothing. The M2 clause would provide $m(B) = 1$ only if $m^\star(C) = 1$, but that antecedent is apparently unavailable in this subcase. Thus, the argument runs aground.

If we use Meyer's positive metavaluation, we see that $m((B \wedge (B \to C)) \to C) = 1$ easily enough. Since (I1) is an axiom of C, it is provable, which suffices for the desired conclusion. The trade-off with the positive metavaluations, however, is that while one can show that many logics with (I1) and related contraction principles are Metacomplete, one cannot combine that with the De Morgran negation principles common to relevant logics. The reason is that R, and some other strong logics, have excluded middle as a theorem for each formula A. Metacompleteness would, then, require that one of A and $\neg A$ be a theorem, for each formula. While that may hold for certain *theories*, it does not (and should not) hold for the *logic*. This issue touches upon an important consequence of Metacompleteness, to which we will now turn.

9.3 Consequences of Metacompleteness

We have shown that a range of logics are Metacomplete and Metasound. We will refer to logics with both properties, that is, those for which theorem 9.5 is provable, as Metacomplete. Metacompleteness has many important consequences. The first important consequence is that Metacomplete logics have the DISJUNCTION PROPERTY.

THEOREM 9.6 (Disjunction property) *Let* L *be Metacomplete. If* $\vdash_{\mathsf{L}} A \vee B$, *then* $\vdash_{\mathsf{L}} A$ *or* $\vdash_{\mathsf{L}} B$.

Proof Let L be Metacomplete. Suppose $\vdash_{\mathsf{L}} A \vee B$. By theorem 9.5, $m(A \vee B) = 1$, so $m(A) = 1$ or $m(B) = 1$. By theorem 9.5 again, we obtain the desired conclusion. □

We can see a reason why extending Meyer's positive metavaluations to include De Morgan negation will be impossible for stronger logics like R. Metacompleteness implies the disjunction property. R has excluded middle, $p \vee \neg p$, as a theorem, but it does not have p or $\neg p$ as theorems. To be sure, there will be prime R-theories with one or the other disjunct, but those non-logical theories properly extend the logic.

The two classes of logics have different features that come out as consequences of Metacompleteness, pointed out by Slaney (1984).

THEOREM 9.7 *Let* L *be an* M1 *logic. Then* L *has no theorems of the form* $\neg(A \to B)$.

Proof Let L be an M1 logic. Suppose L has a theorem of the form $\neg(A \to B)$. Then $m(\neg(A \to B)) = 1$, so $m^\star(A \to B) = 0$. From the definition of M1 metavaluations, this is impossible. □

THEOREM 9.8 *Let* L *be an* M2 *logic. Then* $\vdash_{\mathsf{L}} \neg(A \to B)$ *iff* $\vdash_{\mathsf{L}} A$ *and* $\vdash_{\mathsf{L}} \neg B$.

Proof The proof is left as an exercise for the reader. □

Theorem 9.8 may appear surprising, since it seems in tension with the variable-sharing property. After all, for any two theorems, say $p \to p$ and $\neg\neg(q \to q)$, an M2 logic will have $\neg((p \to p) \to \neg(q \to q))$ as a theorem. The tension, however, is merely apparent. The reason is that a negated implication is not, on its own, an implication.[4] The variable-sharing property holds only for implications, not for formulas claiming a failure of implication. Indeed, it would be strange to require that antecedent and consequent have a substantive meaning connection in an implication that is being claimed not to hold. Negated implications do not feature in any of the standard axiomatizations of relevant logics, and they generally are less studied in the context of relevant logics.[5]

While the fact about negated implications in M2 logics does not violate variable sharing, it is perhaps too close to violating the spirit. M1 logics lack negated implications as theorems, so the problem is avoided for those logics.

9.4 Metacompleteness and (γ)

For a given logic L, L is itself an L-theory. Metavaluations can be adapted to work for theories in general, rather than just logics. For this section, we will drop t from the language, calling the result $\mathfrak{L}^-$.

We will start with adapting Meyer's metavaluations for R. Let T be an R-theory. Following Meyer (1976a), we use the following clauses.

- For $p \in \mathsf{At}$, $m(p) = 1$ iff $p \in T$.
- $m(A \wedge B) = 1$ iff $m(A) = 1$ and $m(B) = 1$.
- $m(A \vee B) = 1$ iff $m(A) = 1$ or $m(B) = 1$.
- $m(\neg A) = 1$ iff $m(A) = 0$ and $\neg A \in T$.
- $m(A \to B) = 1$ iff $A \to B \in T$ and if $m(A) = 1$, then $m(B) = 1$.

Let $m(T) = \{A \in T : m(A) = 1\}$. This set is the METATHEORY of T.[6]

We will now prove some lemmas about the metavaluations on R-theories.

LEMMA 9.9 *Let T be an* R*-theory. Then $m(T) \subseteq T$.*

Proof The proof is by induction on the structure of an arbitrary formula in $m(T)$. The base case is immediate by definition. For the inductive cases, the inductive hypothesis settles the conjunction and disjunction cases. The membership clauses settle the negation and implication cases. □

[4] There is a caveat that in RW, one can define negation as $\to f$. This makes the negated implication itself an implication, although f, being the negation of t, brings with it the same trivialization of the variable-sharing property.

[5] Negated implications are important in connexive logics, which generally do not feature in this book. In connexive logics, the axioms $(A \to \neg B) \to \neg(A \to B)$, $\neg(A \to \neg A)$, and $\neg(\neg A \to A)$ are important. For more on connexive logics, see Omori and Wansing (2019), Francez (2019), and Weiss (2022b), among others.

[6] This sense of "metatheory" is, of course, different from the sense deployed in discussions of Soundness and Completeness.

It is immediate from the definition of a metatheory that it is prime, even when T is not. Further, the metatheory of T is consistent, even when T is not.

LEMMA 9.10 *Let T be an* R*-theory. If* $\neg A \in m(T)$, *then* $A \notin m(T)$.

Proof Suppose $\neg A \in m(T)$. Then $m(A) = 0$, so $A \notin m(T)$. □

LEMMA 9.11 *Let T be a prime, regular* R*-theory. If* $A \notin T$, *then* $\neg A \in T$.

Proof $A \vee \neg A \in T$, as T is regular. Since T is prime, either $A \in T$ or $\neg A \in T$. Suppose $A \notin T$, so $\neg A \in T$. □

LEMMA 9.12 *Let T be a prime, regular* R*-theory. If* $A \notin m(T)$, *then* $\neg A \in T$.

Proof The proof is by induction on the structure of A.

Let A be p. Suppose $p \notin m(T)$. By definition, $p \notin T$. By lemma 9.11, $\neg p \in T$.

Let A be $B \wedge C$. Suppose $B \wedge C \notin m(T)$. So either $B \notin m(T)$ or $C \notin m(T)$. By the inductive hypothesis, $\neg B \in T$ or $\neg C \in T$, whence $\neg B \vee \neg C \in T$. It follows that $\neg(B \wedge C) \in T$.

Let A be $B \vee C$. Suppose $B \vee C \notin m(T)$. So, $B \notin m(T)$ and $C \notin m(T)$. By the inductive hypothesis, $\neg B \in T$ and $\neg C \in T$. It follows that $\neg(B \vee C) \in T$.

Let A be $\neg B$. Suppose $\neg B \notin m(T)$. Either $\neg B \notin T$ or $B \in m(T)$. In the former case, by lemma 9.11, $\neg\neg B \in T$. In the latter case, by lemma 9.9, $B \in T$, so $\neg\neg B \in T$.

Let A be $B \rightarrow C$. Suppose $B \rightarrow C \notin m(T)$. Either $B \rightarrow C \notin T$ or both $B \in m(T)$ and $C \notin m(T)$. In the former case, lemma 9.11 yields $\neg(B \rightarrow C) \in T$. In the latter case, the inductive hypothesis with lemma 9.9 gives $B \in T$ and $\neg C \in T$. Since $\vdash_R (B \rightarrow C) \rightarrow (\neg B \vee C)$, $\neg(B \rightarrow C) \in T$. □

COROLLARY 9.13 *Let T be a prime, regular* R*-theory. If* $A \notin m(T)$, *then* $\neg A \in m(T)$.

Proof Suppose $A \notin m(T)$. Then $\neg A \in T$, by the preceding lemma. It follows from the definition of metavaluation that $\neg A \in m(T)$. □

COROLLARY 9.14 *Let T be a prime, regular* R*-theory.* $A \notin m(T)$ *iff* $\neg A \in m(T)$.

Finally, we show that $m(T)$ is a theory under certain conditions.

LEMMA 9.15 *Let T be a regular, prime* R*-theory. Then* $m(T)$ *is a regular, prime, consistent* R*-theory.*

Proof Let T be a regular, prime R-theory. Primeness and consistency have been established.

We will show that $m(T)$ is closed under detachment, which is to say that if $B \in m(T)$ and $B \rightarrow C \in m(T)$, then $C \in m(T)$. Suppose $B \in m(T)$ and $B \rightarrow C \in m(T)$. The latter yields that if $B \in m(T)$ then $C \in m(T)$, as desired. $m(T)$ is closed under (R2) by the conjunction clause of metavaluation.

Regularity is proven by showing that all the axioms of R receive value 1, having shown that the rules preserve value 1. There are many cases to check. We will do two.

To see that $m(B \to ((B \to C) \to C)) = 1$, suppose $m(B) = 1$. We need to show that $m((B \to C) \to C) = 1$. By lemma 9.9, $B \in T$, so $(B \to C) \to C \in T$. Suppose that $m(B \to C) = 1$. Since $m(T)$ is closed under detachment, $m(C) = 1$, establishing that $m((B \to C) \to C) = 1$. Since T is regular and $B \to ((B \to C) \to C)$ is an axiom, $B \to ((B \to C) \to C) \in T$.

To see that $m(\neg\neg B \to B) = 1$. $\neg\neg B \to B \in T$, by regularity of T. Suppose $m(\neg\neg B) = 1$. By corollary 9.14 twice, $m(B) = 1$, which suffices for the desired conclusion.

That is enough axiom cases. The remainder are left as an exercise.

Finally, to establish that $m(T)$ is an R-theory, we need closure under provable implication of R. This is done by induction on the length of the shortest proof of an R-theorem. Since all axioms get value 1 and the rules of R preserve value 1, all theorems get value 1. Thus, $m(T)$ is a regular R-theory. □

The upshot of these lemmas is that given a regular, prime R-theory T that does not contain some non-theorem A, we can find a regular, prime, consistent R-theory $m(T)$ that does not contain A. This is the basis for the NORMALIZATION THEOREM, which says that there is a reduced model whose base point 0 is a representative of $m(T)$, of the sort used in the canonical model construction, such that A fails there. The consistency ensures that, given the other conditions on R-frames, $0 = 0^*$. This is to say that at the base point, negation is Boolean negation, which is to say that a negation holds at the base point just in case the negated formula fails there.

The normalization theorem provides a reduced frame, called a normal frame, that can be used in a model-based proof of (γ) admissibility for R, which is another important application of metavaluations. The name "(γ)" comes from Ackermann (1956), who presented the precursors of Anderson and Belnap's logic E. Ackermann used the rule (γ), $A, \neg A \vee B \Rightarrow B$. Anderson and Belnap did not include it in their axiomatization of E, and it was, for a long time, an open question whether E admitted that rule. The question of whether a logic or theory admits (γ) is an important one that has driven much interesting work in relevant logics and it is one that we will return to in Section 9.8.

Of the basic relevant logics that we have looked at, many have (γ) admissible. Perhaps the simplest example of a logic in which (γ) is not admissible is the logic of CL.[7] In that logic, $\neg(p \to p) \vee (\neg(p \to p) \to q)$ is valid, as is $p \to p$, but $\neg(p \to p) \to q$ is not, as the logic of CL satisfies the variable-sharing property. Perhaps the most well known failure of (γ) is in the logic Lattice-R, which is R without the distribution axiom (A6).[8] In that logic, the failure of (γ) admissibility is closely related to the lack of the distribution axiom.

A normal frame has the feature that for the base point 0, $0 = 0^*$. The normal frame base point $m(T)$ is supposed to witness the failure of (γ) for R, for some theorems A and $\neg A \vee B$. Suppose $0 \Vdash A$, $0 \Vdash \neg A \vee B$, but $0 \nVdash B$. It follows that $0 \Vdash \neg A$ or $0 \Vdash B$. The latter option is not available by assumption, so $0 \Vdash \neg A$. This implies

[7] I thank Ethan Brauer for raising the question of the (γ) admissibility for CL to me.
[8] Thistlewaite et al. (1988) demonstrate the failure of (γ) admissibility in Lattice-R.

$0^* \not\Vdash A$, which contradicts an assumption. Thus, this is not actually a counterexample, so (γ) is admissible in R.[9]

An alternative metavaluation argument settles the question for R.

THEOREM 9.16 ((γ) admissibility for R) *If* $\vdash_R A$ *and* $\vdash_R \neg A \vee B$, *then* $\vdash_R B$.

Proof This proof follows that of Restall (2000, 101). Suppose $\vdash_R A$ and $\vdash_R \neg A \vee B$, but $\nvdash_R B$. Then $\langle \mathsf{R}, \{B\}\rangle$ is an R-disjoint pair. There is, then, a prime theory T such that $B \notin T$. By lemma 9.15, $m(T)$ is prime, regular, and consistent. Since $\neg A \vee B \in m(T)$ and $m(T)$ is prime, $\neg A \in m(T)$ or $B \in m(T)$. By the consistency of $m(T)$, $A \in m(T)$ only if $\neg A \notin m(T)$. It follows that $B \in m(T)$, so $B \in T$, which is a contradiction. Therefore, $\vdash_R B$. □

The proofs given earlier depended on features of R and R-theories not shared by weaker logics. In particular, they made heavy use of the fact that excluded middle is a theorem. Nonetheless, the normalization techniques can be extended to many of the weaker logics, such as RW or DJ.

The metavaluations given earlier were for R-theories. Slaney (1987a) showed how to adapt metavalautions to theories in logics that lack (I1). The definition of metavaluation is tweaked, in much the way that Meyer's metavaluations were adjusted. We will return to this work in Section 9.7.

Using Slaney's metavaluations, we can show that (γ) is admissible for all M1 and M2 logics.

THEOREM 9.17 ((γ) admissibility for M1 and M2 logics) *Suppose* L *is an* M1 *logic or an* M2 *logic. Then, if* $\vdash_{\mathsf{L}} A$ *and* $\vdash_{\mathsf{L}} \neg A \vee B$, *then* $\vdash_{\mathsf{L}} B$.

Proof Suppose $\vdash_{\mathsf{L}} A$ and $\vdash_{\mathsf{L}} \neg A \vee B$. By theorem 9.6, $\vdash_{\mathsf{L}} \neg A$ or $\vdash_{\mathsf{L}} B$. Since L is a sublogic of TV, we know that $\nvdash_L A \wedge \neg A$, so $\nvdash_L \neg A$. Therefore, $\vdash_{\mathsf{L}} B$. □

A generalization of this theorem is left as an exercise at the end of the chapter.

9.5 Modal Metacompleteness

Metavaluations can be extended to modal logics as well. This is, perhaps, a misleading description, as some of Meyer's early work (1971b) on metavaluations considered modal logics, including E and NR. Meyer used the clause

- $m(\Box B) = 1$ iff $\vdash_{\mathsf{L}} \Box B$ and $m(B) = 1$,

showing that for a range of modal logics L, all theorems of L get value 1. We will see how the notion of metavaluation has been used in the context of relevant modal logics in this section.

Mares and Meyer (1992) use metavaluations to demonstrate the admissibility of (γ) in the logic R4. They define the canonical model for R4, using as K the set of prime R4-theories. For $T_i, T_j \in K$, the canonical accessibility relation is defined as

- ST_iT_j iff $T_i^{-\Box} \subseteq T_j \subseteq T_i^{-\Diamond}$.

[9] The meta-theoretic use of (γ) is a sticking point to some relevant logicians.

They define, for each $T_i \in K$, a metavaluation $v_i(\cdot)$ as follows.

- $v_i(p) = 1$ iff $p \in T_i$, for $p \in$ At.
- $v_i(B \wedge C) = 1$ iff $v_i(B) = 1$ and $v_i(C) = 1$.
- $v_i(B \vee C) = 1$ iff $v_i(B) = 1$ or $v_i(C) = 1$.
- $v_i(B \rightarrow C) = 1$ iff $B \rightarrow C \in T_i$ and if $v_i(B) = 1$ then $v_i(C) = 1$.
- $v_i(\neg B) = 1$ iff $\neg B \in T_i$ and $v_i(B) = 0$.
- $v_i(\Box B) = 1$ iff $\forall j(ST_iT_j \Rightarrow v_j(B) = 1)$.

They define $Tr_i = \{A \in \mathfrak{L}_\Box : v_i(A) = 1\}$. They prove that each Tr_i is a prime, regular, consistent R4-theory. So, every prime, regular R4-theory contains a prime, regular, consistent R4-subtheory. Using these subtheories, they are show that (γ) is admissible for R4.

Seki (2011a) uses a similar technique of giving a metavaluation clause for necessity at a prime, regular theory in the canonical frame for a given logic, although the clauses he initially uses differ from those of Mares and Meyer (1992).

- $v_s(\Box B) = 1$ iff $\Box B \in s$ and $v_s(B) = 1$.
- $v_s(\blacklozenge B) = 1$ iff $\blacklozenge B \in s^*$ or $v_s(B) = 1$.

He shows that for a range of logics extending BX.M plus some disjunctive rules, (γ) is admissible. While Seki obtains Metacompleteness results for logics with excluded middle, they do not have all the usual consequences, such as the disjunction property. Seki goes on to define alternative metavaluations for a range of logics extending BX.M plus some disjunctive rules and (MC) and $(\blacksquare A \wedge \blacklozenge B) \rightarrow \blacklozenge(A \wedge B)$. These alternative metavaluations have the same clauses as those of Mares and Meyer (1992), using an accessibility relation on theories in the clauses rather than membership in a particular theory.

Seki (2013) extends Slaney's metavaluations to $\mathfrak{L}_{\Box\blacklozenge}$. We will focus here on the $\mathfrak{L}_\Box$ extension. In addition to the M1 and M2 distinction, we need to distinguish two classes of modal extensions.

DEFINITION 9.18 (M1s, M2s, M1t, M2t) *An* M1s *or* M2s *logic is an* M1 *logic or an* M2 *logic, respectively, with the addition of* $(\Box\wedge)$ *and* (RM) *as well as zero or more of the following.*

- (K) $\Box(A \rightarrow B) \rightarrow (\Box A \rightarrow \Box B)$.
- (Nec) $A \Rightarrow \Box A$.
- (K$\Diamond$) $\Box(A \rightarrow B) \rightarrow (\Diamond A \rightarrow \Diamond B)$.
- (CoNec) $\Box A \Rightarrow A$.

We will use Ms *logic to refer to an* M1s *or* M2s *logic when the underlying base logic is not the focus.*

An M1t *or* M2t *logic is an* M1 *logic or an* M2 *logic, respectively, plus* $(\Box\wedge)$, (RM), *and the rule* $A \Rightarrow \Diamond A$ *as well as any selection of the modal principles for an* Ms *logic.*

We will use Mt *logic to refer to an* M1t *or* M2t *logic when the underlying base logic is not the focus.*

Metavaluations extend to the modal language with the following clauses.

- $m(\Box A) = 1$ iff $\vdash_{\mathsf{L}} \Box A$ and $m(A) = 1$.
  For Ms logics, $m^\star(\Box A) = 1$.
  For Mt logics, $m^\star(\Box A) = 1$ iff $m^\star(A) = 1$.
- For Ms logics, $m(\blacklozenge A) = 0$.
  For Mt logics, $m(\blacklozenge A) = 1$ iff $m(A) = 1$.
  $m^\star(\blacklozenge A) = 1$ iff $\nvdash_{\mathsf{L}} \neg\blacklozenge A$ or $m^\star(A) = 1$.
- For Ms logics, $m(\Diamond A) = 0$.
  For Mt logics, $m(\Diamond A) = 1$ iff $m(A) = 1$.
  $m^\star(\Diamond A) = 1$ iff $\nvdash_{\mathsf{L}} \neg\Diamond A$ or $m^\star(A) = 1$.

For the $m^\star(\cdot)$ clauses for the conditional, M1s and M1t logics use the M1 clause, and M2s and M2t logics use the M2 clause. The clauses for $\blacklozenge$, it turns out, coincide with the clauses for $\Diamond$, using the latter's standard definition of $\neg\Box\neg$. These clauses are, in fact, lemmas, but the arguments are straightforward.

Seki shows that the Ms and Mt logics are Metacomplete. He goes on to show that Metacompleteness holds when the logics are extended with axioms from four classes of modal axioms. The Ms and Mt logics differ on which classes of axioms can be added preserving Metacompleteness. Of the modal axioms that were highlighted in the previous chapters, (4) and (5) can be added to Ms logics preserving Metacompleteness, and (T), (D), (4), (5), and (B) can be added to Mt logics preserving Metacompleteness. Of course, (K) and (Nec) are included in the definitions of Ms and Mt logics. We will present proofs of these facts below, beginning with Metacompleteness.

THEOREM 9.19 *Let* L *be an* Ms *or* Mt *logic. Then the following hold.*

- *If* $m(A) = 1$, *then* $\vdash_{\mathsf{L}} A$.
- *If* $m^\star(A) = 0$, *then* $\vdash_{\mathsf{L}} \neg A$.

Proof The proof is by simultaneous induction on the structure of A, and it adds two new cases to the earlier proof.

Suppose A is $\Box B$. Suppose $m(\Box B) = 1$. Then $\vdash_{\mathsf{L}} \Box B$, by definition. Suppose that $m^\star(\Box B) = 0$. If L is an Ms logic, this can't happen. Suppose that L is an Mt logic. Then $m^\star(B) = 0$, so $\vdash_{\mathsf{L}} \neg B$, by the inductive hypothesis. It follows that $\vdash_{\mathsf{L}} \neg\Box\neg\neg B$, from the definition of Mt logic, so $\vdash_{\mathsf{L}} \neg\Box B$, as desired.

Suppose A is $\blacklozenge B$. Suppose $m(\blacklozenge B) = 1$. This is impossible if L is an Ms logic. Let L be an Mt logic. Then $m(B) = 1$, so $\vdash_{\mathsf{L}} B$, by the inductive hypothesis. From the definition of Mt logic, it follows that $\vdash_{\mathsf{L}} \blacklozenge B$, as desired. Next, suppose that $m^\star(\blacklozenge B) = 0$. Then by definition, $\vdash_{\mathsf{L}} \neg\blacklozenge B$. □

The second case, while strictly speaking not needed given the stated focus of the section, is instructive and covers the case of $\Diamond$, since the clauses for the two connectives are the same.

THEOREM 9.20 *Suppose* L *is an* M1 *or* M2 *logic. If* L.MZ *is an* Ms *or* Mt *extension of* L, *then if* $\vdash_{\mathsf{L.MZ}} A$, *then* $m(A) = 1$.

Proof As in the proof for the non-modal case, the proof is by induction on the length of proof. We will present only the case for the (K) axiom here.

Assume (K) is in L.MZ. Suppose that $m(\Box(A \to B)) = 1$. From this assumption, $\vdash_{\mathsf{L.MZ}} \Box(A \to B)$. It follows that $\vdash_{\mathsf{L.MZ}} \Box A \to \Box B$. We need to show that $m(\Box A \to \Box B) = 1$. For this, assume $m(\Box A) = 1$, so $\vdash_{\mathsf{L.MZ}} \Box A$, and so $\vdash_{\mathsf{L.MZ}} \Box B$. From the initial assumption, $m(A \to B) = 1$, and by the assumption that $m(\Box A) = 1$, $m(A) = 1$. It follows that $m(B) = 1$, so $m(\Box B) = 1$.

Next, suppose $m^\star(\Box A) = 1$. If L.MZ is an Ms logic, then $m^\star(\Box B) = 1$. If L.MZ is an Mt logic, then $m^\star(A) = 1$. From $m(A \to B) = 1$, it follows that $m^\star(B) = 1$, whence $m^\star(\Box B) = 1$.

Next, suppose $m^\star(\Box(A \to B)) = 1$. If L.MZ is an M1s or M1t logic, then $m^\star(\Box A \to \Box B) = 1$ by the M1 clause for the conditional. If L.MZ is an M2s logic, then assume $m(\Box A) = 1$. By the Ms clause for $\Box$, $m^\star(\Box B) = 1$. If L.MZ is an M2t logic, assume $m(\Box A) = 1$. From $m^\star(\Box(A \to B)) = 1$, we have $m^\star(A \to B)$, so $m(A) = 1$ only if $m^\star(B) = 1$. From $m(\Box A) = 1$ it follows that $m(A) = 1$, so $m^\star(B) = 1$, whence $m^\star(\Box B) = 1$, as desired. □

Finally, we will prove the claims regarding extensions with some common modal axioms, mentioned earlier.

THEOREM 9.21 *Suppose* L *is an* M1 *or* M2 *logic. Then the following hold.*

(i) If L.MZ *is an* Ms *or* Mt *extension of* L *with* (4), *then* $m(\Box A \to \Box\Box A) = 1$.
(ii) If L.MZ *is an* Ms *or* Mt *extension of* L *with* (5), *then* $m(\Diamond A \to \Box\Diamond A) = 1$.
(iii) If L.MZ *is an* Mt *extension of* L *with* (T), *then* $m(\Box A \to A) = 1$.
(iv) If L.MZ *is an* Mt *extension of* L *with* (D), *then* $m(\Box A \to \Diamond A) = 1$.
(v) If L.MZ *is an* Mt *extension of* L *with* (B), *then* $m(A \to \Box\Diamond A) = 1$.

Proof Suppose L is an M1 or M2 logic. For each case, the provability clause for the conditional is satisfied, so we will omit reference to it.

For (i), we need to show that $m(\Box A \to \Box\Box A) = 1$. Assume $m(\Box A) = 1$. By theorem 9.19, $\vdash_{\mathsf{L.MZ}} \Box A$, so $\vdash_{\mathsf{L.MZ}} \Box\Box A$. Since $m(\Box A) = 1$, it follows that $m(\Box\Box A) = 1$. Next, for the Ms case, we have that $m^\star(\Box\Box A) = 1$. For the Mt case, suppose that $m^\star(\Box A) = 1$. It follows that $m^\star(\Box\Box A) = 1$, as desired.

Cases (ii)–(v) are left as exercises for the reader. □

The extant work on metavaluations for relevant modal logics reveals two things. One is that Metacompleteness for the modal extensions does not depend greatly on the choice of base logic, provided that the base logic has an appropriate Metacompleteness result. We can see this, for example, in Seki's (2013) work, where the M1 and M2 distinction does not play a major role. The other thing that the work reveals is that there are many subtleties to metcompleteness for modal extensions, especially when one is considering classes of axioms in the language $\mathfrak{L}_{\Box\blacklozenge}$. It is notable that the clauses for metavaluations used by Mares and Meyer and by Seki for logics with (MC) use an accessibility relation on theories. It appears to be an open question to what extent that technique is useful for logics that do not contain (MC).

9.6 Meyer's Metavaluational Criticism of Classical Logic

Meyer (1976b) presents a criticism of TV as an account of entailment. The criticism uses metavaluations. Let us say a logic is COHERENT when all theorems get value 1 on all metavaluations for the logic. Meyer (1976b, 501) glosses coherence by saying "a logic was called *coherent* provided that it could plausibly be interpreted in its own metalogic." Meyer motivates the CANONICAL METAVALUATION, that of definition 9.1, by saying that a logic is a logic of entailment when the true entailments, formulas of the form $A \rightarrow B$, are just those provable in the logic. If this is not the case, then the "$\rightarrow$" connective does not *mean* entailment, although it might yet *indicate* entailment.

On the canonical metavaluation, we have the following evaluations of theorems of TV.

(C1) $m(((p \rightarrow q) \rightarrow p) \rightarrow p) = 1$.
(C2) $m([((p \rightarrow q) \rightarrow p) \rightarrow p] \rightarrow [p \vee (p \rightarrow q)]) = 1$.
(C3) $m(p \vee (p \rightarrow q)) = 0$.

The evaluation of (C3) can be seen by noting that $m(p) = 0$ and, as $p \rightarrow q$ is not provable, $m(p \rightarrow q) = 0$. Thus, (C3) is invalid according to metavaluations. Consequently, (R1) does not preserve truth, and so the classical implication connective does not mean entailment.

Meyer presents this result as showing that TV is not truth-functional enough, since there are true implications with true antecedents and false consequents. This rings hollow, however, as it is not clear that the canonical metavaluation is a good arbiter of truth-functionality. After all, the clause for the implication is, plausibly, intensional, and the values of the antecedent and consequent do not figure in the evaluation of the whole.

Meyer offers an alternative clause for the conditional that adds a bit of value-functionality.

- $m(A \rightarrow B) = 1$ iff $\vdash_{\mathsf{L}} A \rightarrow B$ and if $m(A) = 1$ then $m(B) = 1$.

He calls this the CANONICAL QUASI-VALUATION. On the canonical quasi-valuation, we still have (C1) come out as 1 and (C3) come out as 0. Because of this, (C2) is evaluated as 0. The validity of (R1) is preserved, but classical theorems whose main connective is "$\rightarrow$" come out as false. The putative entailments of TV are not even all true. By contrast, the positive fragments of intuitionistic logic, many strict conditional logics, R, E, and T are Metacomplete, and so their metavaluations will evaluate their own entailments as true and their rules as truth-preserving.

Meyer's result highlights the tension between the truth-functional, disjunctive nature of the classical material conditional and the non-truth-functional nature of entailment. The metavaluation clause for disjunction is value-functional. In contrast, neither of the clauses for the implication are, as both include a conjunct requiring provability in the logic. It is those clauses that capture the idea that the "$\rightarrow$" connective means entailment, and it is those clauses that are, arguably, what make metavaluations distinctive.

What happens if we evaluate, not (C3), but the result of replacing "$\rightarrow$" with the usual definition of the material conditional? For that, we need the metavaluation clause for negation. Meyer's quasi-valuation clause for negation has a similar intensional character, with the canonical metavaluation clause merely dropping the second conjunct.

- $m(\neg A) = 1$ iff $\vdash_{\mathsf{L}} \neg A$ and $m(A) = 0$.

We can see that $m(p \vee (\neg p \vee q)) = 0$, since $m(p) = 0$, $m(q) = 0$, and $m(\neg p) = 0$, since $\neg p$ is not a theorem of TV. This witnesses a failure of excluded middle for these metavaluations.

Meyer's metavaluations draw a line between the positive fragment of TV and many other logics. The material conditional of TV cannot really mean *entailment*. Many of the (positive fragments of) logics Meyer shows are Metacomplete exhibit fallacies of relevance that would, from the relevantist point of view, disqualify their implications from meaning entailment. Being Metacomplete can, at best, be a necessary condition on a logic for its implication to mean entailment. The search for sufficient conditions seems to require looking elsewhere than metavaluations.

The metavaluational argument presented has its limitations. The metavaluations do not extend to relevant logics with De Morgan negation. As noted earlier, it appears impossible to extend metavaluations to include negation in a way that makes R Metacomplete, since each instance of excluded middle would need to have a disjunct come out with value 1, which would render one or the other disjunct provable. The lack of metavaluations for R with De Morgan negation need not present the same problems for the proponent of R as Meyer's metavaluational points present for the proponent of TV, but there is some tension. If one is truly moved by Meyer's point, then it is plausible that one should want the logic *with negation* to be Metacomplete. Such a desire would push one to some of the weaker, non-contractive logics, which are Metacomplete. Those logics can be interpreted in their own metalogics, in the metavaluational sense.

9.7 Metacompleteness and Reduced Frames

Slaney (1987a) showed that some of the contraction-free logics are complete with respect to their reduced frames. In this section, we will present a simple, albeit limited, method of proving Completeness with respect to reduced frames that uses metavaluations.[10] Say that a frame condition is REDUCIBLE iff it uses only $R, {}^*, \leq$, terms for points, existential quantifiers, universal quantifiers, conjunction, disjunction, and the material conditional. In particular, a frame condition is not reducible if it uses a restricted quantification over N that is not used in an instance of $\leq$. Of the basic conditions on frames, only condition (iii), the upward closure of N under $\leq$, is not reducible, but it holds in all reduced frames. Of the additional conditions listed earlier, (FI8), (FN3), and (FNec) are not reducible.

[10] Standefer (2022a) presents the details of this method in more detail.

LEMMA 9.22 *Suppose* $\langle K, N, R, ^* \rangle$ *is an* L*-frame, all the conditions on* L*-frames are reducible, and there is* $0 \in N$ *such that* $0 \leq a$*, for all* $a \in N$*. Then* $\langle K, N, R, ^* \rangle$ *is a reduced* L*-frame.*

Proof All the conditions on B-frames are satisfied in virtue of being an L-frame. Further, the additional frame conditions hold as the frame is unchanged. Therefore, it is a reduced L-frame. □

THEOREM 9.23 *If* L *is prime and* L *is Complete with respect to* L*-frames, then* L *is Complete with respect to reduced* L*-frames.*

Proof Suppose L is prime and L is Complete with respect to L-frames. Then, the canonical L-frame $\langle K, N, R, ^* \rangle$ is an L-frame and the canonical L-model is a model, by the results of the previous section. Since L is prime, $\mathsf{L} \in K$. In fact, $\mathsf{L} \in N$, and for all $a \in N$, $\mathsf{L} \subseteq a$, so $\mathsf{L} \leq a$. Thus, the canonical L-frame is reduced, with L acting as 0, and the L-frame conditions all hold. □

While the condition (FNec) is not reducible, we can nonetheless use it in the context of reduced frames and maintain Completeness. There is an alternative condition that does not use N but, nonetheless, is equivalent to (FNec): (FrNec) if $0 \leq a$ and Sab then $0 \leq b$. In reduced frames, as we have defined them, $N = \{a \in K : 0 \leq a\}$, (FrNec) is (FNec) with the "$\in N$" portions rewritten using the definition just given. We can then extend the previous theorem to include logics closed under (Nec).

Using metavaluational techniques, we can see that all M1 and M2 logics are complete with respect to their reduced frames. Further, we can see that many Ms and Mt logics are complete with respect to their reduced frames. The latter point is interesting because it answers an open question from Standefer (2021b). After showing that many relevant modal logics were incomplete with respect to their reduced frames, the question was raised whether any relevant modal logics are complete with respect to their reduced frames. The answer is that many logics are. One can extend many of the non-contractive logics with standard modal principles and obtain Completeness with respect to reduced frames. One can even extend the logics with some non-standard modal principles and maintain Completeness, as a consequence of Seki's Metacompleteness results together with his Sahlqvist Completeness results.

This method of proof is somewhat limited, since it only works for relevant modal logics whose base logic is an M1 or M2 logic. Further, the metavaluational method, using Slaney-style metavaluations, does not work for logics with (MC). For those logics, a different technique would be needed, if proof is even possible.

Unfortunately, the addition of (MC) to a logic with the disjunction property can result in a logic that lacks the disjunction property. As an example, take RW.MNec. This is an M2s logic, so it has the disjunction property. The addition of (MC) results in a logic that lacks the disjunction property. It is clear that $\vdash_{RW.MNec} q \vee (p \rightarrow p)$, whence by (Nec), $\vdash_{RW.MNec} \Box(q \vee (p \rightarrow p))$. Using (MC) results in $\Box q \vee \Diamond(p \rightarrow p)$ being a theorem, but neither disjunct is provable. This can be shown using two-element algebras, one of which is shown in Table 9.3.[11] The values are in the set $\{0, 1\}$, with

[11] Both were generated by MaGIC.

Table 9.3. Counterexample to disjunction property

$\to$	0	1	$\neg$	$\Box$	$\Diamond$
0	1	1	1	1	0
1	0	1	0	1	0

1 being the sole designated value, and they are ordered $0 < 1$. For a counterexample to $\Diamond(p \to p)$, set $v(p) = 0$, then $v(\Diamond(p \to p)) = 0$. For a counterexample to the provability of the other disjunct, use the identity interpretation of the modal operators. On that interpretation, $v(q) = v(\Box q) = 0$, so the second disjunct is not provable. Thus, we see that the addition of (MC) can upset the disjunction property.

This method of proving Completeness is quick and comparatively easy, requiring nothing essentially new beyond what was used to prove Completeness with respect to unreduced frames. Slaney (1987a) proved a direct Completeness result for many Metacomplete logics with respect to their reduced frames. Part of the trick of obtaining an appropriate prime theory is identifying the right sort of theory to input into the metavaluation, given that axiom (I1), which was thought to be key to maintaining the validity of (R1) in reduced frames, is absent. For this, we need to define a few kinds of theories, highlighted by Slaney (1987a) and Giambrone (1992).[12]

DEFINITION 9.24 (Varieties of theories) *Let T be an* L*-theory.*

T is DETACHED iff if $A \to B \in T$ and $A \in T$, then $B \in T$.

T is AFFIXED iff if $A \to B \in T$, then both $(C \to A) \to (C \to B) \in T$ and $(B \to C) \to (A \to C) \in T$.

T is TRANSPOSITIVE iff if $A \to B \in T$, then $\neg B \to \neg A \in T$.

T is STABLE iff it is detached, affixed, and transpositive.

T is CLOSED UNDER A RULE iff whenever the premises of the rule are in T, the conclusion is in T.

T is RULE REGULAR iff it is L*-regular and closed under the primitive rules of* L.

T is ORDINARY iff it is rule regular and stable.

T is NORMAL iff it is ordinary, prime, and consistent.

We need these definitions of different theories because, when one moves to weaker logics with more primitive rules, one needs to ensure that theories have the appropriate closure conditions. In particular, if T is detached, then (R1) will hold, and if T is affixed, then (R4) and (R5) will hold. One can see that the the Completeness proof with respect to reduced frames for C works because the axioms in C, namely (I1), (I3), (I4), and (N1), are sufficient to ensure that C-theories are stable.

Slaney (1987a) shows that M1 and M2 logics are complete with respect to their reduced frames. Given a non-theorem A of a logic L, Slaney constructs an ordinary theory T that does not contain L. This theory, however, may not be prime, which is where the metavaluations come in. If L is an M1 or M2 logic, then metavaluations can

[12] Some of these definitions were presented in Chapter 4, but the definitions are repeated here for ease of reference.

be used to obtain $m(T)$, which is a normal theory, and so a prime theory. This theory can act as the base world for the reduced frame.

9.8 The Fifth Indemonstrable: Coda on (γ)

The question of whether a logic or theory is closed under the rule (γ) has driven much interesting work in relevant logics. Indeed, it was one of the earliest questions flagged as needing an answer, in a list compiled by Anderson (1963), and Dunn and Restall (2002) use the issue of (γ) as one of the three main organizing themes of their survey. The question of (γ) admissibility for E and R was settled by Meyer and Dunn (1969), and since, the issue of (γ) admissibility has been solved for many logics and for many theories.[13] One upshot is that most of the commonly studied relevant logics are closed under (γ).[14] We have, however, seen some examples of logics in which (γ) is not admissible. While the examples of failures of (γ) in logics have fallen outside of the range of logics from B to R, some logics in that range fail to have (γ) admissible. Øgaard and Standefer (2025) provide a systematic way to obtain logics between B and R that fail to have (γ) admissible. To illustrate their idea, take a logic L that is between B and E. Let A be any theorem of B, then the logic extending L with $\neg A \vee (B \rightarrow ((B \rightarrow C) \rightarrow C))$ will be a sublogic of R for which γ fails to be admissible.[15]

There can be failures of (γ) admissibility among relevant logics. Indeed, there can be failures of (γ) in the range of the more standard relevant logics, although the standard relevant logics all have (γ) admissible. In Section 13.3, we will see an important *theory* in which (γ) is not admissible. In this section, however, we will step back and briefly look at the reasons why (γ) has the place that it does in relevant logics and whether it should.

The issue of (γ) starts with E, so let us start with why (γ) was not a primitive rule of E. Ackermann took (γ) as a primitive rule, whereas Anderson and Belnap did not. The reason is that they wanted all the primitive rules of their logic to have corresponding valid entailments. The only rules they adopted for E were $A, A \rightarrow B \Rightarrow B$ and $A, B \Rightarrow A \wedge B$. Both of these rules have corresponding valid entailments, namely $(A \wedge (A \rightarrow B)) \rightarrow B$ and $(A \wedge B) \rightarrow (A \wedge B)$. The addition of the entailment form of (γ), $(A \wedge (\neg A \vee B)) \rightarrow B$, to E is disastrous, leading to violations of variable sharing. Indeed, Read (1980, 66) goes so far as to say that rejection of (γ) is "central to the whole conception of relevant logic." Not all relevant logicians, however, share Read's views on (γ).

13 See Mares and Meyer (1992), Restall (2000, 95ff.), Seki (2011a,b; 2012a,b), Weiss (2020b), Kripke (2022), and Dunn (2022). Urquhart (2016b) gives an overview of three methods used to prove that (γ) is admissible in a logic.

14 Jago (2013) raises the question of "how weak a relevant logic needs to be before [(γ)] becomes inadmissible." In many weak relevant logics, (γ) is admissible, and in some strong relevant logics, (γ) is not admissible. The strength of the logic does not determine whether the logic admits (γ) or not. It is an interesting question what features of a logic do, in fact, determine whether (γ) is admissible.

15 The interested reader can verify this fact using MaGIC.

Despite (γ) being unavailable as an axiom, it does seem desirable as an admissible rule. One upshot of having it admissible as a rule, at least in some logics like E and R, is that it will ensure that all classical tautologies in the vocabulary $\{\neg, \wedge, \vee\}$ will be derivable. Some relevant logicians have claimed that one does not need (γ) in ordinary reasoning.[16] Some, such as Mortensen (1983) and Lavers (1987), have suggested that (γ) is acceptable, provided that one is working in a context known to be consistent, or otherwise normal. Arithmetic is, perhaps, an example of such a context, whereas naive set theory is not.

Despite the objections raised to (γ), it does not seem to be quite of a piece with the other paradoxes of implication. For one, it is admissible in many relevant logics. One might still ask, can relevant logicians have (γ) as a primitive rule? The answer turns out to be a qualified "yes."

Robles and Méndez (2010; 2011b) provide ternary-relational models for many relevant logics that have (γ) as a primitive rule, rather than a merely admissible one. They show, further, that many of these logics enjoy the variable-sharing property. Indeed, Ackermann's Π', which is E with the addition of (γ) among its primitive rules, enjoys variable sharing.[17] Øgaard (2019) shows that one can have (γ) as a primitive rule in many logics while maintaining the variable-sharing property. Øgaard further shows that, when considering consequence relations, those logics are *explosive*, not paraconsistent, in the sense that there are formulas A, such that for all B, $\{A, \neg A\} \vdash B$. Relevant logics are typically thought of as paraconsistent, but the story is more complicated if one considers consequence relations rather than just the set of logical truths. Øgaard defines two logics, Æ and M, that are relatives of E and R, respectively, that each have the additional axiom $(\mathsf{t} \wedge \neg \mathsf{t}) \to B$, and he shows that these logics have (γ) admissible along with having many of the virtues of other relevant logics.[18]

While some relevant logicians, such as Read and perhaps Anderson and Belnap, took the rejection of (γ) to be central to the relevant logic enterprise, there are no technical barriers to having it in a relevant logic. The addition of (γ) to certain logics and theories may have undesirable consequences, but that is reason not to add (γ) to those logics and theories. One may want to follow Øgaard in exploring explosive consequence relations for relevant logics. Conversely, if one's philosophical view includes a commitment to paraconsistency in addition to relevance, then one may have reason not to appeal to (γ) as a rule. There is, it seems, philosophical room for a reconsideration of the place of (γ) in the philosophy of relevant logics.

16 For a heated discussion of this claim, see the exchange by Burgess (1981; 1984), Read (1983), and Mortensen (1986). Burgess (2005) provides a more recent discussion of "relevantist" views of (γ) and other issues.

17 Ackermann's logic also had another rule, δ, which was $A \to (B \to C), B \Rightarrow A \to C$. In the context of consequence relations, B was required to be a theorem, and the rule is equivalent to $A \Rightarrow (A \to B) \to B$. For more on δ, see Robles (2013).

18 NB: Øgaard's logic M is a relevant logic, rather than the TV-based modal logic of the same name from Section 6.6.

9.9 Exercises

(1) Complete the proof of lemma 9.4 by verifying the cases for the remaining axioms and rules.

(2) Provide a proof of theorem 9.8.

(3) Complete the proof of lemma 9.15 by verifying the remaining axiom cases.

(4) Complete the proof of theorem 9.21 by doing cases (ii)–(v).

(5) Suppose that L is closed under (γ) and $\vdash_{\mathsf{L}} A \vee \neg A$. Show that for all formulas B that contain only connectives from $\{\neg, \wedge, \vee\}$, $B \in \mathsf{TV}$ iff $B \in \mathsf{L}$.

(6) Suppose L is an Ms logic. Show that L is closed under (CoNec), $\Box A \Rightarrow A$.

(7) Suppose L is an Ms or an Mt logic. Show that if $\vdash_{\mathsf{L}} \Box A \vee \Box B$, then either $\vdash_{\mathsf{L}} A$ or $\vdash_{\mathsf{L}} B$.

(8) Prove that if L is a consistent and prime logic, then (γ) is admissible in L.

10 Connectives

10.1 Introduction

The propositional languages of Part I that we have looked at have contained a fairly standard set of connectives, $\{\mathrm{t}, \neg, \rightarrow, \circ, \wedge, \vee\}$. One can reasonably view these as the *basic connectives* for relevant logics. The modal logics add various *modal operators* to the basic language. There are other connectives that are considered for non-modal logics. In this chapter, we will look at two such connectives, left implication ($\leftarrow$), covered in Section 10.2, and Boolean negation ($\sim$), covered in Section 10.3. Then, in Section 10.4, we will consider constraints on new connectives and some of their general features.

10.2 Implications

In Chapter 3, the ternary relational models use the ternary relation to interpret implication and fusion. To evaluate implications, we start at the first spot of the ternary relation, and to interpret fusions, we start at the third. No connectives start at the second spot. We can, however, introduce such a connective, $\leftarrow$, with the following verification condition.

- $a \Vdash C \leftarrow B$ iff for all $b, c \in K$, if $Rbac$ and $b \Vdash B$, then $c \Vdash C$.

The Heredity lemma holds for the language extended with this connective. The ternary relational frames are subject to the condition that if $Rabc$, $d \leq a$, $e \leq b$, and $c \leq f$, then $Rdef$. Inspection of the cases of the proof of the Heredity lemma reveal that the $d \leq a$ portion is used in the case for $\rightarrow$, and the $c \leq f$ portion is used in the case for $\circ$. The remaining clause is unused in the base language, but the case for $\leftarrow$ uses it, as we will now show.

Proof Suppose that $a \Vdash C \leftarrow B$ and $a \leq b$. Suppose that for some c, d, $Rcbd$, $c \Vdash B$ but $d \nVdash C$. As $a \leq b$, $Rcad$, so $a \nVdash C \leftarrow B$, which contradicts our initial assumption. □

The left implication, $\leftarrow$, is connected to fusion in much the same way as (right) implication. Whereas implication and fusion are connected via the two rules

- $A \rightarrow (B \rightarrow C) \Rightarrow (A \circ B) \rightarrow C$ and
- $(A \circ B) \rightarrow C \Rightarrow A \rightarrow (B \rightarrow C)$,

left implication is connected to fusion via the two rules

- $B \rightarrow (C \leftarrow A) \Rightarrow (A \circ B) \rightarrow C$ and
- $(A \circ B) \rightarrow C \Rightarrow B \rightarrow (C \leftarrow A)$.

The connectives $\{\circ, \rightarrow, \leftarrow\}$ form a family of connectives, interpreted over the same relation. This sort of connection is studied by GAGGLE THEORY, and it yields a host of surprising and illuminating general features of connectives.[1]

In general, $\rightarrow$ and $\leftarrow$ will be non-equivalent connectives. In the presence of (I5), however, the two connectives will be equivalent. It is easiest to see this when considering models. The frame condition (FI5) is that if $Rabc$, then $Rbac$. As a result, any point a such that $a \Vdash A \rightarrow B$ will also have $a \Vdash B \leftarrow A$, and vice versa. In logics without (I5), such as B or T, $\rightarrow$ and $\leftarrow$ will not be equivalent, while in RW and R, they will be equivalent.

10.3 Boolean Negation

Boolean negation ($\sim$) was studied by Meyer and Routley (1973; 1974) and others.[2] Meyer's idea was to consider a negation connective whose verification condition was closer to that of the classical negation condition,

- $a \Vdash {\sim}A$ iff $a \nVdash A$.

To add Boolean negation to a model, however, the model needs to satisfy the condition that if $a \leq b$, then $a = b$. Without this condition, the Heredity lemma will fail. With it, heredity is trivial.

As should be clear from the verification condition, contradictions involving Boolean negation cannot hold anywhere, and excluded middle involving Boolean negation will hold everywhere.

LEMMA 10.1 *At every point a in a model, both*

(i) $a \nVdash A \wedge {\sim}A$, *and*
(ii) $a \Vdash A \vee {\sim}A$.

As a consequence, any class of ternary relational frames will validate some paradoxes of implication.

THEOREM 10.2 *In any class of ternary relational frames, the following will be valid.*

- $A \rightarrow (B \vee {\sim}B)$.
- $(A \wedge {\sim}A) \rightarrow B$.

[1] See Bimbó and Dunn (2008) for more.

[2] While the symbol "¬" is sometimes used for Boolean negation, we are using it for the De Morgan negation. In more recent literature, it is more common to use "¬" for De Morgan negation, although some logicians do use the notation "~" for De Morgan negation and "¬" for Boolean negation.

Proof For the first, note that given (i) of the preceding lemma, there cannot be a counterexample, as there are no points at which the consequent fails. For the second, a counterexample is ruled out as the antecedent does not hold anywhere. □

With Boolean negation, one can define a kind of material implication, $\sqsupset$: $A \sqsupset B$ is $\sim A \vee B$. Unlike the usual material implication of relevant logics using De Morgan negation, it is immediate that *modus ponens* holds for this material implication, $A, A \sqsupset B \Rightarrow B$, as disjunctive syllogism holds for Boolean negation, $A, \sim A \vee B \Rightarrow B$. One can use this fact, as do Meyer and Routley, to define "classical" extensions of relevant logics, which would perhaps be better called "Boolean extensions." They provide a classical form of R by adding

- $\sim\sim A \rightarrow A$ and
- $(A \wedge B) \rightarrow \sim C \Rightarrow (A \wedge C) \rightarrow \sim B$

to axioms for positive R. Similarly, Mares (2000a) defines CE, a classical form of E, using the axiom scheme

- $A \rightarrow B$, whenever $A \sqsupset B$ is a substitution instance of a classical tautology in the vocabulary $\{\wedge, \vee, \sim\}$.

Other Boolean extensions of logics can be defined similarly.

Meyer and Routley showed that adding Boolean negation to R is a conservative extension. This means that R is complete with respect to the class of R-frames where heredity is identity. It turns out that many of the standard relevant logics can be conservatively extended by Boolean negation. Giambrone and Meyer (1989) showed that TW and RW are both conservatively extended by Boolean negation. A notable exception here is E. Mares (2000a) showed that CE is not a conservative extension of E. In particular, the formula

$$\neg\Box(\neg p \vee q) \vee (\neg\Box p \vee \Box q),$$

where $\Box A$ is the defined connective of E, namely $(A \rightarrow A) \rightarrow A$, is a theorem of CE but not of E. Mares shows this by demonstrating that frames for CE permit embeddings of Kripke frames for S4. Over the set N, Mares defines a binary relation S by Sab iff $\exists x \in N, Raxb$. The relation S is reflexive and transitive over N, so any theorem of S4 in the vocabulary $\{\Box, \neg, \wedge, \vee\}$ will hold in CE. Øgaard (2021a) extends Mares's result by axiomatically showing that any sublogic of CE in which excluded middle is valid, $\Box$ distributes over implication, and reasoning by cases holds is not conservatively extended by the addition of Boolean negation.

It is clear that any Boolean negation extension of a relevant logic will violate variable sharing. Limited variable-sharing results for the base language are, nonetheless, available. Øgaard (2021b) shows that in the $\{\neg, \wedge, \vee, \rightarrow\}$-fragment of many base logics, one can have variable sharing, that is, $A \rightarrow B$ is a logical truth only if A and B share an atom. In fact, any sublogic of either the extension of R.MK45NecMC with Boolean negation or the extension of TM with Boolean negation has the variable-sharing property when the formulas under consideration are restricted to the language without Boolean negation.

Conservative extension by Boolean negation in the presence of the Ackermann constant is subtle. Øgaard (2021c) shows that for many logics containing (N3), excluded middle, the addition of Boolean negation will not conservatively extend the base logic. Arguing axiomatically, he shows that any logic between an extension of BBX with t-axioms and an extension of RM will fail to be conservatively extended by the addition of Boolean negation. In particular, the formula $A \rightarrow (\mathsf{t} \vee \neg\mathsf{t})$ will be derivable. This formula is not valid in the $RM3$ algebra of Chapter 5, so the addition of Boolean negation is not conservative. The reader may be struck by the requirement that excluded middle be included in the logics. A natural question is whether the addition of Boolean negation is conservative in logics that lack excluded middle. Øgaard provides a negative answer for many logics. For many logics extending BB, including B, DW, TW, and RW, their strengthening with disjunctive rules has the feature that the addition of Boolean negation is not conservative, as $A \vee ((A \wedge \mathsf{t}) \rightarrow B)$ is valid. That formula, however, is not a theorem of the underlying logics. Øgaard's proof uses disjunctive rules, which are not known to hold in many of the logics weaker than RW. The question of conservative extension for the base logics without disjunctive rules is still open.

The frame condition required for Boolean negation, mentioned earlier, is that $a \leq b$ iff $a = b$. In light of this condition, one can see that there will be issues arising from having excluded middle for De Morgan negation in the logic. The reason is that the frame condition for excluded middle requires that for all $a \in N$, $a^* \leq a$. When one has both excluded middle and Boolean negation, for $a \in N$, $a = a^*$, which has the effect of making De Morgan negation act like Boolean negation, at least over the normal points. In other words, it will not be possible to satisfy contradictions, at least on the normal points. One would then expect there to be effects on theorems involving t. Generally, one would expect Boolean negation to lead to some trouble in classes of frames whose conditions involve the heredity ordering in a substantive, essential way.

10.4 Connectives and Ubiquity

In the preceding chapters and sections, we have looked at some connectives beyond those of the basic relevant language. There is a general question of when a connective is relevant or is compatible with a relevant logic. Standefer (2022c, 930) offers a three-part necessary condition:

(i) New connectives do not generate violations of variable-sharing.
(ii) New connectives do not generate valid weakening-like contexts, $C(A) \rightarrow (B \rightarrow D(A))$, where B shares a variable with none of $C(\cdot)$, $D(\cdot)$, and A.
(iii) The new connective is interpreted on ternary relational models in a way that permits the Heredity lemma to be proved.

The first two conditions are presentation-independent, while the third condition focuses on models. The first two conditions can be read as giving some content to the idea that new connectives should not re-introduce the paradoxes of implication. The third condition seems to be required to ensure that the ternary relational models work for the entirety of the extended language.

This three-part condition is just a necessary condition. There are other potential conditions one could consider adding. One example is the "no loose pieces" principle (NLPP) of Robles and Méndez (2012).[3]

DEFINITION 10.3 (NLPP) *The NLPP says that if A is a logical truth with no conjunctions as ap and no disjunctions as cp, then every atom in A occurs at least once as ap and at least once as cp.*

The formula $(q \rightarrow p) \rightarrow (p \rightarrow p)$ is an example of a formula that violates the NLPP. In particular, the atom q occurs as cp but it does not occur as ap. It is plausible that new connectives should not permit the introduction of loose pieces. Another plausible condition to impose is that the addition of the connective be conservative over the base logic.[4]

Three comments on condition (ii) are in order. First, in practice, some relevant logicians have pointed out the presence of weakening-like contexts as a reason for being uneasy. They are too close to weakening for comfort. Second, in some logics, weakening-like contexts can lead to violations of variable sharing, as illustrated by the following lemma.

LEMMA 10.4 *Suppose* L *satisfies either of the following conditions:*

(1) L *is closed under the rule* $A \rightarrow (B \rightarrow C) \Rightarrow (A \wedge B) \rightarrow C$ *and contains* (N1), *or*
(2) L *contains* (I6).

Then in L, *valid weakening-like contexts will lead to violations of variable-sharing.*

Proof Suppose L satisfies condition (1). Then in L, from $C(A) \rightarrow (B \rightarrow D(A))$ we obtain $C(A) \rightarrow (\neg D(A) \rightarrow \neg B)$ from (N1). The rule then yields $(C(A) \wedge \neg D(A)) \rightarrow \neg B$, but B is assumed not to share any variables with $D(A)$ and $C(A)$.

Suppose L satisfies condition (2). The violation is immediate from the definition of weakening-like context. □

Third, even in logics that do not satisfy the two conditions of the preceding lemma, if $C(A)$ is ever a valid formula, then the weakening-like contexts will give rise to violations of variable-sharing, as the models guarantee *modus ponens* holds.

With the three conditions in place, we can look at some connectives that violate the conditions. We will begin with Boolean negation. Boolean negation leads to violations the first of the conditions. It exhibits a feature that was discussed briefly in Section 8.2.2, namely ubiquity. Boolean negation provides a route to formulas that are ubiquitously true, or ubiquitously false, in all models. If a connective generates

[3] The name comes from Anderson and Belnap (1975, 254–255).
[4] See Belnap's (1962) discussion of adding new connectives to a logic.

formulas that are ubiquitously true or ubiquitously false in all models, then the first condition will be violated, as we can show. First, there are some lemmas to illustrate features of ubiquity.

LEMMA 10.5 *Suppose that A is ubiquitously true in M. Then for all $a \in K$, $a \Vdash B \rightarrow A$.*

Proof Suppose A is ubiquitously true in M. Suppose $Rabc$ and $b \Vdash B$. As A is ubiquitously true, $c \Vdash A$, so $a \Vdash B \rightarrow A$. □

LEMMA 10.6 *Suppose that A is ubiquitously false in M. Then for all $a \in K$, $a \Vdash A \rightarrow B$.*

Proof The proof is similar to that of the preceding lemma. □

These lemmas have the following corollary.

COROLLARY 10.7 *Suppose A is ubiquitously true in M or A is ubiquitously false in M. Then $A \rightarrow A$ is ubiquitously true in M.*

Corollary 10.7 tells us that self-implications of ubiquitously true or ubiquitously false formulas are themselves ubiquitously true. This observation is, then, enough for a theorem, for which we will introduce some terminology. Let us say that a formula A is UBIQUITOUS (in a class $\mathcal{C}$ of frames) iff for all models M (built on a frame in $\mathcal{C}$), A is ubiquitous in M.

THEOREM 10.8 *Suppose that A is ubiquitous in a class $\mathcal{C}$ of frames. Then for all formulas B, $B \rightarrow (A \rightarrow A)$ is ubiquitously true in all models M built on a frame in $\mathcal{C}$.*

Proof From corollary 10.7, $A \rightarrow A$ is ubiquitously true in all such models M. By lemma 10.5, $B \rightarrow (A \rightarrow A)$ is ubiquitously true in any model M built on a frame in $\mathcal{C}$. □

Given such a formula A that does not contain t, we can obtain a violation of variable sharing. Let q be any atom not in A. Then $q \rightarrow (A \rightarrow A)$ will be valid.

Notice that demonstrating that an implication $A \rightarrow B$ is ubiquitously true in a model requires different reasoning from showing that the implication holds in the model. The latter requires showing, for $a \in K$, if $a \Vdash A$, then $a \Vdash B$, while the former requires showing that for all $a, b, c \in K$, if $Rabc$ and $b \Vdash A$, then $c \Vdash B$.

If a formula is ubiquitous in a class of frames, then there will be violations of variable sharing for the logic of that class. Therefore, the relevant logician should reject connectives that give rise to ubiquitous formulas. Let us look at some connectives that yield ubiquitous formulas in the class of all ternary relational frames.

As noted earlier, Boolean negation can be used to obtain ubiquitous formulas. The connective $\mathbb{U}$ of Section 8.2.2 likewise generates ubiquitous formulas. It is not hard, perhaps surprisingly easy, to come up with connectives and verification conditions that will generate ubiquitous formulas.

Some additional connectives, discussed by Standefer (2022c), that lead to violations of the variable-sharing property are the following, given with their proposed verification conditions.

- $a \Vdash B \rhd C$ iff $|B| \subseteq |C|$.
- $a \Vdash B \doteq C$ iff $|B| = |C|$.
- $a \Vdash \#B$ iff $|B| \neq \emptyset$.
- $a \Vdash \mathbb{N}B$ iff $N \subseteq |B|$.

These connectives all generate ubiquitous formulas. They also share the feature that the point of evaluation on the left-hand side of the biconditional does not occur on the right-hand side. This has the effect that if the formula is true anywhere in a model, it will be true everywhere, and similarly if it is false.

Ubiquitous formulas can be obtained using verification conditions that involve the ternary relation as well. The connectives below all satisfy heredity.

- $a \Vdash \mathbb{J}B$ iff for all $b, c \in K$, if $Rabc$, then $c \Vdash B$.
- $a \Vdash \mathbb{M}B$ iff for some $b, c \in K$, $Rbca$ and $c \Vdash B$.

With the former, $\mathbb{J}A \rightarrow A$ is ubiquitously true, and with the latter, $A \rightarrow \mathbb{M}A$ is ubiquitously true.

We will present three more illustrative connectives that violate conditions (i) or (ii) earlier, while also evaluating subformulas at the additional points of the verification condition. For this, define

- Rac as $a \leq c$, and
- $Rac_1, \ldots, c_{n+1}$, for $n \geq 1$, as $\exists x(Rac_1, \ldots, c_{n-1}x$ and $Rxc_nc_{n+1})$.

Using this definition, we can define two R-chain connectives, $\mathbb{R}$ and $\mathbb{Q}$, along with a connective, $\downarrow$, that uses the heredity relation.[5]

- $a \Vdash \mathbb{R}B$ iff for all $c_1, \ldots, c_n \in K$, if $Rac_1, \ldots, c_n$, then for $i \leq n$, $c_i \Vdash B$.
- $a \Vdash \mathbb{Q}B$ iff for some $c_1, \ldots, c_n \in K$, $Rc_1, \ldots, c_na$, and for some $i \leq n$, $c_i \Vdash B$.
- $a \Vdash \downarrow B$ iff for some $b \leq a$, $b \nVdash B$.

The first of these generates weakening-like contexts, such as $\mathbb{R}p \rightarrow (q \rightarrow \mathbb{R}p)$. The latter two generate ubiquitous formulas, such as $\mathbb{Q}(p \rightarrow p)$ and $p \vee \downarrow p$.

The necessary conditions on being a relevant connective are modest and plausible. Nonetheless, they rule out many connectives as being acceptable to the relevant logician. A consequence of the conditions is that connectives that generate ubiquitous formulas must be rejected. Such connectives provide an indication of the limits of the ternary relational models.[6]

[5] The $\downarrow$ connective is studied in the context of models for intuitionistic logic. See López-Escobar (1985).

[6] A natural question is whether a model theory developed with a non-classical logic could accommodate some of the problematic connectives. For an example of such a model theory, see the work of Girard and Weber (2015) and Weber et al. (2016). Evaluation of the suggestion will, however, have to wait for further developments of the theory.

The relevant logician is, it seems, committed to the view that, at the propositional level, no connective, or set of connectives, can generate a formula A that is bound to have the right sort of connection of meaning to every single atom p of the language to guarantee that $p \rightarrow A$ is valid.[7] Every formula must be capable of failing at *some* point in *some* model. We can think of this idea in relation to the implication connective, as a way stating that the paradoxes of implication, in particular (PI1), $A \rightarrow (B \rightarrow B)$, must be rejected. Even logical truths are not exempted from the requirement that they fail somewhere.

Ubiquity avoidance is supported by the anti-suppressive principle discussed by Øgaard (2020a) and Plumwood (2023).[8]

DEFINITION 10.9 (Anti-suppressive principle) *A logic* L *satisfies the* ANTI-SUPPRESSIVE PRINCIPLE *iff for every formula* A*, there are formulas* B *and* C *such that* $\models_{\mathsf{L}} (A \wedge B) \rightarrow C$ *but* $\not\models_{\mathsf{L}} B \rightarrow C$.

The idea behind the anti-suppressive principle is that there are no truths that can, in general, be dropped from the antecedent of some valid implication without rendering the result invalid. Formulas ubiquitously true in all models lead to violations of the anti-suppressive principle. Valid implications $D \rightarrow E$ have the feature that $|D|_M \subseteq |E|_M$, for all models M. If A is ubiquitously true in all models, then for any D, $|A|_M \cap |D|_M = |D|_M$. If A is ubiquitously true and $(A \wedge B) \rightarrow C$ is valid, then $|A \wedge B|_M \subseteq |C|_M$, but $|A \wedge B|_M = |A|_M \cap |B|_M = |B|_M \subseteq |C|_M$. Thus, $B \rightarrow C$ is valid as well.

Ubiquity avoidance is also supported by the joint force principle considered by Plumwood and by Øgaard.

DEFINITION 10.10 (Joint force principle) *A logic* L *satisfies the* JOINT FORCE PRINCIPLE *iff for every formula* A*, there are formulas* B *and* C *such that* $\models_{L} (A \wedge B) \rightarrow C$ *but* $\not\models_{\mathsf{L}} B \rightarrow C$ *and* $\not\models_{\mathsf{L}} A \rightarrow C$.

The idea behind the joint force principle is that for every formula, some formula can be combined with it to get new consequences unavailable to either on their own. Formulas that are ubiquitously false in all models lead to violations of this principle. Let A be ubiquitously false in all models. Then, for any formulas B and C, $\models_{\mathsf{L}} (A \wedge B) \rightarrow C$ and also $\models_{\mathsf{L}} A \rightarrow C$. Nothing can be paired with such an A that results in a consequence not already implied by A on its own.

The anti-suppressive principle and the joint force principle can be stated in forms that are presentation-independent. One might, then, consider adding them as conditions on connectives for relevant logics by requiring that the extension of a logic with a new connective must obey these conditions. The additional conditions would not make a difference. Øgaard (2020a) has shown that the variable-sharing property, condition

[7] Things get trickier once propositional quantification is permitted into the language. The formula $\exists p.p$ is implied by every atom, indeed every formula, and $\forall p.p$ similarly implies every formula. The relevance intuitions behind the variable-sharing property seem clearest, and strongest, at the propositional level with no quantification.

[8] This is Øgaard's formulation of the principle, although we have changed the provability turnstile to the validity turnstile, as we are discussing models.

(i) earlier, already entails both the anti-suppressive and joint force principles, while they do not on their own entail (i).

10.5 Exercises

(1) The fission connective, $\oplus$, has been described as an intensional disjunction, much like fusion is an intensional conjunction. In RW, we can define fission in two ways, for which we will use the symbols $\oplus$ and $+$.
- $(A \oplus B) \leftrightarrow \neg(\neg A \circ \neg B)$.
- $(A + B) \leftrightarrow (\neg A \rightarrow B)$.

Show that in RW, $(A \oplus B) \leftrightarrow (A + B)$. Conclude that an analog of (γ) holds for $\oplus$: $\neg A, A \oplus B \Rightarrow B$.

(2) Using the definitions in the previous problem, provide verification conditions for $\oplus$ and $+$ in ternary relational models. In logics weaker than T, $\oplus$ and $+$ are no longer equivalent. Using either algebras or ternary relational models, show that the two definitions of fission are not equivalent in the weaker logics.

(3) Suppose that the $\star$ connective is introduced with the following verification condition.
- $a \Vdash \star B$ iff $a^* \Vdash B$.

To ensure that the Heredity lemma is provable, we impose the condition that if $a \leq b$, then $a^* \leq b^*$. Show that the addition of $\star$ to B yields violations of variable sharing.

(4) Show that the $\mathbb{N}$ connective defined in this chapter yields violations of variable sharing. Further, show that $\mathbb{N}A$ is not equivalent to $\mathsf{t} \rightarrow A$.

(5) Show that the joint force principle follows from the variable-sharing property.

(6) Consider the binary connective $\multimap$ with the following verification condition.
- $a \Vdash B \multimap C$ iff for all $b \in K$, if $a \leq b$ and $b \Vdash B$, then $b \Vdash C$.

Show that the addition of this connective to B is a conservative extension. Then, show that the addition of this connective results in violations of variable sharing.

Part III

Quantification

11 Quantifiers

11.1 Introduction

In this third part of the book, we will consider extensions of the base logics with quantifiers. Before getting to first-order quantifiers, however, we will start with propositional quantifiers, a topic that provides a natural midway point between propositional logic and first-order logic. While proof-theoretically the quantifiers are fairly straightforward, their frame theory is surprising. The surprise comes with difficulties arising from the standard Tarskian verification condition. There have been two alternative proposals, one by Fine (1988) and one by Mares and Goldblatt (2006). The focus in this chapter will be on the latter, since it also connects neatly with the models for propositional quantifiers, developed by Goldblatt and Kane (2009).[1] Both of these papers on relevant quantification use the method of general, or admissible proposition, frames, also explored by Goldblatt (2011) in the context of TV-based modal logics.

Quantifiers, both first-order and propositional, have a long history in relevant logics. Anderson (1960) considered a first-order extension of E. Anderson (1972) considered extensions of E and R with propositional quantifiers, and Meyer (1973) considered alternative axiomatizations of these logics. Routley and Meyer (1973) provided frames for propositional quantifiers, although they did not prove completeness, which, as we will see below, turned out to be impossible, given the work of Kremer (1993). Anderson et al. (1992) begin with two chapters on quantification, starting with propositional quantification before proceeding to first-order.

The issues with the Tarskian verification condition were identified by Fine (1989), who showed that a first-order extension of R was incomplete with respect to the straightforward constant-domain extensions of ternary relational frames for it. Fine (1988) proposed an alternative frame theory, for which he proved Completeness results. That account has been criticized by Brady (2017b), and it has been defended by Logan (2019). Goldblatt (2009) proved a conservative extension result using Fine's models. Brady (1988; 1989a) provided his own theory of relevant quantification, based on his *content semantics*. Neither Fine's nor Brady's approaches have been adapted to propositional quantification.

In this chapter, we will begin with axiomatic logics for propositional quantification (Section 11.2) and the frames for them (Section 11.3), before proceeding to axiomatics

[1] For a nice exposition of Fine's approach to quantifiers, see Logan (2024).

for first-order quantification (Section 11.5) and the first-order frames (Section 11.6). While the Mares–Goldblatt frames were first developed for first-order quantification before being adapted to propositional quantification, this way of proceeding allows us to introduce the new machinery for the frames in a way that avoids some of the extra complications of first-order quantification, namely the addition of terms and a domain of individuals.[2] After introducing the frames for propositional quantification, we will present proofs of Soundness and Completeness (Section 11.4), and after introducing the frames for first-order quantification we will present proofs of Soundness and Completeness (Section 11.7). The latter proofs largely build on the core ideas of the former proofs. Finally, in Section 11.8, we will discuss the standard Tarskian verification conditions for the quantifiers and issues of incompleteness.

11.2 Axioms for Propositional Quantification

To begin, we will add the universal propositional quantifier, $\forall$, to the language to obtain the language $\mathfrak{L}_{\forall p}$. We will distinguish a set of propositional variables, $\mathsf{Var} = \{p_1, \ldots, p_n, \ldots\}$, although we will still use $q, r, s, \ldots$ to refer to them, and a set of propositional constants $\mathsf{Con} = \{c_1, \ldots, c_n, \ldots\}$, such that $\mathsf{At}{=}\mathsf{Var} \cup \mathsf{Con}$. Occurrences of propositional variables will be free or bound, using the usual definitions of these terms. A formula is CLOSED if it does not contain any occurrences of free variables. The existential quantifier, $\exists p$, will be defined as $\neg\forall p\neg$.

The substitution of a formula B for an atom p_n in A, $A[p_n/B]$, is obtained by replacing all free occurrences of p_n in A with B, provided that no free occurrence of p_n is in the scope of a quantifier that binds an atom in B.

Given the base logic B, we will obtain the propositional quantifier extension $\mathsf{B}^{\forall p}$, by adding the following axioms.

(PQ1) $\forall p_n A \rightarrow A[p_n/B]$.
(PQ2) $A \rightarrow B \Rightarrow A \rightarrow \forall p_n B$, provided p_n is not free in A.

This is the basic axiom system for propositional quantification over B. There are some additional axioms that can be added. One can extend the base logic with any of the axioms and rules for relevant logics from Chapter 2. One can also add the quantificational axiom

- (ECp) $\forall p_n(A \vee B) \rightarrow (\forall p_n A \vee B)$, provided p_n is not free in B.

This is the propositional quantificational analog of (MC) from the modal logic chapters. It is called the axiom of EXTENSIONAL CONFINEMENT, following Mares and Goldblatt (2006). The extension L of B with (PQ1), (PQ2), any of the base logic axioms considered in Chapter 2, or (ECp) will be the logic $\mathsf{L}^{\forall p}$.

Following Goldblatt and Kane (2009), we will prove a few lemmas about any such logic.

[2] The reader that follows up on some references in Goldblatt and Kane (2009) will find references to Mares and Goldblatt (2006).

LEMMA 11.1 *The following are provable in* $\mathsf{L}^{\forall p}$.

(i) $(\forall p_n A \wedge \forall p_n B) \to \forall p_n(A \wedge B)$.
(ii) $A \to \forall p_n A$, *if* p_n *is not free in* A.
(iii) $\forall p_n(A \to B) \to (\forall p_n A \to \forall p_n B)$.

Proof For the first, $(\forall p_n A \wedge \forall p_n B) \to A$, by (PQ1) and (A2). Then by (R2) and (A3), we have $(\forall p_n A \wedge \forall p_n B) \to (A \wedge B)$, whence $(\forall p_n A \wedge \forall p_n B) \to \forall p_n(A \wedge B)$ by (PQ2).

The second follows from (PQ2).

The third is obtained from (PQ1) using (R6), (PQ2), and (R7), followed by (PQ1) again. □

Note that Goldblatt and Kane (2009) do not have fusion in their logics, so (iii) given earlier is an optional additional axiom.

LEMMA 11.2 *The logic* $\mathsf{L}^{\forall p}$ *is closed under the following rules.*

(i) $A \Rightarrow \forall p_n A$.
(ii) $A \to B[p_n/c] \Rightarrow A \to \forall p_n B$, *if* c *is not in* A *or* B, *and* p_n *is not free in* A.
(iii) $A[p_n/c] \Rightarrow \forall p_n A$, *if* c *is not in* A.
(iv) $B \leftrightarrow C \Rightarrow A[p_n/B] \leftrightarrow A[p_n/C]$, *where* A *is a formula with at most one free variable,* p_n *and* B *and* C *are closed formulas.*

Proof For (i) suppose A is provable. Then by (R8), we have $\mathsf{t} \to A$, so by (PQ2) we have $\mathsf{t} \to \forall p_n A$. Thus, we have $\forall p_n A$, from (A8) and (R1).

For (ii), suppose $A \to B[p_n/c]$ is provable, p_n is not free in A, and c does not occur in A or B. Take such a proof and replace all occurrences of c by a new variable q not appearing the proof. The result is a proof of $A \to B[p_n/q]$. The desired conclusion then follows by (PQ2).

For (iii), use the replacement technique of (ii) on the proof of (i).

For (iv), the proof is by induction on the complexity of A. In all the cases except where A is of the form $\forall p_m D$, the result is straightforward. If A is $\forall p_m D$, there are two cases, $m = n$ and $m \neq n$. In the former, there is no substitution, since p_n is not free in A. Suppose we are in the latter case. In that case, $(\forall p_m D)[p_n/B] = \forall p_m D[p_n/B]$ and $(\forall p_m D)[p_n/C] = \forall p_m D[p_n/C]$. By the inductive hypothesis, $D[p_n/B] \leftrightarrow D[p_n/C]$. By (i) given earlier and part (iii) of lemma 11.1, we obtain the desired conclusion. □

That is enough of axiom chopping for now. Let us turn to the frames for propositional quantifiers.

11.3 Mares–Goldblatt Frames for Propositional Quantifiers

To define frames for propositional quantifiers, we will build on the propositional frames of Section 3.4. In addition to a set Prop of propositions on a ternary relational frame, we need an additional set, PropFun, that will be instrumental in interpreting the quantifiers.

Given a ternary relational frame $\langle K, N, R, ^*\rangle$ we let Prop be a model-like set of propositions, as in definition 3.14. For $S \subseteq \wp(K)$, set

$$\bigsqcap S =_{Df} \bigcup \left\{ X \in \mathsf{Prop} : X \subseteq \bigcap S \right\}.$$

From this definition, $\bigsqcap S \in \mathcal{H}$, although it may not be in Prop without further conditions. It will be the case that $\bigsqcap S \subseteq \bigcap S$, and, in general, $\bigsqcap S \in \mathsf{Prop}$ is not sufficient for $\bigcap S \in \mathsf{Prop}$.

For uniformity with the first-order quantificational sections, we will define the sets Var and Con of propositional variables and constants, respectively, as arbitrary infinite subsets of At such that $\mathsf{t} \in \mathsf{Con}$. An assignment $f\colon \mathsf{Var} \mapsto \mathsf{Prop}$ can be viewed as a function $\omega \mapsto \mathsf{Prop}$, as the set Var is countable, so an assignment f can be written as $\langle f(0), f(1), \ldots\rangle$. Given an assignment $f\colon \mathsf{Var} \mapsto \mathsf{Prop}$ and $X \in \mathsf{Prop}$, define

$$f[n/X] =_{Df} \langle f(0), \ldots, f(n-1), X, f(n+1), \ldots\rangle.$$

We will say that two assignments f and g are n-VARIANTS, $f \asymp_n g$ iff $f(m) = g(m)$, where $m \neq n$. As an example, $f \asymp_n f[n/X]$. From the definition, it is straightforward that, for all $n \in \omega$, $\asymp_n$ is reflexive and symmetric.

DEFINITION 11.3 (Propositional function) *Given a propositional frame F, written as $\langle K, N, R, ^*, \mathsf{Prop}\rangle$, a* PROPOSITIONAL FUNCTION *on F is a function $\phi\colon \mathsf{Prop}^{\mathsf{Var}} \mapsto \mathsf{Prop}$.*

Given an assignment $f\colon \mathsf{Var} \mapsto \mathsf{Prop}$, define operations on propositional functions ϕ and ψ as follows.

$$\begin{aligned}
(\phi \cap \psi)f &=_{Df} (\phi f) \cap (\psi f).\\
(\phi \cup \psi)f &=_{Df} (\phi f) \cup (\psi f).\\
(\phi \Rrightarrow \psi)f &=_{Df} (\phi f) \Rrightarrow (\psi f).\\
(\phi \otimes \psi)f &=_{Df} (\phi f) \otimes (\psi f).\\
(-\phi)f &=_{Df} -(\phi f).\\
(\forall_n \phi)f &=_{Df} \bigsqcap_{X \in \mathsf{Prop}} (\phi f[n/X]).
\end{aligned}$$

DEFINITION 11.4 (Propositional quantifier frame, closed set of propositional functions) *A* PROPOSITIONAL QUANTIFIER FRAME *for the base logic* L *is a sextuple $\langle K, N, R, ^*, \mathsf{Prop}, \mathsf{PropFun}\rangle$ where* Prop *is a model-like set of propositions and* PropFun *is a set of propositional functions obeying the following conditions.*

FTee $\phi_{\mathsf{t}} \in \mathsf{PropFun}$, *where* $\phi_{\mathsf{t}}(f) = N$.
FEval $\phi_n \in \mathsf{PropFun}$, *where* $\phi_n(f) = f(n)$.
FImp *If* $\phi, \psi \in \mathsf{PropFun}$, *then* $\phi \Rrightarrow \psi \in \mathsf{PropFun}$.
FFus *If* $\phi, \psi \in \mathsf{PropFun}$, *then* $\phi \otimes \psi \in \mathsf{PropFun}$.
FConj *If* $\phi, \psi \in \mathsf{PropFun}$, *then* $\phi \cap \psi \in \mathsf{PropFun}$.
FDisj *If* $\phi, \psi \in \mathsf{PropFun}$, *then* $\phi \cup \psi \in \mathsf{PropFun}$.
FNeg *If* $\phi \in \mathsf{PropFun}$, *then* $-\phi \in \mathsf{PropFun}$.
FAll *If* $\phi \in \mathsf{PropFun}$, *then* $\forall_n \phi \in \mathsf{PropFun}$, *for* $n \in \omega$.

A set of propositional functions that obeys the conditions given earlier apart from (FEval) *is said to be* CLOSED. *If in addition, the set of propositional functions obeys* (FEval), *then it is said to be* FULLY CLOSED.

DEFINITION 11.5 *A* PROPOSITIONAL QUANTIFIER MODEL M *is a pair of a propositional quantifier frame* $\langle K, N, R, ^*, \mathsf{Prop}, \mathsf{PropFun}\rangle$ *and a valuation function* $V\colon \mathsf{Con} \mapsto \mathsf{Prop}$ *obeying the following condition.*

FMod *For all* $c \in \mathsf{Con}$, *the constant function* $\phi_c \in \mathsf{PropFun}$, *where* $\phi_c(f) = V(c)$, *for all assignments* $f \in \mathsf{Prop}^\omega$.

The condition (FMod) ensures that there is a propositional function for the value of each propositional constant.

The verification, or satisfaction, conditions for formulas are defined relative to a point a in a model M and an assignment f. We will write $|A|_{M,f} = \{a \in K : f,a \Vdash A\}$, usually omitting the M to reduce clutter slightly.

- $f,a \Vdash p_n$ iff $a \in f(n)$.
- $f,a \Vdash c$ iff $a \in V(c)$.
- $f,a \Vdash \mathsf{t}$ iff $a \in N$.
- $f,a \Vdash \neg B$ iff $f,a^* \nVdash B$.
- $f,a \Vdash B \wedge C$ iff $f,a \Vdash B$ and $f,a \Vdash C$.
- $f,a \Vdash B \vee C$ iff $f,a \Vdash B$ or $f,a \Vdash C$.
- $f,a \Vdash B \to C$ iff $\forall b,c \in K$(if $Rabc$ and $f,b \Vdash B$, then $f,c \Vdash C$).
- $f,a \Vdash B \circ C$ iff $\exists b,c \in K(Rbca$ and $f,b \Vdash B$ and $f,c \Vdash C)$.
- $f,a \Vdash \forall p_n B$ iff $\exists X \in \mathsf{Prop}\left(a \in X \text{ and } X \subseteq \bigcap_{P \in \mathsf{Prop}} |A|_{f[n/P]}\right)$.

The final clause can be rewritten using the notion of an n-variant.

- $f,a \Vdash \forall p_n B$ iff $\exists X \in \mathsf{Prop}\left(a \in X \text{ and } X \subseteq \bigcap_{g \asymp_n f} |A|_g\right)$.

These can be rewritten in a more algebraic, truth-set form, for a given model M.

- $|c|_f = V(c)$.
- $|p_n|_f = f(n)$.
- $|\mathsf{t}|_f = N$.
- $|\neg B|_f = -|B|_f$.
- $|B \wedge C|_f = |B|_f \cap |C|_f$.
- $|B \vee C|_f = |B|_f \cup |C|_f$.
- $|B \to C|_f = |B|_f \Rightarrow |C|_f$.
- $|B \circ C|_f = |B|_f \otimes |C|_f$.
- $|\forall p_n B|_f = \prod_{P \in \mathsf{Prop}} |B|_{f[n/P]} = \forall_n |B|_f$.

From the earlier observation about $\bigsqcap S$, in addition to previous results, it follows that each of these truth sets is a proposition.

Every formula defines a propositional function, relative to a propositional quantifier model M. Define these propositional functions as follows.

- ϕ_c is given by FMod.
- $\phi_{p_n} = \phi_n$.
- ϕ_{t} is given by FTee.
- $\phi_{\neg B} = -\phi_B$.
- $\phi_{B \wedge C} = \phi_B \cap \phi_C$.
- $\phi_{B \vee C} = \phi_B \cup \phi_C$.
- $\phi_{B \to C} = \phi_B \Rrightarrow \phi_C$.
- $\phi_{B \circ C} = \phi_B \otimes \phi_C$.
- $\phi_{\forall_{p_n} B} = \forall_n \phi_B$.

These propositional functions relate to truth sets as follows.

LEMMA 11.6 *For any propositional quantifier model M and assignment f, for any formula A,* $\phi_A(f) = |A|_f$.

Proof The proof is by the induction on the construction of A.

The base cases are straightforward. If A is c, then $\phi_c(f) = V(c) = |c|_f$. If A is p_n, then $\phi_{p_n} = f(n) = |p_n|_f$. If A is t, then $\phi_{\mathsf{t}}(f) = N = |\mathsf{t}|_f$.

If A is $\forall p_n B$, then $\phi_{\forall_{p_n} B}(f) = \forall_n \phi_B(f)$, so by the inductive hypothesis, $\forall_n \phi_B(f) = \forall_n |B|_f$. It then follows that $\forall_n \phi_B(f) = |\forall p_n B|_f$ by definition, so $\phi_{\forall p_n B}(f) = |\forall p_n B|_f$.

The inductive hypothesis takes care of the remaining cases. □

For any propositional quantifier model, for every formula of the language, there is a propositional function in PropFun that interprets that formula. In particular, this means that for any formula, B, the universal closure of B, namely $\forall p_{n_1} \ldots \forall p_{n_k} B$, will be interpreted as a proposition in Prop.

Next, we will define counterexample, holding in a model, and validity.

DEFINITION 11.7 (Counterexample, holding, validity) *A* COUNTEREXAMPLE *to a formula A is a pair of a propositional quantifier model M and an assignment f such that for some* $a \in N$, $f, a \not\Vdash A$.

A formula A HOLDS *in a model M relative to f iff* $\langle M, f \rangle$ *is not a counterexample to A.*

A formula A is VALID IN A MODEL *M iff A holds in M relative to every f.*

A formula A is VALID ON A PROPOSITIONAL QUANTIFIER L-FRAME *F iff A is valid on M, for every model M built on F.*

A formula A is VALID IN A CLASS *C of propositional quantifier* L*-frames iff A is valid on F for every* $F \in C$*. When this holds, write* $\models_C A$.

Before we move to Soundness and Completeness, there are a few more basic facts to verify concerning the propositional quantifier models.

LEMMA 11.8 *Let M be a propositional quantifier model and let f and g be assignments. Let A be a formula and suppose $f(n) = g(n)$ for all p_n occurring freely in A. Then, $|A|_f = |A|_g$.*

Proof The proof is by induction on the construction of A. We will do a couple of cases. Suppose A is p_n. Then $|A|_f = f(n)$. By the assumption that $f \asymp_n g$, $f(n) = g(n)$, so $|A|_f = |A|_g$. The other base cases are straightforward.

Suppose A is $\forall p_n B$. Then, $|\forall p_n B|_f = \forall_n Bf = \bigsqcap_{P \in \mathsf{Prop}} |B|_{f[n/P]}$. By the inductive hypothesis, $|B|_{f[n/P]} = |B|_{g[n/P]}$, so $\bigsqcap_{P \in \mathsf{Prop}} |B|_{f[n/P]} = \bigsqcap_{P \in \mathsf{Prop}} |B|_{g[n/P]} = |\forall p_n B|_g$.

The remaining cases are handled by the inductive hypothesis. □

The next basic fact to verify about the models concerns substitution.

LEMMA 11.9 (Semantic substitution lemma) *Let M be a propositional quantifier model and let f be an assignment. Let A and B be formulas and $p_n \in \mathsf{Var}$ such that B is free for p_n in A. Then, $|A[n/B]|_f = |A|_{f[n/|B|_f]}$.*

Proof For the proof, see Goldblatt and Kane (2009, 84–85). □

Next is the Heredity lemma.

LEMMA 11.10 (Heredity lemma) *For any propositional quantifier model M and any assignment f, if $a \leq b$ and $f, a \Vdash A$, then $f, b \Vdash A$.*

Proof This follows from lemma 11.6, since $|A|_f \in \mathsf{Prop} \subseteq \mathcal{H}$. □

From this the Verification lemma follows as usual.

LEMMA 11.11 (Verification lemma) *Let M be a propositional quantifier model and f an assignment. For any formulas A and B, if $A \rightarrow B$ holds in M relative to f, then $\forall a \in K$, $f, a \Vdash A$ only if $f, a \Vdash B$.*

Finally, we will record one fact about validity.

LEMMA 11.12 *Suppose that $\models_{\mathcal{C}} A \rightarrow B$. Let M be a model built on a frame $F \in \mathcal{C}$. Then for all assignments f, $|A|_f \subseteq |B|_f$.*

Those are all the basic facts about propositional quantifier models we will prove here. In the next section, we turn to Soundness and Completeness.

11.4 Soundness and Completeness for Propositional Quantification

In this section, we will prove Soundness and Completeness of $\mathsf{L}^{\forall p}$ with respect to $\mathsf{L}^{\forall p}$-frames. As usual, the Soundness part is easy and is proved by induction on the length of proof. The Completeness direction is harder and proceeds via canonical model construction.

THEOREM 11.13 (Soundness) *For any formula A, if $\vdash_{\mathsf{L}^{\forall p}} A$, then $\models_{\mathsf{L}^{\forall p}} A$.*

Proof Since most of the axioms and rules have been covered in previous Soundness proofs, we will only cover the new ones.

For (PQ1), suppose $f,a \Vdash \forall p_n A$, so there is $X \in \mathsf{Prop}$ such that $a \in X$ and $X \subseteq \bigcap_{g \asymp_n f} |A|_g$. Suppose that B is free for p_n in A. By lemma 11.9, $|A[p_n/B]|_f = |A|_{f[n/|B|_f]}$. Since $f \asymp_n f[n/|B|_f]$, it follows that $X \subseteq |A|_{f[n/|B|_f]}$, so $f[n/|B|_f],a \Vdash A$, whence, by lemma 11.9, $f,a \Vdash A[p_n B/]$ as desired.

For (PQ2), suppose $A \to B$ is provable and p_n is not free in A. By the inductive hypothesis, $\vDash_{\mathsf{L}^{\forall p}} A \to B$. Suppose $f,a \Vdash A$. Since p_n is not free in A, for all $g \asymp_n f$, $g,a \Vdash A$, by lemma 11.8. It then follows that $g,a \Vdash B$, for all $g \asymp_n f$. From lemma 11.12 and $\vDash_{\mathsf{L}^{\forall p}} A \to B$, $|A|_g \subseteq |B|_g$, for all such assignments g. By lemma 11.8 again, $|A|_f \subseteq |B|_g$, for all $g \asymp_n f$, so $|A|_f \subseteq \bigcap_{g \asymp_n f} |B|_g$. As $|A|_f \in \mathsf{Prop}$ and $a \in |A|_f$, $f,a \Vdash \forall p_n B$, as desired. □

There was one additional axiom considered in Section 11.2, (ECp). This axiom has an additional frame condition.

(FECp) $\bigsqcap_{X \in \mathsf{Prop}} (\phi(f) \cup \psi(f[n/X])) \subseteq (\phi \cup \forall_n \psi)(f).$

We can prove Soundness for logics with the axiom with respect to frames obeying the frame condition.

LEMMA 11.14 *If $\mathcal{C}$ is a class of frames obeying the frame condition* (FECp), *then* $\vDash_{\mathcal{C}}$ (ECp).

Proof For (ECp), suppose $f,a \Vdash \forall p_n(A \vee B)$, where p_n is not free in B. So, $a \in (\phi_{\forall p_n A} \cup \phi_B)(f) = \bigsqcap_{X \in \mathsf{Prop}} (\phi_A \cup \phi_B)(f[n/X])$. Since p_n is not free in B, we have $\bigsqcap_{X \in \mathsf{Prop}} (\phi_A \cup \phi_B)(f[n/X]) = \bigsqcap_{X \in \mathsf{Prop}} (\phi_A(f[n/X]) \cup \phi_B(f))$. By (ECp), $\bigsqcap_{X \in \mathsf{Prop}} (\phi_A(f[n/X]) \cup \phi_B(f)) \subseteq (\phi_{\forall p_n A} \cup \phi_B)(f)$, so $f,a \Vdash \forall p_n A \vee B$, as desired. □

That suffices for Soundness. Next, we turn to Completeness. As in previous chapters, the proof uses a canonical model construction. Much of the proof follows the pattern of what has been done before. In particular, for a given logic $\mathsf{L}^{\forall p}$, the definitions of $\mathsf{L}^{\forall p}$-regular, prime $\mathsf{L}^{\forall p}$-theories, R, and * match those from before. The Pair extension lemma extends to the present context as well. The new additions to the canonical frame are the definitions of Prop and PropFun. For PropFun, we need to define the result of substituting representative formulas for propositions into formulas based on assignments, A^f, which itself requires a few lemmas to justify. Let us start with Prop.

- $\|A\| = \{a \in K : A \in a\}$.
- $\mathsf{Prop} = \{\|A\| \subseteq K : A \in \mathfrak{L}_{\forall p}$ and A is a closed formula$\}$.

We will tie propositions in the canonical $\mathsf{L}^{\forall p}$-frame to provable implication.

LEMMA 11.15 *For any closed formulas A and B, $\|A\| \subseteq \|B\|$ iff $\vdash_{\mathsf{L}^{\forall p}} A \to B$.*

Proof For the left-to-right direction, assume $\|A\| \subseteq \|B\|$ and assume $\nvdash_{\mathsf{L}^{\forall p}} A \to B$. Then $\langle\{A\},\{B\}\rangle$ is an $\mathsf{L}^{\forall p}$-disjoint pair. So there is a prime $\mathsf{L}^{\forall p}$-theory a such that $A \in a$ but $B \notin a$. This contradicts the initial assumption, so $\vdash_{\mathsf{L}^{\forall p}} A \to B$.

For the right-to-left direction, suppose $\vdash_{\mathsf{L}\forall p} A \to B$. Suppose $a \in ||A||$, so $A \in a$. As a is a theory, $B \in a$, so $a \in ||B||$, as desired. □

COROLLARY 11.16 *For any closed formulas A and B, $||A|| = ||B||$ iff $\vdash_{\mathsf{L}\forall p} A \leftrightarrow B$.*

Let $B_1, \ldots, B_k, \ldots$ be an enumeration of the closed formulas of $\mathfrak{L}_{\forall p}$. For a proposition $X \in \mathsf{Prop}$, let B_X be the first B in the enumeration such that $B \in X$. Then set $A[p_n/X] =_{=Df} A[p_n/B_X]$, so that we can talk about substituting a proposition into a formula via substituting its representative formula. This substitution is unique in the following sense.

LEMMA 11.17 *Let A be a formula with at most p_n free and let B and C be closed formulas. if $||B|| = ||C||$, then $||A[p_n/B]|| = ||A[p_n/C]||$.*

Proof Suppose $||B|| = ||C||$. By corollary 11.16, $\vdash_{\mathsf{L}\forall p} B \leftrightarrow C$. By lemma 11.2, $\vdash_{\mathsf{L}\forall p} A[p_n/B] \leftrightarrow A[p_n/C]$, so $||A[p_n/B]|| = ||A[p_n/C]||$, as desired. □

The preceding lemma then justifies the following definition.

DEFINITION 11.18 *Let A be a formula of $\mathfrak{L}_{\forall p}$ and $f \in \omega \mapsto \mathsf{Prop}$. A^f is defined as*

$$A^f = A[p_0/f(0), p_1/f(1), \ldots, p_n/f(n), \ldots],$$

where this is the simultaneous substitution of the propositions for the variables.

Define $A^{f\backslash n}$ as

$$A^{f\backslash n} = A[p_0/f(0), p_1/f(1), \ldots, p_n/p_n, \ldots].$$

In light of the finitude of A, this can be written instead as

$$A^f = A[p_{i_1}/f(i_1)] \ldots [p_{i_n}/f(i_n)],$$

where $p_{i_1}, \ldots, p_{i_n}$ are all the variables free in A. In terms of the assumed enumeration of closed formulas given earlier, this can be written as

$$A^f = A[p_{i_1}/B_{f(i_1)}] \ldots [p_{i_n}/B_{f(i_n)}].$$

We will record four facts about these substitutions.

LEMMA 11.19 *Let A be a formula and f an assignment. The following are true.*

- *If A is closed, then $A^f = A$.*
- *A^f is a closed formula.*
- *$A^{f\backslash n}[p_n/X] = A^{f[n/X]}$.*
- *$(\forall p_n A)^f = \forall p_n(A^{f\backslash n})$.*

Proof These are more or less immediate from the definitions. □

We can now resume the definition of the canonical frame.

- Define $\phi_A \colon \mathsf{Prop}^\omega \mapsto \mathsf{Prop}$ to be $\phi_A(f) = ||A^f||$.
- $\mathsf{PropFun} = \{\phi_A : A \in \mathfrak{L}_{\forall p}\}$.

The canonical model takes the valuation V to be $V(c) = ||c||$, for $c \in \mathsf{Con}$.

Next, we need to verify that Prop is model-like and that PropFun is fully closed.

LEMMA 11.20 Prop *is model-like.*

Proof This is left as an exercise for the reader. □

To prove that PropFun is fully closed, it will be useful to have a couple of lemmas.

LEMMA 11.21 *Let* $\forall p_n A$ *be a closed formula and* a *a prime* $\mathsf{L}^{\forall p}$*-theory. Then the following are equivalent.*

(i) $a \in ||\forall p_n A||$.
(ii) There is an $X \in$ Prop *such that* $a \in X$ *and* $X \subseteq ||A[p_n/P]||$*, for all* $P \in$ Prop.

Proof To prove that (i) implies (ii), suppose $a \in ||\forall p_n A||$. Let $X = ||\forall p_n A||$. Then $a \in X$. $X \subseteq ||A[p_n/P]||$, for each $P \in$ Prop from lemma 11.12 and (PQ1).

To prove that (ii) implies (i), suppose there is an $X \in$ Prop such that $a \in X$ and $X \subseteq ||A[p_n/P]||$, for all $P \in$ Prop. By definition, there is a closed formula B such that $||B|| = X$, so $B \in a$. Let c be a constant not occurring in B or A, and let $||c|| = P \in$ Prop. From the assumptions and lemma 11.12, $B \rightarrow A[p_n/c]$ is provable, so by lemma 11.2, $B \rightarrow \forall p_n A$ is provable. From lemma 11.12, $||B|| \subseteq ||\forall p_n A||$, whence $a \in ||\forall p_n A||$. □

COROLLARY 11.22 *Let* $\forall p_n A$ *be a closed formula. Then*

$$||\forall p_n A|| = \bigsqcap_{P \in \mathsf{Prop}} ||A[p_n/P]||.$$

LEMMA 11.23 *For all formulas* A, $\forall_n \phi_A = \phi_{\forall p_n A}$.

Proof By definition, $\forall_n \phi_A(f) = \bigsqcap_{P \in \mathsf{Prop}} \phi_A(f[n/P])$. By the definition of PropFun,

$$\bigsqcap_{P \in \mathsf{Prop}} \phi_A(f[n/P]) = \bigsqcap_{P \in \mathsf{Prop}} ||A^{f[n/P]}||.$$

By lemma 11.19, □

$$\bigsqcap_{P \in \mathsf{Prop}} ||A^{f[n/P]}|| = \bigsqcap_{P \in \mathsf{Prop}} ||A^{f\setminus n}[p_n/P]||.$$

By the preceding corollary,

$$\bigsqcap_{P \in \mathsf{Prop}} ||A^{f\setminus n}[p_n/P]|| = ||\forall p_n(A^{f\setminus n})||.$$

By lemma 11.19, $||\forall p_n(A^{f\setminus n})|| = ||(\forall p_n A)^f||$, and by definition of PropFun, $||(\forall p_n A)^f|| = \phi_{\forall p_n A}(f)$, as desired.

LEMMA 11.24 PropFun *is fully closed.*

Proof There are several cases to verify. Most are straightforward and covered by Goldblatt and Kane (2009). The details are left to the reader. As a hint, the case of (FAll) is handled by the preceding lemma. $\phi_{\forall p_n A} \in$ PropFun, so $\forall_n \phi_A \in$ PropFun, for all $n \in \omega$. □

LEMMA 11.25 PropFun *obeys* (FMod).

Proof By lemma 11.19, $||c^f|| = ||c||$, $\phi_c(f) = ||c|| = V(c)$. □

If $\mathsf{L}^{\forall p}$ contains (ECp), we need to verify that the frame obeys the additional frame condition.

LEMMA 11.26 *Suppose* $\mathsf{L}^{\forall p}$ *contains* (ECp). *Then the canonical frame obeys* (FECp).

Proof Suppose $a \in \prod_{X \in \mathsf{Prop}} (\phi(f) \cup \psi(f[n/X]))$. For some formulas A and B, $\phi = \phi_A$ and $\psi = \phi_B$, so $a \in \prod_{X \in \mathsf{Prop}} (\phi_A(f) \cup \phi_B(f[n/X]))$. Then we have

$$\prod_{X \in \mathsf{Prop}} (\phi_A(f) \cup \phi_B(f[n/X])) = \prod_{X \in \mathsf{Prop}} ||A^f \vee B^{f[n/X]}||.$$

Since A^f is closed, $||A^f \vee B^{f[n/X]}|| = ||(A^f \vee B)^{f[n/X]}||$. It then follows that $||(A^f \vee B)^{f[n/X]}|| = ||(A^f \vee B)^{f\setminus n}[p_n/X]||$. Using these identities, we have

$$\prod_{X \in \mathsf{Prop}} (\phi_A(f) \cup \phi_B(f[n/X])) = ||\forall p_n(A^f \vee B)^{f\setminus n}||.$$

Since A^f is closed, $||\forall p_n(A^f \vee B)^{f\setminus n}|| = ||\forall p_n(A^f \vee B^{f\setminus n})||$. Then, $||\forall p_n(A^f \vee B)^{f\setminus n}|| = ||\forall p_n(A^f \vee B^{f\setminus n})||$, and using (PQ3) and lemma 11.12,

$$||\forall p_n(A^f \vee B^{f\setminus n})|| \subseteq ||(A^f \vee \forall p_n B^{f\setminus n})||.$$

As

$$||(A^f \vee \forall p_n B^{f\setminus n})|| = \phi_A(f) \cup \forall_n \phi_B(f) = (\phi_A \cup \forall_n \phi_B)(f),$$

it follows that $\prod_{X \in \mathsf{Prop}} (\phi_A(f) \cup \phi_B(f[n/X])) \subseteq (\phi_A \cup \forall_n \phi_B)(f)$, so the frame satisfies (FECp), as desired. □

The final lemma to prove for Completeness is the Truth lemma.

LEMMA 11.27 (Truth lemma) *For any formula* A, *for any assignment* f, $|A|_f = ||A^f||$.

Proof The proof is by induction on the construction of A. The base cases are handled by the definition of Prop, (FMod), and (FTee) together with lemma 11.19.

The cases of the propositional connectives are handled as in previous proofs.

Suppose A is $\forall p_n B$. Then $|\forall p_n B|_(f) = \forall n |B|_f$ by definition. By the inductive hypothesis, $\forall n |B|_f = \forall n \phi_B(f)$. By lemma 11.23, $\forall n \phi_B(f) = \phi_{\forall p_n B}(f)$, so $|\forall p_n B|_f = \phi_{\forall p_n B}(f)$, as desired. □

With that, we can consider the Completeness theorem proved.

THEOREM 11.28 (Completeness) *For all formulas* A, *if* $\models_{\mathsf{L}^{\forall p}} A$, *then* $\vdash_{\mathsf{L}^{\forall p}} A$.

The mechanics of the Mares–Goldblatt frames have been exhibited in the context of propositional quantifiers. The same basic idea carries over to first-order quantifiers, to which we will now turn.

11.5 Syntax and Axioms for First-Order Quantifiers

In the move to first-order logic, we have to introduce terms and predicates. Terms will be either constants or variables. The language will have a countably infinite set of variables $\mathsf{Var} = \{x, y, z, x_1, \ldots\}$ and a set of constants $\mathsf{Con} = \{c, d, c_1, \ldots\}$, which may be empty and is at most countably infinite.[3] We will use $s, t, s_1, \ldots$ for arbitrary terms. The set of terms will be Terms.

Predicates, $F, G, \ldots,$ will be n-ary, where $n \geq 1$. Atomic formulas will be n-ary predicates followed by n terms, such as $Fs_1, \ldots, s_n$. When the internal structure of an atom is not salient to the point at hand, $p, q, r \ldots$ may be used.

The language $\mathfrak{L}_\forall$ will be obtained by adding a singulary universal quantifier, $\forall$, to $\mathfrak{L}$. The existential quantifier, $\exists$, will be defined from the universal as $\exists x A =_{Df} \neg\forall x \neg A$. We will use the standard definitions for free and bound variables in a formula. The notation $A[x/t]$ means that t is substituted for all free occurrences of x in A, provided that t is free for x in A. Let us turn now to axioms.

The axioms for the zero-order fragment of a base relevant logic L are the axioms of L. Note that the axioms permit instances using the new atomic formulas with predicates. The additional axiom and rule needed for the quantifiers are

(Q1) $\forall x A \rightarrow A[x/t]$, provided t is free for x in A, and
(Q2) $A \rightarrow B \Rightarrow A \rightarrow \forall x B$, where x is not free in A.

The logic QL will be L with the addition of (Q1) and Q2. There is an additional quantifier axiom that one might add, the axiom of extensional confinement (EC) $\forall x(A \vee B) \rightarrow A \vee \forall x B$, where x is not free in A. QL with the addition of (EC) is $\mathsf{QL.EC}$.[4]

We will now state a few lemmas concerning QL. The proofs will be left to the reader, who can consult the analogous lemmas from Section 11.2 for hints.

LEMMA 11.29 *The following are provable in* QL.

(*i*) $(\forall x A \wedge \forall x B) \rightarrow \forall x(A \wedge B)$.
(*ii*) $A \rightarrow \forall x A$, *if* x *is not free in* A.
(*iii*) $\forall x(A \rightarrow B) \rightarrow (\forall x A \rightarrow \forall x B)$.

LEMMA 11.30 *The logic* QL *is closed under the following rules.*

(*i*) $A \Rightarrow \forall x A$.
(*ii*) $A \rightarrow B[x/c] \Rightarrow A \rightarrow \forall x B$, *if* c *is not in* A *or* B, *and* x *is not free in* A.
(*iii*) $A[x/c] \Rightarrow \forall x A$, *if* c *is not in* A.

Let us turn to the frames for these logics.

[3] There is some repurposing of terminology and symbols from the previous sections. Context should make it clear what is meant.

[4] Mares and Goldblatt (2006) use RQ for what we call QR.EC, but we think that the new nomenclature is a bit easier to remember and in line with the nomenclature for modal logics in previous chapters.

11.6 Mares–Goldblatt Frames for First-Order Quantifiers

To obtain a frame for first-order quantifiers, we simply add to a propositional quantifier frame a non-empty domain of individuals. Note that in this context, Var and Con are subsets of Terms rather than of At.

DEFINITION 11.31 (Mares–Goldblatt frame) *A* MARES–GOLDBLATT QL-FRAME *is a septuple* $\langle K, N, R, ^*, I, \mathsf{Prop}, \mathsf{PropFun}\rangle$, *where* $\langle K, N, R, ^*\rangle$ *is an* L*-frame, I is a non-empty domain of individuals,* Prop *is model-like, and* PropFun *is closed.*

DEFINITION 11.32 (Valuation, assignment) *A* VALUATION *V on a Mares–Goldblatt frame is a function such that*

- $V(c) \in I$ *for all* $c \in \mathsf{Con}$, *and*
- $V(F)\colon I^n \mapsto \mathsf{Prop}$, *for F an n-ary predicate.*

An ASSIGNMENT *f for a Mares–Goldblatt frame is a function such that* $f(x_j) \in I$, *for* $j \in \omega$.

As before, it will be useful to view assignments as sequences,

$$f = \langle f(0), f(1), f(2), \ldots \rangle,$$

and defining substitutions on them as

$$f[n/d] = \langle f(0), \ldots, f(n-1), d, f(n+1), \ldots \rangle,$$

where $d \in I$.

Following Mares and Goldblatt (2006, 168), given an assignment f and a valuation V on a Mares–Goldblatt frame, we can define some additional functions.

- $Vf(s) = Vfs$, for $s \in \mathsf{Terms}$.
- $Vfc = V(c)$, for $c \in \mathsf{Con}$.
- $Vfx = f(x)$, for $x \in \mathsf{Var}$.
- $V(Fs_1, \ldots, s_n)(f) = V(F)(Vfs_1, \ldots, Vfs_n)$.

This allows us to view an assignment V and an atomic formula $Fs_1, \ldots, s_n$ as determining a function $I^\omega \mapsto \mathsf{Prop}$. This notation will be useful later.

DEFINITION 11.33 (Mares–Goldblatt model) *A* MARES–GOLDBLATT MODEL *M is a pair of a Mares–Goldblatt frame F and a valuation function V such that for all atomic formulas A,* $V(A) \in \mathsf{PropFun}$.

Since the set of variables is countable, it will be treated as ordered in a given way. The notion of an n-variant is adapted straightforwardly, and where y is x_n, we may write $f \asymp_y g$ for $f \asymp_n g$.

The verification conditions on Mares–Goldblatt models are the same as those for propositional quantifier models, swapping the universal propositional quantifier for the universal quantifier over objects. One difference emerges with the universal quantifier clause.

- $f, a \Vdash \forall x B$ iff $\exists X \in \mathsf{Prop}\left(a \in X \text{ and } X \subseteq \bigcap_{i \in I} |A|_{f[x/i]}\right)$.

This clause can be rewritten using the notion of x-variants.

- $f,a \Vdash \forall x B$ iff $\exists X \in \mathsf{Prop}\left(a \in X \text{ and } X \subseteq \bigcap_{g \asymp_x f} |A|_g\right)$.

In this clause, the assignments are in I^ω, rather than in Prop^ω. The more algebraic version of the conditions can be adapted, with the following minor adjustments.

- $|Fs_1, \ldots, s_n|_f = V(Fs_1, \ldots, s_n)(f)$.
- $|\forall x_n A|_f = \bigsqcap_{g \asymp_n f} |A|_g = \forall_n |A|_f$.

The equations then support the following lemma.

LEMMA 11.34 *Let M be a Mares–Goldblatt model, A an arbitrary formula, and $f \in I^\omega$. Then $|A| \in \mathsf{PropFun}$ and $|A|_f \in \mathsf{Prop}$.*

For the propositional function $|A|$, we will write ϕ_A.

The definitions for counterexample, holding, and validity match those of Section 11.3. The Heredity lemma holds for Mares–Goldblatt models, as do the Verification lemma and Semantic substitution lemma. Similarly, whenever two assignments $f, g \in I^\omega$ agree on the variables free in A, then $|A|_f = |A|_g$.

11.7 Soundness and Completeness for Mares–Goldblatt Frames

The arguments of this section are largely similar to those of §11.4.

THEOREM 11.35 (Soundness theorem) *For all formulas A, if $\vdash_{QL} A$, then $\models_{QL} A$.*

Proof The proof is by induction on the structure of the proof of A. We will only do the new cases.

For (Q1), suppose that $f,a \Vdash \forall x B$. Then $a \in |\forall x B|_f \in \mathsf{Prop}$. From the verification condition for the universal quantifier, $|\forall x B|_f \subseteq |B|_g$, for $g \asymp_x f$. Let $i = Vf(s)$, where s is free for x in B. Then, $|\forall x B|_f \subseteq |B|_{f[n/i]}$, so by the Semantic substitution lemma, $|\forall x B|_f \subseteq |B[x/s]|_f$. Thus, $f,a \Vdash B[x/s]$, as desired.

For (Q2), suppose that $\models_{QL} A \rightarrow B$, where x is not free in A. Suppose that $f,a \Vdash A$. Since x is not free in A, $g,a \Vdash A$, for all $g \asymp_x f$. From the assumption, it follows that $g,a \Vdash B$, for all such g. Therefore, $|A|_f = |A|_g \subseteq |B|_g$, so $|A|_f \subseteq \bigsqcap_{g \asymp_x f} |B|_g = |\forall x B|_f$. It follows that $f,a \Vdash \forall x B$, as desired. □

Completeness is proved via the canonical model method. The canonical frame for QL is similar in many ways to that of the canonical frame for $\mathsf{L}^{\forall p}$. One difference concerns constants, here being terms. If the language does not already contain a countably infinite supply of constants, then they are added to the language. Next, we will define the new parts of the canonical QL-frame.

- $I = \mathsf{Con}$.
- $||A|| = \{a \in K : A \in a\}$, where A is a closed formula.

- $\mathsf{Prop} = \{\|A\| : A \text{ is a closed formula}\}$.
- For each formula A, $\phi_A(f) = \|A^f\|$.
- $\mathsf{PropFun} = \{\phi_A : A \in \mathfrak{L}_\forall\}$.

To extend this to a QL-model, we add the following clauses for the valuation.

- $V(c) = c$.
- $V(F)(c_1, \ldots, c_n) = \|Fc_1, \ldots, c_n\|$.

We will record a few lemmas, whose proofs are largely the same as in Section 11.4.

LEMMA 11.36 *For any closed formulas A and B, $\|A\| \subseteq \|B\|$ iff $\vdash_{QL} A \to B$.*

LEMMA 11.37 Prop *is model-like.*

LEMMA 11.38 PropFun *is closed.*

LEMMA 11.39 *If the formula $\forall x A$ is closed, then the following are equivalent.*

- $a \in \|\forall x A\|$
- *There is $X \in$ Prop such that $a \in X$ and $X \subseteq \|A[x/c]\|$ for all $c \in I$*

COROLLARY 11.40 *If the formula $\forall x A$ is closed, then* $\|\forall x A\| = \bigsqcap_{c \in \mathsf{Con}} \|A[x/c]\|$.

LEMMA 11.41 (Truth lemma) *For all formulas A, $|A| = \phi_A$. Consequently, for $f \in I^\omega$, $|A|_f = \phi_A(f)$.*

These pieces are enough for Completeness.

THEOREM 11.42 (Completeness) *For all formulas A, if $\models_{QL} A$, then $\vdash_{QL} A$.*

Proof Following Mares and Goldblatt (2006), we will give a slightly different proof than in other cases. Suppose that $\models_{QL} A$. Then A holds at all regular points in the canonical model, so A^f is in all regular prime QL-theories, for any $f \in I^\omega$. In particular, this holds for an assignment f that assigns only constants not occurring in A. This implies $\|\mathsf{t}\| \subseteq \|A^f\|$, which yields that $\mathsf{t} \to A^f$ is a QL-theorem, when $\mathsf{t} \to \forall x_1, \ldots, \forall x_n A$ is a theorem as well. It then follows that $\forall x_1, \ldots, \forall x_n A$ is a theorem, and so A is a theorem by (Q1). □

There is one lingering issue for this section, namely logics with (EC). The model condition for (EC) is the following.

(FEC) $X \backslash Y \subseteq \bigcap_{g \asymp_x f} \phi(g)$ only if $X \backslash Y \subseteq \forall_n \phi(f)$

for all $\phi \in$ PropFun, all $X, Y \in$ Prop, and all $f \in I^\omega$.

LEMMA 11.43 *The axiom* (EC) *is valid any class of frames obeying* (FEC).

Proof Let M be a model on a frame obeying (FEC). Suppose that $f, a \Vdash \forall x(A \vee B)$, where x is not free in A. So there is $X \in$ Prop such that $a \in X$ and $X \subseteq \bigcap_{g \asymp_x f} |\phi_A \cup \phi_B|_g$.

$$\bigcap_{g \asymp_x f} (\phi_A \vee \phi_B)(g) = \bigcap_{g \asymp_x f} (\phi_A(g) \cup \phi_B(g)) = \bigcap_{g \asymp_x f} (\phi_A(f) \cup \phi_B(g)).$$

As $\bigcap_{g \asymp_x f} (\phi_A(f) \cup \phi_B(g)) = \phi_A(f) \cup \bigcap_{g \asymp_x f} \phi_B(g)$, it then follows that $X\backslash|A|_f \subseteq \bigcap_{g \asymp_x f} \phi_B(g)$. From (FEC), we have $X\backslash|A|_f \subseteq \forall_n \phi_B(f)$, so $f, a \Vdash A \vee \forall x B$, as desired.

□

For Completeness, we will show that the canonical model for QL.EC obeys the required condition.

THEOREM 11.44 *The canonical frame for* QL.EC *obeys* (FEC).

Proof Suppose that $X\backslash Y \subseteq \bigcap_{g \asymp_{x_n} f} \phi(g)$. Since $X, Y \in$ Prop, there is a formula A such that $Y = ||A||$, and as $\phi \in$ PropFun, there is a B such that $\phi = \phi_B$. Rewriting our assumption, we then have $X\backslash||A|| \subseteq \bigcap_{g \asymp_{x_n} f} \phi_B(g) = \bigcap_{g \asymp_{x_n} f} ||B^g||$. For every g such that $g \asymp_{x_n} f$, there is $c \in I$ such that $g = f[n/c]$, so $X\backslash||A|| \subseteq ||B^{f[n/c]}||$. This, in turn, implies $X \subseteq ||A|| \cup ||B^{f[n/c]}||$, for each $c \in I$. Let c be a constant not occurring in A or B. Since A is closed, $||A|| \cup ||B^{f[n/c]}|| = (||A|| \cup ||B||)^{f[n/c]}$. Since $X \in$ Prop, there is a C such that $||C|| = X$. Using the lemma 11.36, $X \subseteq (||A|| \cup ||B||)^{f[n/c]}$ implies $C \rightarrow (A \vee B)^{f[n/c]}$, for some c not occurring in A or B. From the Semantic substitution lemma, we then have $C \rightarrow (A \vee B)^{f\backslash n}[n/c]$. By lemma 11.30, $C \rightarrow \forall x_n((A \vee B)^{f\backslash n})$, which is $C \rightarrow \forall x_n(A \vee B^{f\backslash n})$, so by (EC), $C \rightarrow (A \vee \forall x_n B^{f\backslash n})$. By lemma 11.36, $X \subseteq ||A|| \cup ||\forall x_n B^{f\backslash n}||$. Using the definition of PropFun and the analog of lemma 11.23, we then have $X \subseteq ||A|| \cup \phi_{\forall x_n B}(f\backslash n)$, but as $\phi_{\forall x_n B}(f\backslash n) = \forall_n \phi_B(f\backslash n) = \forall_n \phi(f)$, we have $X\backslash||A|| \subseteq \forall_n \phi_B(f)$. Substituting assumed identities, we then have the desired $X\backslash Y \subseteq \forall_n \phi(f)$.

□

With the adequacy of the frames established, we will close this section by looking at the philosophical interpretation of the framework offered by Mares (2009). Recall that Mares (2004; 2021) interprets the points in the frames as situations that carry information, which is used in his theory of situated inference. The verification conditions for a connective state the conditions under which certain kinds of information are available in a situation. A situation verifies a universal claim if it carries *general information* that entails every instance of that claim. On this interpretation, the non-Tarskian verification condition used earlier is appropriate, whereas the Tarskian verification condition, to be discussed in the next section, is not, as in many cases lots of information about particulars is not the same as general information. Let us now turn to the Tarskian verification conditions for the quantifiers.

11.8 Tarskian Conditions and Incompleteness

The verification condition for quantifiers is different from the clause usually used in first-order classical logic, namely,

- $f, a \Vdash \forall x B$ iff for all $g \asymp_x f$, $g, a \Vdash B$.

Let us call this the TARSKIAN CONDITION. Indeed, this was the condition used by Routley and Meyer (1973) for early work on quantificational extensions of relevant logics.[5] The frames of Sylvan and Meyer didn't use a set of propositions, but rather only added a domain of individuals to the basic ternary relational frames. These frames, $\langle K, N, R, ^*, I\rangle$, will be called Q-FRAMES, where $\langle K, N, R, ^*\rangle$ is a ternary relational frame and I is a non-empty set of individuals. A Q-frame is a Q-frame for L when the underlying ternary relational frame obeys the frame conditions for L. We'll note that the logic $\mathsf{QR.EC}$ is sound for the class of all Q-frames for R.

One might wonder how the Tarskian condition relates to the non-Tarskian condition adopted in the earlier sections. First, Mares and Goldblatt (2006, 171) prove two sufficient conditions for the non-Tarskian condition to coincide with the Tarskian condition. One condition is that I is finite. The other condition is that the frame is FULL in the sense that $\mathsf{Prop} = \mathcal{H}$. When the frame is full, then if $S \subseteq \mathcal{H}$, $\bigcap S \in \mathsf{Prop}$ and $\bigcap S = \bigsqcap S$. When frames are not full, one may have $\bigsqcap S \subsetneq \bigcap S$, and indeed, one can have $\bigsqcap S \in \mathsf{Prop}$ while $\bigcap S \notin \mathsf{Prop}$.

Inspection of the canonical frame construction for $\mathsf{L}^{\forall p}$ and QL will reveal that the Prop sets in those frames are not full. In virtue of corollaries 11.22 and 11.40, we see that for closed formulas, the quantifiers are evaluated in a Tarskian way, and, because of the details of how assignments work in the canonical models, all formulas are evaluated as closed formulas relative to an assignment. In the canonical model, the proposition that entails all the instances of B will simply be the proposition that $\forall x B$.

Let us pause for a moment to consider the existential quantifier, which is defined in terms of negation and the universal quantifier. Unpacking the definition and using the corresponding verification clauses, we get the following clause for the existential quantifier.

- $f, a \Vdash \exists x A$ iff for all $X \in \mathsf{Prop}$ such that $a^* \in X$, there is a $b \in X$ and $g \asymp_x f$ such that $b^* \in |A|_g$.

Following Dunn (1993), Mares and Goldblatt (2006, 170) suggest using the "compatibility" understanding of the Routley star. A proposition X is compatible with a point a just in case $a^* \in X$, and two propositions X and Y are compatible iff there is some $b \in K$ such that $b \in X$ and $b^* \in Y$. The verification clause then says that $\exists x A$ holds at a iff every proposition compatible with a is compatible with some instance of A. When the universal quantifier obeys the Tarskian condition, however, this ends up simplifying to the corresponding Tarskian condition for the existential.

- $f, a \Vdash \exists x A$ iff for some $g \asymp_x f$, $g, a \Vdash A$.

The simplifications resulting from the Tarskian clause for the universal quantifier extend to the existential as well.

Given all this, why would one opt for the non-Tarskian condition, rather than the Tarskian condition? The reason is that the Tarskian condition results in incompleteness with respect to Q-frames for R, as shown by Fine (1989). The formula Fine gives that

[5] Priest (2008, ch. 24) uses the Tarskian condition as well.

is valid in all Q-frames for R that is not a theorem of QR.EC is large: It is an implication whose antecedent is

$$(p \rightarrow \exists x Ex) \wedge \forall x\big((p \rightarrow Fx) \vee (Gx \rightarrow Hx)\big)$$

and whose consequent is

$$\big(\forall z((Ez \wedge Fz) \rightarrow q) \wedge \forall y((Ey \rightarrow q) \vee Gy)\big) \rightarrow \big(\exists x Hx \vee (p \rightarrow q)\big).^6$$

Let us call this formula (F), for Fine. For a proof of its validity, see *Entailment volume II* (1992, 232–233).

The proof of the validity of (F) does not use much that is distinctive of R-frames. In fact, Fine points out that really it only depends on (I4), but the formula and argument can be adapted to work with (I3) instead. This means that the incompleteness extends to QRW.EC, QTW.EC, QT.EC, QC.EC, and QE.EC. It is worth also noting that while QRM-frames are defined with a single domain for every point, Fine's argument proceeds without that assumption, permitting domains at different points to vary.

Fine (1989, 214–225) provides a countermodel showing that (F) is not a theorem of QR.EC, a summary of which is provided by Anderson and Belnap (1992, 233–235). The construction is involved and uses infinite domains, and we leave it to the interested reader to investigate further. Fine's incompleteness result holds for some strong logics, but, as far as we know, it has not been shown to hold for weaker logics. Completeness with the Tarskian condition may yet be available for some relevant logics. One can give completeness proofs for the Q-models with the Tarskian condition, using tableaux systems, as is done by Priest (2008, ch. 24) and Weiss (2020b).

One upshot of the disparity between the Mares–Goldblatt verification clause for quantifiers and the Tarskian one is that the Tarskian clause results in more validities. This means that for purposes of showing that a formula does not hold in the logics valid on the Mares–Goldblatt frames, we can, in many cases of interest, use the Tarskian clause, or alternatively consider full Mares–Goldblatt frames. While this does not work in general, as guaranteed by the incompeleteness result given earlier, it can simplify finding counterexamples for some formulas of interest. This is illustrated by Dunn in the appendix to his (1987) as well as by Badia and Tedder (2018) in showing that elimination of quantifiers fails for some relevant theories.

Kremer (1993) shows that the propositional quantification, when interpreted with the analogous Tarskian clause, there called the primary interpretation of the quantifiers, is not axiomatizable. It is, in fact, isomorophic to second-order classical logic. In footnote 1 of his paper, Kremer says that completeness can be had by using a secondary interpretation of those quantifiers, one that restricts quantification to a subset of all the propositions, an idea suggested also by Routley and Meyer (1973, 236–237). While Kremer (1993, 347) takes his results to "bode ill" for relevant logics, we are less sure, since as Kremer notes, the secondary interpretation has a good proof-theoretic

[6] Fine (1989, 212) remarks, "The reader may be somewhat mystified by the above formula, which just appears out of the blue. It arose naturally, however, from a certain attempted method for proving completeness. Unfortunately, it would take too long to describe the method or to retrace the steps by which the formula was found."

motivation, and it appears to have a connection to the Mares–Goldblatt framework. Kremer's, and related unaxiomatizability and incompleteness results, do, however, deserve further philosophical discussion.

Goldblatt and Kane (2009, 95ff.) show that in full propositional quantifier frames, many propositional quantificational logics are incomplete. In terms of the base relevant logics of interest for this book, any sublogic L of $\mathsf{E}^{\forall p}$ is incomplete with respect to the full class of propositional quantifier L-frames. To explain the incompleteness, let us define two formulas.

- Exm is $\forall p(p \vee \neg p)$.
- Atm is $\exists q(q \wedge \forall r((q \rightarrow r) \vee (r \rightarrow q)))$.

Goldblatt and Kane show that the formula $(Exm \wedge \mathfrak{t}) \rightarrow Atm$ is valid over all full frames for a wide range of logics. This formula is not, however, valid over the class of propositional frames for those logics, when Prop is permitted not to be full. This shows that the possibility that Prop is not full is essential to the completeness of these logics.

Finally, Badia (2019) shows that the Tarskian interpretation of the propositional quantifiers leads to incompactness as well. A logic is COMPACT iff for any set of formulas X, if there is not a model, in the class of models for the logic, for every formula in X, then there is some finite $Y \subseteq X$ such that there is not a model for all the formulas in Y. A logic is incompact iff it is not compact. Badia shows that a particular fragment of the language $\mathfrak{L}_{\forall p}$ generates incompactness over the class of full propositional frames for $\mathsf{B}^{\forall p}$. The fragment in question is the $\forall_1$ fragment, which contains all formulas equivalent to a formula of the form $\forall p_1 \ldots \forall p_n A$, where A is quantifier-free. The class of formula is axiomatizable, but it is incompact. While we are not considering strong Completeness results here, namely Completeness permitting assumptions, it is worth noting that Badia's result implies failure of strong Completeness.

11.9 Exercises

(1) Provide a proof for lemma 11.20, showing that in the canonical frame, Prop is model-like.

(2) Provide a proof for lemma 11.24, showing that in the canonical frame, PropFun is fully closed.

(3) Provide proofs for lemmas 11.29 and 11.30.

(4) State and prove the Semantic substitution lemma for the first-order quantifiers.

(5) Show that (F) is valid in the class of QR.EC frames using the Tarskian verification condition.

(6) Prove that in models where both I and Prop are finite, $|\forall x A|f = \bigcap_{g \asymp_x f} |A|g$. Then show that in models where Prop is full that $|\forall x A|f = \bigcap_{g \asymp_x f} |A|g$. Conclude that QL is not complete with respect to frames where either Prop and I are finite or Prop is full.

12 Identity

12.1 Introduction

In this chapter, we will add a binary relation of identity, $=$, to the language $\mathfrak{L}_\forall$. We will then develop the models for identity in the Mares–Goldblatt framework.[1] As we will see, there are some hurdles to modeling identity while also holding on to basic relevant intuitions and insights.

To begin, we will address the question of why new models for identity are needed. It will be worth looking at the problems with the naive modeling. Suppose that we require that "$=$" be interpreted as the finest equivalence relation:

$$a, f \Vdash s = t \text{ iff } Vf(s) = Vf(t).$$

In that case, $t = t$ will be ubiquitously true in all models, as there will be no way to force the two occurrences of "t" to denote different objects. This implies that for any formula A, $A \rightarrow t = t$ will be valid. At the propositional level, we do not have logical truths being implied by everything, so it seems like self-identities should not be implied by everything either. This, then, is the problem: We want to have some logical truths involving identity, in particular

- $t = t$,
- $s = t \rightarrow t = s$, and
- $(s = t \wedge t = u) \rightarrow s = u$,

but we want to avoid arbitrary formulas implying these. There are plausible axiom systems that achieve this, and it would be good to have models for them. Coming up with these models will provide some additional insight into how the ternary relational models work, and they will help indicate some of their limitations.

One of the distinctive applications of identity in relevant logic is Dunn's RELEVANT PREDICATION.[2] There is an intuitive difference between saying of Mount Taranaki that it is in New Zealand and saying of Taroko Mountain that it is such that Mount Taranaki is in New Zealand. The predicate "is such that Mount Taranaki is in New Zealand" does not really involve its object, as opposed to "is in New

[1] This chapter generalizes the results of Standefer (2021a), which focused on adding identity to R and some of its quantificational extensions.

[2] For more on relevant predication, see Dunn (1987; 1990a,b) and Kremer (1989; 1997).

Zealand," which is a paradigm of an object-involving predicate. Dunn defines relevant predication as

$$(\rho x A(x))t =_{Df} \forall x(x = t \to A(x)).$$

Relevantly predicating $A(x)$ of t is saying that being identical to t *implies* having the property $A(x)$. True relevant predications establish more of a connection between their subject and predicate than standard predications. While Taroko Mountain may be such that Mount Taranaki is in New Zealand, something being Taroko Mountain plausibly falls short of implying the location of Mount Taranaki.

Of importance for the theory of relevant predication is the fact that there are different formulations of the substitution axiom for identity, a conjunctive form,

$$(A(s) \wedge s = t) \to A(t),$$

and an iterated implication form,

$$A(s) \to (s = t \to A(t)).$$

The iterated implication form, it turns out, does not sit well with relevance intuitions, as argued by Dunn (1987). Hence, we will focus on the conjunctive form here. Mares (1992c) shows that allowing substitution into the scope of a conditional in the conjunctive substitution axiom has serious consequences for the theory of relevant predication, and so he considers a restricted form of the conjunctive substitution axiom.[3] Kremer (1999) discusses motivations for different axiomatizations of relevant identity. Mares (2019) explores connections between the relevant biconditional and identity. Øgaard (2020b) looks at substitution of identicals in relevant logics. As is apparent from this work, there is much to explore regarding identity in relevant logics.

There has been some work on alternative models for quantified relevant logics with and without identity. The pioneering work of Mares (1992c) provided models for a range of relevant logics with identity and quantifiers, building on the work of Fine (1974; 1988). Apart from Fine's models for quantifiers, Brady (1988; 1989a) has given a kind of algebraic semantics for relevant logics with quantifiers, but there has not been an extension of Brady's semantics to deal with identity.

Priest (2008, ch. 24) develops constant-domain models for quantified relevant logics, as well as the extension with identity, although Priest uses the Tarskian truth condition for the universal quantifier. Priest provides a tableau system that is Sound and Complete for his models, rather than a Hilbert-style axiom system like the other authors mentioned earlier.

Finally, Ferenz (2019) has, independently, developed an approach to incorporating identity into Mares–Goldblatt models, different from the one pursued here. Ferenz is motivated more by Kremer's concerns over indiscernibility and issues of relevant predication, and so his conditions for identity in models are weaker than the ones considered below. A detailed comparison of Ferenz's approach and the one of this

[3] The restricted and unrestricted forms of the substitution axiom are, respectively, (I3) and (I4) below.

chapter will be left to the interested reader. As with most things in relevant logics, alternative approaches are welcome and often illuminating.

The remainder of the chapter will proceed as follows. In Section 12.2, we will begin by defining models for zero-order logics and then present the axioms and models for identity in the context of zero-order logics. In Section 12.3, we will add propositions and propositional functions from the Mares–Goldblatt frames for quantification. Completeness for the identity logics so far covered is presented in Section 12.4. In Section 12.5, we add quantifiers to the language and we can then provide conditions for stronger substitution axioms in the context of quantifiers. We sketch a proof of Completeness for the strong logic of identity and quantification. Finally, in Section 12.6, we discuss relevant predication, an idea due to Dunn.

12.2 Zero-Order Logics

We will begin by defining the zero-order logics, since these have not been defined before. The formulas are built up as follows, where Var is a countably infinite set of variables, Con a countable set of constants, $\mathsf{Var} \cap \mathsf{Con} = \emptyset$, and $\mathsf{Terms} = \mathsf{Var} \cup \mathsf{Con}$. This is the language of Chapter 11 with the quantifiers removed. As in that chapter, we treat the set of variables as coming with a canonical ordering, so that we can interchangeably use functions on ω as functions on variables.

Frames and models for the zero-order language can be obtained with some minor modifications on the frames and models of Chapter 11. We will provide the definitions of frames and models for B, which can be adjusted for other logics in the usual ways.

DEFINITION 12.1 (Zero-order ternary relational frame) *A ZERO-ORDER TERNARY RELATIONAL FRAME G is a quintuple $\langle K, N, {}^*, R, I\rangle$, where $\langle K, N, {}^*, R\rangle$ is a ternary relational frame and the domain $I \neq \emptyset$.*

DEFINITION 12.2 (Zero-order ternary relational model) *A ZERO-ORDER TERNARY RELATIONAL MODEL M is a pair $\langle F, V\rangle$, where F is a zero-order ternary relational frame and V is an interpretation function such that*

- $V(P)\colon I^n \mapsto \wp(K)$,
- $V(c) \in I$, *for* $c \in \mathsf{Con}$, *and*
- *for* $a, b \in K$, *if* $a \leq b$ *and* $a \in V(P)(i_1, \ldots, i_n)$, *then* $b \in V(P)(i_1, \ldots, i_n)$.

A model M is BUILT ON a frame F iff M is $\langle F, V\rangle$, for some V.

An ASSIGNMENT $f\colon \omega \mapsto I$ is a function from the variables to the domain, also written I^ω below.

The verification clauses are the same as from Chapter 11. The definitions of counterexample and validity are carried over as well.

With the models for the underlying language defined, we can move on to the logics of identity. We will proceed slowly with the logic of identity, adding axioms in stages.

The axioms for identity are surprisingly modular in the sense that there is a lot of control of how much substitution is sanctioned.

We will begin with adding the identity predicate to a zero-order language, extending the language with a binary predicate for identity, $=$. We will show how to augment the zero-order ternary relational frames to obtain frames and models that are adequate; crucially, they do not make self-identities ubiquitously true. Beginning with the zero-order models allows us to factor out aspects of the models that are there solely to deal with quantifiers. The interactions of identity and quantification add some complications, which we will see below, but the core ideas of the models are already present at the zero-order level.

To begin, we will consider some constraints on interpretations of the identity predicate. These constraints also formed the basis of the Fine models for identity developed by Mares (1992c).

(1) The interpretation of identity needs to be reflexive at the regular points, but it need not be reflexive at other points. This will ensure that $t = t$ is valid while $p \rightarrow t = t$ is not.
(2) It needs to be symmetric and transitive at every point.
(3) There needs to be relations between the interpretation of identity at a point b and at b^*.
(4) The valuations need to reflect identities in some way to ensure that substitutions work as required.

These constraints will translate to conditions on the frames and models. In particular, the first three will be conditions on frames, while the fourth will be a condition on models. The use of model conditions appears unavoidable, as we need to deal directly with the valuations, rather than the frame parts.

To satisfy these constraints, we will use a family of relations, $\approx_a$, for $a \in K$, where $\approx_a \subseteq I \times I$. These relations will satisfy the constraints and we will require that the interpretation functions respect the relations in a certain way.

DEFINITION 12.3 (Identity frames) *A* ZERO-ORDER IDENTITY FRAME F *is sextuple* $\langle K, N, {}^*, R, I, \{\approx_a\}_{a \in K}\rangle$, *where the initial quintuple is a zero-order ternary relational frame, and the family of relations obeys the following conditions, for all* $a, b \in K$ *and* $i, j, k \in I$.

(ID1) *If* $a \leq b$, *then* $\approx_a \subseteq \approx_b$.
(ID2) *If* $a \in N$, *then* $(i, i) \in \approx_a$.
(ID3) *If* $(i, j) \in \approx_a$, *then* $(j, i) \in \approx_a$.
(ID4) *If* $(i, j) \in \approx_a$ *and* $(j, k) \in \approx_a$, *then* $(i, k) \in \approx_a$.
(ID5) *If* $(i, j) \in \approx_a$ *and* $(j, k) \in \approx_{a^*}$, *then* $(i, k) \in \approx_{a^*}$.

A few comments are in order. The first condition ensures that the Heredity lemma is provable. The second ensures that reflexivity of identity is valid. The third is for symmetry, and the fourth is for transitivity. The fifth ensures that substitution in negated identities works out.[4]

In order to define the models, we will need two subsidiary definitions.

DEFINITION 12.4 (Weakly symmetric) *A point a is WEAKLY SYMMETRIC in (i,j) iff if $Vf(u) = i$ and $Vf(v) = j$, then for each predicate P,*

$$a \in V(Pt_1,\ldots,t_n)(f) \text{ iff } a \in V(Ps_1,\ldots,s_n)(f),$$

where

- *if $s_\ell \notin \{u,v\}$, then $t_\ell = s_\ell$, and*
- *if $s_\ell \in \{u,v\}$, then $t_\ell = u$ or $t_\ell = v$.*

Next, we need to define the agreement required between the interpretation and the indexed relations, which we will call conformity.

DEFINITION 12.5 (Conformity) *A valuation V CONFORMS to a zero-order identity frame iff for all $f \in I^\omega$, $s,t \in$ Terms, $a \in K$,*

$$a \in V(=)(Vf(s), Vf(t)) \text{ iff } (Vf(s), Vf(t)) \in \approx_a.$$

With that definition in hand, we can give the definition of identity models.

DEFINITION 12.6 (Identity models) *A zero-order identity model M is a pair $\langle F, V\rangle$, where F is a zero-order identity frame, V is a valuation conforming to F and obeying the conditions from the zero-order models, as well as the following two conditions.*

(ID6) *If $(i,j) \in \approx_a$, then a^* is weakly symmetric in (i,j).*

(ID7) *If $Vf(u) = i$, $Vf(v) = j$, $a \in V(Pt_1,\ldots,t_n)(f)$, and $(i,j) \in \approx_a$, then $a \in V(Ps_1,\ldots,s_n)$, where*
   - *if $t_\ell \notin \{u,v\}$, then $s_\ell = t_\ell$, and*
   - *if $t_\ell \in \{u,v\}$, then $s_\ell \in \{u,v\}$.*

Condition (ID6), with (ID5), captures a maximality intuition about the Routley star, which says that a^* "asserts what [the point a] does not deny."[5] Together they yield the soundness of substitution in the scope of a negation. Condition (ID7) ensures that substitution of identicals works for atomic formulas.

Using the definition of conformity, we can derive an equivalent verification condition for the identity predicate.

- $f,a \Vdash s = t$ iff $(Vf(s), Vf(t)) \in \approx_a$.

In other words, an identity $s = t$ holds at a if and only if the objects denoted by "s" and by "t" are in the relation $\approx_a$. This is useful since it allows us to reason directly with the indexed relations, which provide good intuitions about identity.

For the remainder of this chapter, we will be concerned with validity with respect to a class of models.

[4] If one is using an incompatibility relation, rather than the Routley star, to interpret negation, this condition will need to be changed.

[5] See Dunn (1993, 332).

DEFINITION 12.7 (Validity) *A model M is a counterexample to a formula A iff there is $a \in N$ such that for some assignment $f \in I^{\omega}$, $f, a \not\Vdash A$.*

A formula is valid in a class of models iff no model in that class is a counterexample to the formula.

A formula A is valid, $\models_{\mathsf{B}^=} A$, iff A is valid in the class of all zero-order identity models.

The Heredity lemma can be proven for zero-order identity models.

LEMMA 12.8 (Heredity lemma) *If $a \leq b$ and $f, a \Vdash A$, then $f, b \Vdash A$.*

Proof The new case is for $s = t$, and it is immediate from condition (ID1). □

Let us turn from models to axioms. To axiomatize the target logic, add the following axioms to B.

(I1) $t = t$.
(I2) $s = t \rightarrow t = s$.
(I3) $(A \wedge s = t) \rightarrow A'$, where A is $\{\rightarrow, \circ\}$-free, and A' is the result of substituting one or more occurrences of s for t in A.

The resulting logic will be called $\mathsf{B}^=$. We can obtain other logics by taking another zero-order logic L and adding the earlier identity axioms to obtain $\mathsf{L}^=$. We will focus on $\mathsf{B}^=$ here.

One form of the transitivity of identity is obtained via the third axiom.[6] We will record this in a lemma.

LEMMA 12.9 $(s = t \wedge t = u) \rightarrow s = u$ *is provable in* $\mathsf{B}^=$.

Let us call this form the conjunctive form of transitivity. There are other forms of transitivity that one might want, namely the iterated implication forms: $A \rightarrow (s = t \rightarrow A')$ and $s = t \rightarrow (A \rightarrow A')$. Both of these have an air of paradox, looking similar to the undesirable $A \rightarrow (B \rightarrow A)$ and $B \rightarrow (A \rightarrow A)$. In weaker base logics, such as those lacking permutation, the two iterated implication axioms are not equivalent, and in logics that lack contraction, they need not entail the conjunctive form. We will return to discussion of the iterated implication forms of the transitivity of identity later. For now, we will note that they require additional frame conditions to validate.

The notion of proof and theorem of $\mathsf{B}^=$ will be defined in the usual way. We will write $\vdash_{\mathsf{B}^=} A$ iff there is a $\mathsf{B}^=$-proof of A. With this definition in hand, we can turn to the proof of the soundness of $\mathsf{B}^=$ with respect to zero-order identity models.

THEOREM 12.10 (Soundness) *If $\vdash_{\mathsf{B}^=} A$, then $\models_{\mathsf{B}^=} A$.*

Proof The proof is by induction on the length of the proof of A. Most of the cases are standard, and the essentially new cases are those for the axioms (I1)–(I3).

Axiom (I1) is immediate from (ID2).

Axiom (I2) is covered by (ID3).

6 See Mares (1992c, 12).

For (I3), we argue by induction on the structure of A, where A is $\{\rightarrow, \circ\}$-free. We will suppose that $Vf(s) = i$ and $Vf(t) = j$.

Subcase: A has the form $Pt_1, \ldots, t_n$. Assume $f, a \Vdash Pt_1, \ldots, t_n$ and $f, a \Vdash s = t$, so $a \in V(Pt_1, \ldots, t_n)(f)$ and $(i, j) \in \approx_a$. By (ID7), it follows that $a \in V(Ps_1, \ldots, s_n)(f)$, where $s_\ell = t_\ell$, if $t_\ell \notin \{s, t\}$ and otherwise s_ℓ may be either of s or t. So, $f, a \Vdash A'$, as desired.

Subcase: $t = u$. Assume $f, a \Vdash t = u$ and $f, a \Vdash s = t$. Then $(Vf(t), Vf(u)) \in \approx_a$ and $(Vf(s), Vf(t)) \in \approx_a$. By (ID4), $(Vf(s), Vf(u)) \in \approx_a$, so $f, a \Vdash s = u$, as desired. The other possibilities for this subcase are similar.

Subcase: $\neg Pt_1, \ldots, t_n$. Assume $f, a \Vdash \neg Pt_1, \ldots, t_n$ and $f, a \Vdash s = t$, so $f, a^* \nVdash Pt_1, \ldots, t_n$. So, $a^* \notin V(Pt_1, \ldots, t_n)(f)$. From the assumption it follows that $(i, j) \in \approx_a$, and with (ID6) we can conclude that a^* is weakly symmetric in (i, j). It then follows that $a^* \notin V(Ps_1, \ldots, s_n)(f)$, where $s_i = t_i$, except possibly if s_i is s. Therefore, $f, a \Vdash \neg Ps_1, \ldots, s_n$, which is A', as desired.

Subcase: $\neg(t = u)$. Assume $f, a \Vdash \neg(t = u)$ and $f, a \Vdash s = t$. Then $(i, j) \in \approx_a$ and $f, a^* \nVdash t = u$. Let us suppose $Vf(u) = k$, so $(j, k) \notin \approx_{a^*}$. Suppose $f, a \nVdash \neg(s = u)$. Then $f, a^* \Vdash s = u$. So, $(i, k) \in \approx_{a^*}$, which together with $(j, i) \in \approx_a$, from (ID3), yields $(j, k) \in \approx_{a^*}$ from (ID5). This, however, implies that $f, a^* \Vdash t = u$, and so $f, a \nVdash \neg(t = u)$, contradicting an assumption. Therefore, $f, a \Vdash \neg(s = u)$.

The remaining subcases, $\neg\neg B$, $B \wedge C$, $\neg(B \wedge C)$, $B \vee C$, and $\neg(B \vee C)$ are handled by the inductive hypothesis.

Since the desired conclusion is obtained in all subcases, we conclude that axiom (I3) is valid.

□

Next, we turn to Completeness. For the proof of Completeness, we will adapt the canonical model construction of Chapter 4. In addition to the standard parts of the frame, whose definitions are carried over from the earlier chapter, we add the binary relations for identity defined as follows.

- $(s, t) \in \approx_a$ iff $s = t \in a$.

We need to show that the frame, so defined, satisfies all the frame and model conditions required, (ID1)–(ID7). Let us start with the frame conditions.

LEMMA 12.11 *The canonical frame for* B$^=$ *obeys the frame conditions* (ID1)–(ID5).

Proof For (ID1), suppose $a \leq b$ and $(s, t) \in \approx_a$. This implies that $s = t \in a$, and so $s = t \in b$ since $a \leq b$ implies $a \subseteq b$. Therefore, $(s, t) \in \approx_b$.

For (ID2), let $a \in N$. Since $s = s$ is an axiom, $s = s \in a$, so $(s, s) \in \approx_a$, as desired.

For (ID3), suppose $(s, t) \in \approx_a$. Then $s = t \in a$. It then follows, by axiom (I2), that $t = s \in a$, so $(t, s) \in \approx_a$.

For (ID4), Suppose $(s, t) \in \approx_a$ and $(t, u) \in \approx_a$. Then $s = t \wedge t = u \in a$, and since $(s = t \wedge t = u) \rightarrow s = u$ is a theorem, by lemma 12.9, $s = u \in a$. Thus $(s, u) \in \approx_a$.

For (ID5), suppose $(s, t) \in \approx_a$ and $(t, u) \in \approx_{a^*}$. For reductio, assume $(s, u) \notin \approx_{a^*}$. It follows that $s = u \notin a^*$, so $\neg(s = u) \in a$. From axiom (I3) and $s = t \in a$, $\neg(t = u) \in a$. This implies $t = u \notin a^*$, so $(t, u) \notin \approx_{a^*}$, contradicting the assumption. Thus, $(s, u) \in \approx_{a^*}$.

□

LEMMA 12.12 *The canonical model for* $\mathsf{B}^=$ *obeys the model conditions* (ID6)–(ID7) *and conforms to the canonical frame.*

Proof For (ID6), suppose $Vf(u) = s$, $Vf(v) = t$, $(s,t) \in \approx_a$, and $a^* \in V(Pt_1, \ldots, t_n)(f)$. Suppose that $a^* \notin V(Ps_1, \ldots, s_n)(f)$, where the s_ℓ's and t_ℓ's satisfy the conditions for weak symmetry in s,t. So, $\neg(Ps_1, \ldots, s_n)^f \in a$. By assumption, $Vf(u) = s$, $Vf(v) = t$, and $(s,t) \in \approx_a$, so $u = v \in a$. By (I3), we then have $(\neg Pt_1, \ldots, t_n)^f \in a$, so $(Pt_1, \ldots, t_n)^f \notin a^*$, contradicting the assumption. Therefore, $a^* \in V(Ps_1, \ldots, s_n)(f)$.

For the other direction in weak symmetry, suppose $Vf(u) = s$, $Vf(v) = t$, $(s,t) \in \approx_a$, $a^* \notin V(Pt_1, \ldots, t_n)(f)$, and $a^* \in V(Ps_1, \ldots, s_n)(f)$, where the s_ℓ's and t_ℓ's satisfy the conditions for weak symmetry in s,t. So, $(\neg Pt_1, \ldots, t_n)^f \in a$. By assumption, $Vf(u) = s$, $Vf(v) = t$, and $(s,t) \in \approx_a$, so $(u = v)^f \in a$. By (I3), we then have $(\neg Ps_1, \ldots, s_n)^f \in a$, so $(Ps_1, \ldots, s_n)^f \notin a^*$, contradicting the assumption. Therefore, $a^* \notin V(Ps_1, \ldots, s_n)(f)$.

For (ID7), suppose $Vf(u) = s$, $Vf(v) = t$, $(s,t) \in \approx_a$, and $a \in V(Pt_1, \ldots, t_n)(f)$. Suppose that $a \notin V(Ps_1, \ldots, s_n)(f)$, where the s_ℓ's and t_ℓ's satisfy the conditions of (ID7). Then $(Pt_1, \ldots, t_n)^f \in a$ and $(u = v)^f \in a$, but $(Ps_1, \ldots, s_n)^f \notin a$. As $(u = v \wedge Pt_1, \ldots, t_n)^f \in a$, by (I3), $(Ps_1, \ldots, s_n)^f \in a$, which is a contradiction. Therefore, $a \in V(Ps_1, \ldots, s_n)(f)$, as desired.

For conformity, note that $a \in V(=)(Vf(s), Vf(t))$ iff $(s = t)^f \in a$, which is equivalent to $(Vf(s), Vf(t)) \in \approx_a$.

□

LEMMA 12.13 (Satisfaction lemma) *For all formulas A, for all assignemnts f, $A^f \in a$ iff $f, a \Vdash A$.*

Proof The proof is by induction on the structure of A. The only new case is the identity case.

$$\begin{array}{lll} (s = t)^f \in a & \text{iff} & s^f = t^f \in a \\ & \text{iff} & (Vf(s), Vf(t)) \in \approx_a \\ & \text{iff} & a \in V(=)(Vf(s), Vf(t)) \\ & \text{iff} & f, a \Vdash s = t. \end{array}$$

The transition from line 1 to 2 is by definition, the transition from 2 to 3 is by conformity, and the transition from 3 to 4 is the standard evaluation clause for a predicate.

□

THEOREM 12.14 (Completeness) *For closed formulas A, if $\models_{\mathsf{B}^=} A$, then $\vdash_{\mathsf{B}^=} A$.*

Proof Suppose that it is not the case that $\vdash_{\mathsf{B}^=} A$. Construct the canonical model for $\mathsf{B}^=$. As A is not a theorem, there is a regular, prime theory a such that $A \notin a$. From the satisfaction lemma, we then have $a \nVdash A$, as A is closed. Therefore, it is not the case that $\models_{\mathsf{B}^=} A$.

□

Thus, our models are adequate for the logic $\mathsf{B}^=$. The identity axioms and frame conditions appear to be fairly independent of the underlying logic. One can add the identity axioms (I1)–(I3) to other base logics and obtain sound and complete modelings, provided one restricts to a class of ternary relational frames satisfying appropriate frame conditions.

Focus on the zero-order models has allowed us to highlight the mechanics of the features of the models pertaining primarily to identity. There is one intermediate class of models we will look at before adding quantifiers. That class of models adds some of the Mares–Goldblatt machinery of propositional functions. We will use this to provide a model condition for the general form of (I3), one that does not include a restriction on the connectives in A.

The axioms for $\mathsf{B}^=$ are sound and complete with respect to the zero-order identity models. The substitution axiom, (I3), has a restriction, which one may want to drop. One can validate the unrestricted axiom, provided an additional model condition is adopted. Stating that condition will require the additional machinery of the Mares–Goldblatt frames for quantified relevant logics, so we will turn to that now.

12.3 Mares–Goldblatt Proposition Frames

In order to drop the restriction on the substitution axiom, we will add sets of admissible propositions and propositional functions to our zero-order models. We will need to do this anyway once we add quantifiers, so it is not too much work. We will use the propositions to provide a condition appropriate for the unrestricted version of (I3), dropping the restriction that A be $\{\to, \circ\}$-free:

(I4) $(A \wedge s = t) \to A'$, where A' is the result of substituting one or more occurrences of s for t in A.

Let us call the addition of this axiom to $\mathsf{B}^=$ the logic $\mathsf{B}^=_{sub}$. Much as before, we can obtain $\mathsf{L}^=_{sub}$, for a stronger logic L, by adding the identity axioms (I1)–(I4) to L.

We will define propositional zero-order identity frames.

DEFINITION 12.15 (Propositional zero-order identity frame) *A PROPOSITIONAL ZERO-ORDER IDENTITY FRAME is an octuple,*

$$\langle K, N, {}^*, R, I, \{\approx_a\}_{a \in K}, \mathsf{Prop}, \mathsf{PropFun}\rangle,$$

where the first six components comprise a zero-order identity ternary relational frame, Prop *is a non-empty, model-like set of hereditary subsets of K, and* PropFun *is a non-empty set propositional functions from assignments to* Prop, *satisfying the following closure conditions.*

(FTee) $\phi_N \in \mathsf{PropFun}$, *where* $\phi_N(f) = N$, *for all* $f \in I^\omega$.
(FImp) *If* $\phi, \psi \in \mathsf{PropFun}$, *then* $\phi \Rightarrow \psi \in \mathsf{PropFun}$.
(FConj) *If* $\phi, \psi \in \mathsf{PropFun}$, *then* $\phi \cap \psi \in \mathsf{PropFun}$.
(FDisj) *If* $\phi, \psi \in \mathsf{PropFun}$, *then* $\phi \cup \psi \in \mathsf{PropFun}$.
(FNeg) *If* $\phi \in \mathsf{PropFun}$, *then* $-\phi \in \mathsf{PropFun}$.
(FFuse) *If* $\phi, \psi \in \mathsf{PropFun}$, *then* $\phi \otimes \psi \in \mathsf{PropFun}$.

In other words, PropFun satisfies the conditions to be closed except for the universal quantifier condition.

DEFINITION 12.16 (Propositional zero-order identity model) *A PROPOSITIONAL ZERO-ORDER IDENTITY MODEL M is a pair $\langle F, V\rangle$, where F is a propositional zero-order identity frame and V is an admissible valuation that conforms to F and satisfies conditions* (ID6) *and* (ID7).

The Heredity lemma follows from the fact that the closure conditions on Prop and PropFun preserve the property of being hereditary. As in the previous section, the indexed relations affect the available valuations. The propositional function for the identity predicate is determined by the indexed relations via conformity, as shown by the following.

$$\begin{aligned} (Vf(s), Vf(t)) \in \approx_a \quad &\text{iff} \quad a \in V(=)(Vf(s), Vf(t)) \\ &\text{iff} \quad a \in V(s=t)(f) \\ &\text{iff} \quad a \in \phi_{s=t}(f). \end{aligned}$$

Thus, we can move freely between the indexed relations and the propositional function for the identity predicate.

In order to show the soundness of (I4), we will define some additional concepts and some notation.

DEFINITION 12.17 (Conflation) *For $f, g \in I^\omega$, $f \sim^i_j g$ iff for all $n \in \omega$,*

- *$f(n) = g(n)$, or*
- *$f(n) \neq g(n)$ and either*
  (i) $f(n) = i$ and $g(n) = j$, or
  (ii) $f(n) = j$ and $g(n) = i$.

For a given frame, for $a \in K$, say a is (i, j)-conflating iff for all $\phi \in$ PropFun, for all $f, g \in I^\omega, (f \sim^i_j g \Rightarrow (a \in \phi(f) \Leftrightarrow a \in \phi(g)))$.[7]

DEFINITION 12.18 (Strong zero-order identity frame) *A STRONG ZERO-ORDER IDENTITY FRAME is a zero-order identity frame that additionally satisfies the condition*

(FFullSub) *if $(i, j) \in \approx_a$ then a is (i, j)-conflating.*

Finally, we define two more concept to prove Soundness.[8]

DEFINITION 12.19 (Skeleton, assignment substitutions) *Let $B[x_1/t_1, \ldots, x_n/t_n]$ be the result of simultaneously substituting the term t_i for x_i in B, provided t_i is free for x_i.*

[7] This terminology is based on that of Ripley (2018). I thank Lloyd Humberstone for suggesting the term "conflation" in this context.

[8] A skeleton is basically a *matrix* from Priest (2016, 17), without the requirement that variables appear in increasing order. I thank Ellie Ripley for pointing out Priest's term to me.

A SKELETON of a formula A is a formula B, whose unbound terms are only free variables, $x_1, \ldots, x_n$, each of which occurs only once, such that there are terms $t_1, \ldots, t_n$ such that $A = B[x_1/t_1, \ldots, x_n/t_n]$.

DEFINITION 12.20 (Assignment substitutions) *For $f \in I^\omega$ and $j \in I$, define $f[j/n]$ as*

$$f[j/n] = \langle f0, f1, \ldots f(n-1), j, f(n+1), \ldots \rangle$$

and $f[j_1/n_1, \ldots, j_{m+1}/n_{m+1}]$ as

$$f[j_1/n_1, \ldots, j_{m+1}/n_{m+1}] = (f[j_1/n_1, \ldots, j_m/n_m])[j_{m+1}/n_{m+1}],$$

where the n_i are distinct

These are all the pieces we need for Soundness. We will proceed to list some facts about skeletons and substitution.

LEMMA 12.21 *If A' is the result of substituting zero or more occurrences of term s for a term t in A, provided s is free for t in A, then there is a formula B that is a skeleton for both A and A'.*

LEMMA 12.22 (Semantic substitution lemma) *Suppose B is a skeleton for A, where $B[x_{n_1}/t_1, \ldots, x_{n_m}/t_m] = A$. Suppose that $Vf(t_i) = j_i$, for each i such that $1 \leq i \leq m$. Then*

$$f[j_1/n_1, \ldots, j_m/n_m], a \Vdash B \text{ iff } f, a \Vdash A.$$

COROLLARY 12.23 *Suppose B is a skeleton for A, where $B[x_{n_1}/t_1, \ldots, x_{n_m}/t_m] = A$. Suppose that $Vf(t_i) = j_i$, for each i such that $1 \leq i \leq m$. Then*

$$\phi_B(f[j_1/n_1, \ldots, j_m/n_m]) = \phi_A(f).$$

Those are the pieces we need for Soundness, so let us turn to its proof.

THEOREM 12.24 *If $\vdash_{\mathsf{B}^=_{sub}} A$ then $\models_{\mathsf{B}^=_{sub}} A$.*

Proof We will present just the case for the new axiom, (I4).

Suppose that $f, a \Vdash A$ and $f, a \Vdash s = t$. Suppose $Vf(s) = i$ and $Vf(t) = j$, so it follows that $(i, j) \in \approx_a$. By the condition (FFullSub), a is (i, j)-conflating. Suppose that $f, a \nVdash A'$. By lemma 12.21, A and A' have a common skeleton, B, and for simplicity, we will assume that A and A' differ on exactly one term occurrence, meaning that

- $A = B[x_{n_1}/t_1, \ldots, x_{n_m}/t_m, x_{n_{m+1}}/t]$, and
- $A' = B[x_{n_1}/t_1, \ldots, x_{n_m}/t_m, x_{n_{m+1}}/s]$.

Let g be the assignment $f[j_1/n_1, \ldots, j_m/n_m]$, where $Vf(t_i) = j_i$, for $1 \leq i \leq m$. As $f, a \Vdash A$, by lemma 12.22, $g[i/n_{m+1}], a \Vdash B$. By lemma 12.22, $g[j/n_{m+1}], a \Vdash B$ iff $f, a \Vdash A'$, so $a \notin \phi_B(g[j/n_{m+1}])$ while $a \in \phi_B(g[i/n_{m+1}])$. Since by definition we have $g[i/n_{m+1}] \sim^i_j g[j/n_{m+1}]$, we have reached a contradiction with the fact that a is (i, j)-conflating. Therefore, $f, a \Vdash A'$, as desired.

□

So the logic $\mathsf{B}^{=}_{sub}$ is Sound for the class of all strong zero-order identity models. We can use alternative base logics L to get identity logics, $\mathsf{L}^{=}_{sub}$, that are Sound with respect to appropriate classes of strong zero-order identity models. We just need to use base logics that are sound with respect to an appropriate class of ternary relational frames.

With Soundness out of the way, we can turn to Completeness.

12.4 Completeness for Zero-Order Propositional Models

The proof of Completeness uses the techniques developed in the previous chapter. We will provide the definitions, changes, and additions needed to adapt the arguments of Section 12.2.

- For closed formulas A, $\|A\| = \{a \in K : A \in a\}$.
- $\mathsf{Prop} = \{\|A\| : A$ is a closed formula$\}$.
- For each formula A, $\phi_A : I^\omega \mapsto \mathsf{Prop}$ is defined as $\phi_A(f) = \|A^f\|$.
- $\mathsf{PropFun} = \{\phi_A : A$ is a formula$\}$.
- For n-ary predicate letters P, $V(P)(c_1, \ldots, c_n) = \|Pc_1, \ldots, c_n\|$.

From the work of Section 12.2, we know that the canonical frame satisfies conditions (ID1)–(ID7). That leaves only the frame condition (FFullSub) to verify. For that, it will be easiest to introduce a definition and a lemma.

DEFINITION 12.25 ((s,t)-variants) *Suppose that two formulas A and B have a common skeleton C, $A = C[x_1/t_1, \ldots, x_n/t_n]$, and $B = C[x_1/s_1, \ldots, x_n/s_n]$. A and B are (s,t)-VARIANTS iff for all i such that $1 \leq i \leq n$, either $t_i = s_i$, or $t_i \neq s_i$ but both $t_i \in \{s,t\}$ and $s_i \in \{s,t\}$.*

LEMMA 12.26 *Suppose that A and B are (s,t)-variants. Then, $\vdash_{\mathsf{B}^{=}_{sub}} (A \wedge s = t) \to B$.*

Proof Suppose that A and B are (s,t)-variants. Then they have a common skeleton C such that $A = C[x_1/t_1, \ldots, x_n/t_n]$, and $B = C[x_1/s_1, \ldots, x_n/s_n]$. By repeated use of (ID4) and the affixing rules, we can derive $(A \wedge s = t) \to B$. □

With that lemma in hand, we can prove that the canonical frame obeys the condition (FFullSub).

LEMMA 12.27 *The canonical frame for $\mathsf{B}^{=}_{sub}$ obeys condition* (FFullSub).

Proof Suppose $(i,j) \in \approx_a$. Then $i = j \in a$. Suppose a is not (i,j)-conflating, so there is a $\phi \in \mathsf{PropFun}$ such that for some $f, g \in I^\omega$ such that $f \sim^i_j g$, either

- $a \in \phi(f)$ and $a \notin \phi(g)$, or
- $a \notin \phi(f)$ and $a \in \phi(g)$.

Suppose $a \in \phi(f)$ and $a \notin \phi(g)$. From the definition of $\mathsf{PropFun}$, there is a formula A such that $\phi = \phi_A$. Without loss of generality, we can assume that A is a skeleton of some formula. Then A^f and A^g differ in so far as the latter has zero or more occurrences of i where the former has j and zero or more occurrences of j where the former has i. Then A^f and A^g are (i, j)-variants. By the previous lemma, $\vdash_{\mathsf{R}^=_{sub}}$ $A^f \wedge i = j \rightarrow A^g$. Since $a \in \phi_A(f)$, $a \in ||A^f||$, so $A^f \in a$. So, $A^f \wedge i = j \in a$. We then have $A^g \in a$. This implies $a \in ||A^g||$, which implies $a \in \phi_A(g)$, contradicting the assumption.

The argument for the other disjunct is similar. Therefore, we conclude the canonical model obeys (FFullSub).

□

The canonical frame then satisfies all the required conditions. That the canonical model does as well follows from the results of Section 12.2 and Chapter 11. So, we can claim Completeness.

THEOREM 12.28 (Completeness for $\mathsf{B}^=_{sub}$) *For closed formulas A, if $\models_{\mathsf{B}^=_{sub}} A$, then $\vdash_{\mathsf{B}^=_{sub}} A$.*

We have seen how to model identity in ways that permit substitution of identicals outside the scope of implication and fusion as well as within their scope. These are the logics $\mathsf{B}^=$ and $\mathsf{B}^=_{sub}$. Next, we turn to identity together with quantification.

12.5 Models for Identity and Quantification

In this section, we will present models for the quantified logics with identity, both with the axioms (I1)–(I3) and with the axioms (I1)–(I4). We will prove Soundness and Completeness. We will begin with the logic with only (I1)–(I3).

DEFINITION 12.29 (First-order identity frame, model) *A first-order identity frame, or $\mathsf{QB}^=$-frame, F is a zero-order propositional identity frame obeying the following additional condition.*

(FAll) *If $\phi \in$ PropFun, then $\forall_n \phi \in$ PropFun, for each $n \in \omega$.*

A $\mathsf{QB}^=$-model M is pair $\langle F, V \rangle$ where F is a first-order identity frame, and V is an admissible valuation that conforms to F and obeys the following additional conditions.

(FUnivSub) *If $a \in (\phi_{s=t} \cap \phi_{\forall x_n A})(f)$, then there is $X \in$ Prop such that $a \in X$ and $X \subseteq \bigcap_{g \sim n} \phi_{A'}(g)$, where A is $\{\rightarrow, \circ\}$-free and A' is the result of substituting one or more occurrences of s for t in A.*

(FNegUnivSub) *If $a \in \phi_{s=t}(f)$ and $a^* \in \phi_{\forall x_n A'}(f)$, then there is $X \in$ Prop such that $a^* \in X$ and $X \subseteq \bigcap_{g \sim_n x} \phi_A(g)$, where A is $\{\rightarrow, \circ\}$-free and A' is the result of substituting one or more occurrences of s for t in A.*

The definitions of a model being a counterexample to formula A, a formula A being valid in a class of models, and being valid, $\models_{\mathsf{QB}^=} A$, are straightforward adjustments of the earlier definitions.

The new conditions, (FUnivSub) and (FNegUnivSub), are used in the proof of Soundness. As we will see, they are needed to bridge a crucial gap arising from the use of the non-Tarskian verification condition for the quantifier rather than the Tarskian one. That suggests that these conditions can be omitted when interest is restricted to either finite models or full models.

Next, I will prove soundness for $\mathsf{QB}^=$ with respect to the class of all $\mathsf{QB}^=$-models.

THEOREM 12.30 (Soundness) *If* $\vdash_{\mathsf{QB}^=} A$, *then* $\models_{\mathsf{QB}^=} A$.

Proof We will focus on the two new subcases for the induction for (I3): when A has the form $\forall x_n B$ or the form $\neg\forall x_n B$. In both cases we can assume that s and t are distinct from the quantified variable, otherwise the substitution would be vacuous.

Subcase: $\forall x B$. Suppose that $f,a \Vdash \forall x B$ and $f,a \Vdash s = t$. From the assumption, $a \in (\phi_{s=t} \cap \phi_{\forall x_n B})(f)$. It follows that by (FUnivSub), there is $X \in \mathsf{Prop}$ such that $a \in X$ and $X \subseteq \bigcap_{g \sim_n x} |B'|_{Vg}$. This means that $f,a \Vdash \forall x_n B'$, as desired.

Subcase: $\neg\forall x_n B$. Suppose that $f,a \Vdash \neg\forall x_n B$ and $f,a \Vdash s = t$. Assume $f,a \nVdash \neg\forall x_n B'$. It follows that $f,a^* \Vdash \forall x_n B'$, so $a \in \phi_{s=t}(f)$ and $a^* \in \phi_{\forall x_n B'}(f)$. By (FNegUnivSub), there is $X \in \mathsf{Prop}$ such that $a^* \in X$ and $X \subseteq \bigcap_{g \sim_n x} |B|_{Vg}$. This implies that $f,a^* \Vdash \forall x_n B$, so $f,a \nVdash \neg\forall x_n B$, which contradicts an assumption. Thus, we conclude $f,a \Vdash \neg\forall x_n B'$, as desired.

□

As illustrated in the earlier proof, the two conditions, (FUnivSub) and (FNegUnivSub), are key to the new subcases for the inductive argument. Without them, there is no guarantee of the Soundness of substitution in the scope of a quantifier. This suggests a further restriction on (I3), namely requiring that A be $\{\forall, \rightarrow, \circ\}$-free, because the universal quantifier is the only one that appeals to Prop. This might be a plausible restriction if one thinks of the universal quantifier as being more like a kind of necessity operator rather than a generalized conjunction. If it is akin to a modal operator, then it is plausible that there would be a degree of opacity that could interfere with substitution in its scope. We will, however, not explore this idea here.

The two conditions are worth further comment. They are needed because we are evaluating the universal quantifier using a non-Tarskian verification clause. If one attempts the Soundness argument without the conditions, one can show that all the instances work out. In the Mares–Goldblatt framework, verification of all the instances can, in general, fall short of verifying a universal, which requires the existence of a proposition in Prop entailing all the instances. These additional conditions are needed to guarantee the existence of such a proposition.

Given what we have done earlier, the Completeness proof is straightforward. The only new part to the proof, given what has been in the previous sections and chapters, is to verify that the canonical model satisfies the two new conditions, (FUnivSub) and (FNegUnivSub). For this proof, we will note a lemma, that universals imply their instances.

LEMMA 12.31 *In the canonical model for* QB$^=$, $||(\forall x_n A)^f|| \subseteq \bigcap_{g \sim_n x} \phi_A(g)$.

LEMMA 12.32 *The canonical model for* QB$^=$ *satisfies the conditions* (FUnivSub) *and* (FNegUnivSub).

Proof For (FUnivSub), suppose $a \in (\phi_{s=t} \cap \phi_{\forall x_n A})(f)$, for $\{\rightarrow, \circ\}$-free A. We need to show that there is $X \in$ Prop such that $a \in X$ and $X \subseteq \bigcap_{g \sim_n x} \phi_{A'}(g)$. Since $a \in (\phi_{s=t} \cap \phi_{\forall x_n A})(f)$, $a \in ||(s = t \wedge \forall x_n A)^f||$, so $(s = t \wedge \forall x_n A)^f \in a$. From (I3), it follows that $(\forall x_n A')^f \in a$. By lemma 12.31,

$$||(\forall x_n A')^f|| \subseteq \bigcap_{g \sim_n x} \phi_{A'}(g).$$

As $(\forall x_n A')^f \in a$, $a \in ||(\forall x_n A')^f||$, so $||(\forall x_n A')^f||$ is the desired member of Prop.

For (FNegUnivSub), suppose $a \in \phi_{s=t}(f)$ and $a^* \in \phi_{\forall x_n A'}(f)$, for $\{\rightarrow, \circ\}$-free A'. From the assumption, $(s = t)^f \in a$ and $(\forall x_n A')^f \in a^*$. Suppose that $a^* \notin ||(\forall x_n A)^f||$. Then, $(\forall x_n A)^f \notin a^*$. So, $(\neg \forall x_n A)^f \in a$. By (I3), it follows that $(\neg \forall x_n A')^f \in a$. But then, $(\forall x_n A')^f \notin a^*$, so $a^* \notin ||(\forall x_n A')^f||$, which implies $a^* \notin \phi_{\forall x_n A'}(f)$, contradicting an assumption. Therefore, $a^* \in ||(\forall x_n A)^f||$. By lemma 12.31,

$$||(\forall x_n A)^f|| \subseteq \bigcap_{g \sim_n x} \phi_A(g).$$

Thus, $||(\forall x_n A)^f||$ is the desired member of Prop.

□

With the previous lemma in hand, we conclude with Completeness for QB$^=$.

THEOREM 12.33 (Completeness) *For all formulas A, if* $\models_{QB^=} A$, *then* $\vdash_{QB^=} A$.

Proof The proof is essentially the same as that of Mares and Goldblatt (2006), so we omit it. □

Conditions (FUnivSub) and (FNegUnivSub) are somewhat unappealing, given that their formulation mixes syntactic and model-based concepts. While it would be good to have an equivalent way to state them that appeals only to propositional functions, we will note that they can be replaced for logics with (I4). In models for QB$^=_{sub}$, (FFullSub) suffices, and we do not need to adopt (FUnivSub) and (FNegUnivSub) separately. This leads to another definition.

DEFINITION 12.34 (Strong first-order identity models) *A first-order identity model satisfying* (FFullSub) *is a* STRONG *first-order identity model.*

The logic QB$^=_{sub}$ is Sound and Complete with respect to the class of all strong first-order identity models.

THEOREM 12.35 *For all formulas A,* $\models_{QB^=_{sub}} A$ *iff* $\vdash_{QB^=_{sub}} A$.

Proof For the Soundness direction, the proof of the new case, (I4), is essentially the same as in theorem 12.24.

For the Completeness direction, the proof that the canonical model satisfies (FFullSub) is essentially the same as in the previous section. The rest of the proof follows from what we have done in earlier sections, so we omit it. □

Thus, we have a range of identity logics, with and without quantifiers, and models that are adequate for them. The models use a family of binary relations to interpret identity. This technique is based on the work of Mares (1992c), who provides models for identity in the context of Fine models for quantification.

The use of the binary relations in connection with the Mares–Goldblatt framework is somewhat out of the spirit of the framework. It would be more in the spirit of the framework if models for identity could be defined that only use conditions on Prop and PropFun. We will, however, leave that for future work.

To close this section, we will comment on one other feature of the models: In the Mares–Goldblatt models, the domain, I, does not vary from point to point.[9] Nonetheless, once identity is around, there is a sense in which there are varying domains. Let us say that the domain I of a frame is the GLOBAL DOMAIN. With identity, we can define, for each point $a \in K$,

$$[i]_a = \{j \in I : (i, j) \in \approx_a\},$$

and then the LOCAL DOMAIN, I_a, of each point $a \in K$ is defined as

$$I_a = \{[i]_a : i \in I \text{ and } \exists j \in I((i, j) \in \approx_a)\}.$$

The local domain of a is, roughly, the set of things that a "thinks" are distinct. The local domains may not exhaust I. Further, they can vary from point to point, even as one goes up the heredity ordering.[10] As one goes up the heredity ordering, points may recognize new objects that were not previously in the local domain. They may also conflate objects that were previously regarded as distinct. This may provide the basis for relevant logicians to develop a view on contingent identity and its associated puzzles.[11]

12.6 Relevant Predication

The models and logics of identity are only the start of the story. Many important concepts in logic and philosophy are defined using the identity predicate. Just to take one example, functions crucially use identity, whether to specify the value of a function, for example, $f(s) = t$, or to say when a relation is functional, that is, if Hst and Hsu, then $t = u$. Functions are a prime example of an important area of logic

[9] Mares (2009) shows how to deal with varying domains in Mares–Goldblatt frames.

[10] Varying local domains occur also with models for intuitionistic logic with identity. See van Dalen (2002), especially pp. 35–37 and 61–64, for more on this.

[11] See Schwarz (2013) for an overview of contingent identity.

that needs to be explored against the backdrop of relevant logics. We will not explore functions here. Instead, we will turn to relevant predication, which we discussed in Section 12.1.

Relevant predication is a way of predicating some property F of an object t in a way that is stronger than merely saying Ft. Recall that the relevant predication $(\rho x A(x))t$ is defined as

$$(\rho x A(x))t =_{Df} \forall x(x = t \rightarrow A(x)).$$

A reason we have opted for the conjunctive form of substitution is that the iterated implicational forms have bad consequences for relevant predication, as noted by Dunn (1987).[12] In particular, there are two bad consequences. First, where s and t do not occur freely in A, the implicational forms of substitution, $A \rightarrow (s = t \rightarrow A)$ and $s = t \rightarrow (A \rightarrow A)$, are worryingly close to implicational paradoxes, particularly when s and t do not occur in A.

The second bad consequence that provides reason for avoiding the implicational forms is that Dunn wants to use $\forall y(A(y) \rightarrow (\rho x A(x))y)$ as the characteristic axiom for a predicate $A(x)$ being a relevant predicate, which is to say determining a relevant property. Not all predicates are relevant predicates; rather, only the ones for which that axiom holds are. Adopting the implicational forms of substitution will trivialize the concept of being a relevant predicate, as every predicate will be such. Rather than adopt the implicational form of substitution in general, we can adopt the conjunctive form and then form theories in which certain predicates are relevant predicates. We will say that a predicate A for which $\forall y(A(y) \rightarrow (\rho x A(x))y)$ holds, whether as a matter of logic or adopted as an axiom, is a predicate that determines a relevant property.

Before we go further, it is worth asking what relevant predication is supposed to do. Relevant predication is supposed to express a close connection between a property and an object, something more involved than (merely) having a property necessarily. Using the logical machinery of relevant logics, Dunn shows that it captures a natural distinction between

(1) Socrates is wise; and,
(2) Socrates is such that Reagan is wise.

(1) is formalized as a regular predication, Ws, while the latter is formalized as $\forall x(x = s \rightarrow Wr)$. The latter can fail, even if the consequent were a necessary truth. Dunn (1990a,b) explores the prospects of using relevant predication to formalize notions of intrinsic and essential properties. Kremer (1997) critically evaluates relevant predication and argues in favor of understanding it in terms of a distinguished class of properties. Although we cannot get into the details of Dunn's and Kremer's analyses, their work suggests that relevant predication has much potential for formalizing concepts of metaphysics.

[12] Dunn was focused on R as the base logic, so one of the implicational forms of substitution yielded the other.

Let us turn to some logical issues concerning relevant predication. We start with a broadly logical comment. The form of a relevant predication is connected to a kind of logical factorization. In classical logic, the following equivalences are valid.

- $Fc \leftrightarrow \forall x(x = c \rightarrow Fx)$.
- $Fc \leftrightarrow \exists x(x = c \wedge Fx)$.

These are connected to restricted quantification, which, as we will see in Chapter 13, presents many of its own challenges. While these two biconditionals hold in classical logic, they are *not* valid in the relevant logics we have been studying. The first equivalence is of particular importance here, since its right-hand side, $\forall x(x = c \rightarrow Fx)$, features in the definition of relevant predication. The invalidity of this formula gives relevant predication its interest.

Now that we have models for identity and quantification, we can give verification conditions for ρ. Unpacking the definitions gives the following verification condition.

- $a \Vdash (\rho x A(x))t$ iff $\exists X \in \mathsf{Prop}\left(a \in X \text{ and } X \subseteq \bigcap_{g \sim_x f} (\phi_{x=t}(g) \Rightarrow \phi_{A(x)}(g))\right)$.

Let us make the simplifying assumption that $t \in \mathsf{Con}$, so $Vg(t) = Vf(t)$, and we will assume that $Vf(t) = j$, for a particular $j \in I$. Unpacking the definition a little more, where $g(x) = i$, $\phi_{x=t}(g) \Rightarrow \phi_{A(x)}(g)$ is the set

$$\{a \in K : \forall b, c \in K(\text{if } Rabc \text{ and } (i, j) \in \approx_b, \text{then } g, c \Vdash A(x))\}.$$

While the verification condition for a relevant predication may be fairly involved, seeing what it takes for a relevant predication to hold should indicate that relevant predication is very demanding, requiring a high degree of coordination among different relations across different points.

Studies of relevant predication have mostly been carried out in the stronger logics, particularly R. It is worth looking at some features of relevant predication to see how it works in weaker logics. To that end, we will examine some of the facts about relevant predication proved by Dunn (1987) in the context of R to generalize them to other logics. First, for all relevant logics we consider, $A(t) \rightarrow (\rho x A(x))t$, provided that A determines a relevant property. This is immediate from the definition of determining a relevant property.

Next, $(\rho x A(x))t \rightarrow A(t)$ holds provided that L contains permutation or obeys the rule $A \rightarrow (B \rightarrow C), B \Rightarrow A \rightarrow C$. Note that from the definition of relevant predication, $(\rho x A(x))t \rightarrow (t = t \rightarrow A(t))$ is an instance of universal instantiation. Given either permutation or the rule mentioned, $(\rho x A(x))t \rightarrow A(t)$ then follows. For logics that lack these, such T and its sublogics, relevant predication of a term need not imply that the predicate holds of that term. The failure of relevant predication to imply regular predication limits its intuitive usefulness in logics that lack permutation.

For any of the logics we consider, if A and B determine relevant properties, then so do $A \wedge B$ and $A \vee B$. The reason is that, at the propositional level, if we have $C_1 \rightarrow (D \rightarrow E_1)$ and $C_2 \rightarrow (D \rightarrow E_2)$, it then follows that $(C_1 \wedge C_2) \rightarrow (D \rightarrow E_1)$ and $(C_1 \wedge C_2) \rightarrow (D \rightarrow E_2)$ are derivable, so $(C_1 \wedge C_2) \rightarrow (D \rightarrow (E_1 \wedge E_2))$ is derivable. Similar reasoning holds for disjunction in place of conjunction. In contrast, $\neg A$ determines a relevant property, where A does, in logics with permutation and contraposition. The reason is that, at the propositional level, we can reason as follows. From $A(y) \to (x = y \to A(x))$ we can get $x = y \to (A(y) \to A(x))$ via permutation, whence $x = y \to (\neg A(x) \to \neg A(y))$. Then we can use permutation again to obtain $\neg A(x) \to (x = y \to \neg A(y))$, which then yields the desired result by universal quantification twice.

For all the relevant logics L, if $A \to B$ is a theorem of L, then so is $(\rho x A(x))t \to (\rho x B(x))t$. Suppose that $A \to B$ is a theorem. From the prefixing rule, $(x = t \to A) \to (x = t \to B)$ is a theorem, so $\forall x((x = t \to A) \to (x = t \to B))$ is as well. The desired theorem, $(\rho x A(x))t \to (\rho x B(x))t$, then follows from distribution of the universal quantifier. For logics that contain the suffixing and prefixing axioms, this can be strengthened to the theoremhood of $(A \to B) \to ((\rho x A(x))t \to (\rho x B(x))t)$, as shown by Dunn (1987, 363).

Finally, identity determines some relevant properties as a matter of logic. We will simply list them, as the proofs that these are theorems are a matter of unpacking definitions.

- $(\rho x(x = t))t$.
- $(\rho x(t = x))t$.
- $\forall y(\rho x(x = x))y$.

Going beyond monadic properties, Dunn defines a relevant binary predication of a pair as

$$(\rho xy A(x, y))(s, t) =_{Df} \forall x \forall y((x = s \wedge y = t) \to A(s, t))).$$

Determining a relevant binary property is defined as $\forall u \forall v(A(u, v) \to (\rho xy A(x, y))(u, v))$. Using this definition, identity determines a relevant binary property, as a matter of logic. The proof uses the conjunctive form of transitivity of identity. The identity predicate determines relevant properties, just as a matter of logic. This seems sensible, as what could be more relevant to an object than being that object?

We can see that many, though not all, of the features of relevant predication that Dunn highlighted were due to the use of R as the base logic. Some of the features fail in weaker logics. For the logician that wants to use relevant predication with some of the weaker logics, theories will need additional axioms to make up some of the strength. Of course, relevant predication may simply be something more useful to the logician that prefers R. This leads to a broader question of whether there are any alternative conceptions of predication to be had for the weaker logics, whether definable from the available logical resources or defined in terms of features of the models.

13 Varia

13.1 Introduction

In this chapter, we will briefly look at some topics germane to earlier chapters that did not otherwise fit. These topics would require full chapters, if not books, to cover in appropriate detail. Rather than provide full coverage, we will provide the reader a brief overview of the topics. We will begin with alternative proof systems for relevant logics (Section 13.2). Next, we will look at two applications of quantified relevant logics: arithmetic (Section 13.3) and naive set theory (Section 13.4). There has long been an interest in using relevant logics for non-classical foundations for mathematics, and these two applications are the most developed attempts.[1] Then, in Section 13.5 we will survey two strategies for adding restricted quantification to relevant logics. Next, we will, in Section 13.6, look briefly at quantified relevant modal logics, bringing together themes from Parts 2 and 3 of the book. Finally, we will conclude in Section 13.7 with some directions for future research.

13.2 Proof Systems

Throughout the book, for the proof systems for relevant logics, we have used *axiom systems*, which have their virtues and vices. Their virtues include having simple extensions to modal and quantified logics, with many axioms being neatly associated with frame conditions. Alternative proof systems have been developed, although these are often, though not always, focused on the basic vocabulary, rather than the modal and quantificational extensions. Fully developing the details of even a few of the alternative proof systems would take chapters on their own, but we will here briefly describe two natural deduction systems. References will be provided for these and other systems, for the interested reader to follow up.

We begin with indexed natural deduction systems, first developed by Anderson and Belnap (1975).[2] These systems are Fitch–Jaśkowski-style natural deduction systems where each formula is indexed with a subscripted set α of natural numbers. In this

[1] Mares (2012) canvasses three different ways to develop the philosophy of mathematics with relevant logics.

[2] Anderson and Belnap (1962) presented the indexed natural deduction system for the $\{\rightarrow\}$-fragment of E. The extension to the full vocabulary is presented in *Entailment*.

style of natural deduction, there are nested subproofs, where subproofs begin with an assumption, or hyp step, and the indices on hyp steps are singletons containing the number depth of the subproof. Here are the rules for the conditional and for conjunction. The rule $\rightarrow$I has the condition that $k \in \alpha$.

$$\begin{array}{ll} \quad A_{\{k\}} & \text{hyp} \\ \quad \vdots & \\ \quad B_\alpha & \\ A \rightarrow B_{\alpha \setminus \{k\}} & \rightarrow\text{I} \end{array} \qquad \begin{array}{ll} A_\alpha & \\ A \rightarrow B_\beta & \\ B_{\alpha \cup \beta} & \rightarrow\text{E} \end{array}$$

$$\begin{array}{ll} A_\alpha & \\ B_\alpha & \\ A \wedge B_\alpha & \wedge\text{I} \end{array} \qquad \begin{array}{ll} A \wedge B_\alpha & \\ A_\alpha & \wedge\text{E} \end{array} \qquad \begin{array}{ll} A \wedge B_\alpha & \\ B_\alpha & \wedge\text{E} \end{array}$$

Brady (1984b) provides natural deduction systems for a range of relevant logics and shows that these are equivalent to their axiomatic formulations, in the sense that $A_\emptyset$ is provable in the natural deduction system for L if and only if A is provable in the axiomatic system for L.[3] Mares and Standefer (2017) adapt these systems to provide indexed natural deduction systems for relatives of E, and Standefer and Brady (2018) provide an alternative indexed proof system for E.

While the indexed systems of Anderson, Belnap, and Brady are flexible and user-friendly, they lack some of the nice features one often wants in a proof system. Part of the reason for this is the presence of a distribution rule, permitting one to infer from $A \wedge (B \vee C)_\alpha$ to $(A \wedge B) \vee (A \wedge C)_\alpha$. Without that rule, the distribution axiom is not derivable. A challenge for proof systems for relevant logics is permitting the derivation of distribution using the conjunction and disjunction rules, without postulating a special, separate rule.[4]

Next are sequent-style natural deduction systems.[5] Proofs in these natural deduction systems are trees where each node is a sequent, $X \succ A$. The antecedent of the sequent is a structure, and, in addition to connective rules, there are typically structural rules for manipulating the structures, such as the following.

$$\frac{X; Y \succ A}{Y; X \succ A}\, CI \qquad \frac{X; X \succ A}{X \succ A}\, WI \qquad \frac{X; (Y; Z) \succ A}{(X; Y); Z \succ A}\, B.$$

For comparison, we will include the connective rules for the conditional and conjunction.

[3] Standefer (2021c) provides tree-style natural deduction systems for Brady's systems and shows how to translate between the two proof formats.

[4] Cf. Brady (2006a).

[5] The brief presentation here follows that of Restall (2000), but other presentations have been offered, by Read (1988) and by Slaney (1990), for example.

$$\frac{X; A \succ B}{X \succ A \rightarrow B} \rightarrow I \qquad \frac{X \succ A \rightarrow B \quad Y \succ A}{X; Y \succ B} \rightarrow E$$

$$\frac{X \succ A \quad X \succ B}{X \succ A \wedge B} \wedge I \qquad \frac{X \succ A \wedge B}{X \succ A} \wedge E \qquad \frac{X \succ A \wedge B}{X \succ B} \wedge E.$$

Many of the sequent-style natural deduction systems permit derivation of distribution, without postulating a special rule for it. This is managed by adding additional structure that can be used in the antecedents along with more structural rules.[6]

The sequent-style natural deduction system for a logic is typically shown equivalent to its axiomatic counterpart by using a translation.[7] Define a translation τ by setting

- $\tau(A) = A$, for formulas A, and
- $\tau(X; Y) = \tau(X) \circ \tau(Y)$.

The equivalence is that for any structure X and formula A, $X \succ A$ is derivable in the natural deduction system for L iff $\tau(X) \rightarrow \tau(A)$ is derivable in the axiom system for L.[8]

There are a wide variety of other proof systems for relevant logics. Sequent systems, also called Gentzen systems, have been used to study relevant logics since the early days of the field, as illustrated by Anderson and Belnap (1975). Restall (2000) and Bimbó (2014) provide excellent overviews of more recent work in the area.

13.3 Arithmetic

One of the early applications of relevant logics, especially R, was arithmetic. For this, we add to the language $\mathfrak{L}_{\forall=}$ the function symbols $'$ (successor), $+$ (addition), $\times$ (multiplication), and a constant 0. This language will be $\mathfrak{L}_{Arith}$. We will adopt the usual abbreviations for numerals: 1 for $0'$, 2 for $0''$, and so on. Let us turn to arithmetic theories.

Meyer (1976c; 2021a; 2021b) formulated and explored a relevant arithmetic, $\mathsf{R}^{\#}$, by adding to QR the following axioms and rule, adapting them from familiar axiomatizations of Peano arithmetic (PA).

(#1) $x = y \leftrightarrow x' = y'$.
(#2) $x' \neq 0$.
(#3) $x + y' = (x + y)'$.
(#4) $x \times y' = (x \times y) + x$.
(#5) $x = y \rightarrow (x = z \rightarrow y = z)$.
(#6) $x + 0 = x$.

6 See Dunn (1973), Belnap et al. (1980), Read (1988), Slaney (1990), or Restall (2000) for more on the technique of adding structure to the proof systems for relevant logics.

7 Øgaard (2021d) provides a detailed comparison of the sequent systems of Restall and the Hilbert systems given earlier.

8 See Restall (2000, ch. 4.1).

(#7) $x \times 0 = 0$.
(#8) $A(0), \forall x(A(x) \rightarrow A(x')) \Rightarrow \forall x A(x)$.

Friedman and Meyer (1992) trade the final rule for an axiom scheme.

The arithmetic $\mathsf{R}^\#$ can reproduce many of the standard results of PA. It proves all true arithmetical equations and it proves the negation of all false arithmetical equations. In $\mathsf{R}^\#$, addition and multiplication are commutative and associative. One can define the usual ordering on natural numbers, $y \leq z =_{Df} \exists x(x + y = z)$, and the ordering obeys many of the standard features familiar from PA. In fact, $\mathsf{R}^\#$ proves every theorem of PA in the following sense: A is theorem of PA iff $A \vee \neg\mathsf{t}$ is a theorem of $\mathsf{R}^\#$.[9] As A and $A \vee \neg\mathsf{t}$ are equivalent in PA, this implies undecidability and negation-incompleteness for $\mathsf{R}^\#$.

Meyer showed that there are finite models of $\mathsf{R}^\#$. In particular, using the $RM3$ matrix, he showed that all the axioms get designated values, the rules preserve designated values, but $0 = 1$ is not designated. Therefore, the theory is not trivial. The rub is that the model is inconsistent in the sense that there is a formula A such that A and $\neg A$ are both designated. Meyer and Mortensen (1984) showed that the model construction can be generalized to provide inconsistent models for different modular arithmetics.[10] Slaney (2022) engages in a systematic study of De Morgan algebras that can be used to model strengthenings of $\mathsf{R}^\#$. The matrix $C4$, presented in Chapter 5, plays an important role in classifying the models. Restall (2011) explores frame semantics for substructural arithmetics, including relevant arithmetics, providing some contrast with the algebraic models.

The relevant arithmetic $\mathsf{R}^\#$ has important differences with PA, as one might expect. Two of these, explored by Meyer, will be mentioned here in addition to the finite models. While $\mathsf{R}^\#$ has non-standard models, Meyer and Mortensen (2021) showed that it has non-standard models that use rational numbers as the non-standard numbers. These models are further explored by Ferguson and Ramirez-Camara (2021). Another important feature of $\mathsf{R}^\#$ is that cancellation of division fails: It is not the case that $n \times k = m \times k \rightarrow n = m$ is provable. Meyer (2008) shows that cancellation of division appeals to a small amount of weakening, noting that $(2 \times 1 = 2 \times 0) \rightarrow 1 = 0$ is refuted by an $RM3$ model.

Meyer and Mortensen formulate a stronger arithmetic theory, $\mathsf{R}^{\#\#}$, which adds the ω-rule to $\mathsf{R}^\#$:

$$A(0), A(1), A(2), \ldots \Rightarrow \forall x A(x).$$ [11]

Meyer (1998) claims $\mathsf{R}^{\#\#}$ as the "true" relevant arithmetic, and he shows it is closed under (γ), using metavaluational techniques.

[9] See theorem 8 of Meyer (2021a, 280ff.).

[10] Cf. the finite arithmetics axiomatized by Tedder (2015).

[11] Logan and Leach-Krouse (2021) explore alternative formulations of relevant arithmetics that strengthen $\mathsf{R}^\#$ while falling short of $\mathsf{R}^{\#\#}$. They use a non-standard definition of a term, which allows them to build much of the arithmetic into the closed terms themselves. A cost of this is that it becomes undecidable whether a given formula is an axiom.

Meyer and Mortensen (1984) show, using finitary means, that neither $\mathsf{R}^{\#}$ nor $\mathsf{R}^{\#\#}$ prove any equations of the form $0^{(n)} = 0^{(m)}$, for distinct n and m.[12] While PA, too, has no theorems of the mentioned form, that fact cannot be shown using finitary means.

A major question concerning $\mathsf{R}^{\#}$ was whether it admitted (γ), and so contained all the theorems of PA. Friedman and Meyer (1992) showed that the answer to this question was negative, exhibiting a positive theorem of PA that is not a theorem of $\mathsf{R}^{\#}$. This is done using a result, due to Meyer (1986), that the positive fragment of $\mathsf{R}^{\#}$ is conservatively extended by the addition of De Morgan negation. This is sufficient for showing that $\mathsf{R}^{\#}$ does not contain all the theorems of PA and, further, that (γ) fails.[13] Friedman and Meyer show that $\mathsf{R}^{\#\#}$, however, contains all the theorems of PA. In fact, they show that for a formula A that does not contain $\to$, if A is true in the standard model of arithmetic, A is a theorem of $\mathsf{R}^{\#\#}$. The theory $\mathsf{R}^{\#\#}$ is still incomplete as both $0 = 2 \to 0 = 1$ and its negation fail to be theorems.

Negation-incompleteness seems like a flaw in a putative theory of *true* arithmetic. Meyer and Restall (1999) develop a theory of arithmetic to remedy this. They start with $\mathsf{E}^{\#\#}$, trading the axioms of R in $\mathsf{R}^{\#\#}$ for those of E.[14] Then, using metavaluational techniques, Meyer and Restall are able to define an extension, $\mathsf{TE}^{\#\#}$, that is negation-complete and, in particular, contains true negated implications, such as $\neg(0 = 2 \to 0 = 1)$, as theorems.[15]

Dunn (1979b) explored Robinson arithmetic, arithmetic without induction as a rule or axiom, over R. Dunn actually looked at two versions, one where the natural numbers start at 0 and one in which they start at 1. The former, obtained from $\mathsf{R}^{\#}$ by dropping induction and adding an additional axiom, $\forall x(0 \neq x \to \exists y(x = y'))$, not provable in $\mathsf{R}^{\#}$, collapses into classical Robinson arithmetic. The latter, obtained from $\mathsf{R}^{\#}$ by dropping induction, changing the axioms involving 0, and adding the additional axiom, does not collapse into classical Robinson arithmetic.

It is worth pausing to see why the presence of 0 results in a collapse to classical Robinson arithmetic, for which Dunn (1979b) provides the details. The key fact is that for any term s, $0 \times s = 0$. Given the other axioms that make 0 special, one can obtain implications that show that in Dunn's Robinson arithmetic, $0 = 0$ acts like t, in the sense that $A \leftrightarrow (0 = 0 \to A)$ is provable, for all formulas A. Further, one can show that $0 = 1 \to A$ is also provable. Both of these hold in $\mathsf{R}^{\#}$, but the theorem that is not derivable in $\mathsf{R}^{\#}$ is $0 \neq 0 \to 0 = 1$, for which there is a three-valued countermodel. Using these implications, one goes from $A \to (0 = 0 \to A)$, to $A \to (\neg A \to 0 \neq 0)$. Using that with $0 \neq 0 \to 0 = 1$, one obtains $A \to (\neg A \to B)$, which results in TV when added to R.

12 Weber (2021b) explores the philosophical significance of this result.

13 Meyer et al. (1984) explore more theories for which (γ) fails.

14 The formulation of E used is actually that of Ackermann, who took (γ) as a primitive rule, although it can be dropped in the context of $\mathsf{E}^{\#\#}$, as noted by Meyer and Restall.

15 Badia and Tedder (2018) show that negation-completeness fails for a range of relevant theories, not just arithmetic. Their techniques, however, do not apply to theories closed under the ω-rule.

There has not been, as far as we are aware, any study of Robinson arithmetic over weaker base relevant logics.[16] There has, however, been some study of Peano arithmetic over weaker base logics, such as Brady (2012; 2021) using his logic MC, Slaney et al. (1996) using linear logic, or Restall (1994, ch. 11) using a variety of non-contractive substructural logics, including relevant logics.

13.4 Naive Set Theory

One other area that has attracted a lot of attention is naive set theory. A lot of work has been done in this area, often of a highly technical nature, so we will give only a broad overview.

The language of naive set theory adds to $\mathfrak{L}_\forall$ a binary membership relation, $\in$, and set abstract terms, $\{x : A\}$, for each variable x and each formula A, which need not contain x free. Naive set theory takes as its axioms every instance of the set abstraction principle:[17]

$$\forall x(A(x) \leftrightarrow x \in \{y : A(y)\}).$$

As is well known, the addition of the naive comprehension axioms to classical logic result in inconsistency, and so triviality. The addition of these axioms to R also results in triviality, via Curry's paradox.[18]

We will rehearse Curry's paradox here, following the presentation of Meyer et al. (1979). Let C be defined as the set $\{x : x \in x \to (x \in x \to p)\}$, for an arbitrary p. We then argue as follows.

(1) $C \in C \leftrightarrow (C \in C \to (C \in C \to p))$, as an instance of abstraction.
(2) $C \in C \to (C \in C \to (C \in C \to p))$, from (1).
(3) $(C \in C \to (C \in C \to p)) \to (C \in C \to p)$, as an instance of contraction.
(4) $C \in C \to (C \in C \to p)$, from (2) and (3).
(5) $(C \in C \to (C \in C \to p)) \to C \in C$, from (1).
(6) $C \in C$, from (4) and (5).
(7) p, from (6) and (4), twice.

In the definition of C, p was arbitrary, so similar arguments result in every formula being a theorem.

There are two responses one can have to Curry's paradox. One is to think that naive set abstraction is to blame, and so reject it while not changing the logic. More popular among relevant logicians is holding on to naive set abstraction and blaming the logical principles. Contraction principles, such as (I1) and (I7), play an important role in the trivializing argument. Different principles that can feature in trivializing arguments

[16] There have been studies of Robinson arithmetic over weak substructural logics, which may include weakening, such as Restall (1994, ch. 11) and Badia et al. (2022).

[17] Weber (2010a) distinguishes set abstraction, which is the principle presented, from set comprehension, $\exists y \forall x(A(x) \leftrightarrow x \in y)$. The differences will not matter for this section.

[18] See Shapiro and Beall (2021) for more on Curry's paradox.

have been explored by a number of logicians, including Restall (1993), Rogerson and Butchart (2002), Rogerson and Restall (2004), Rogerson (2007), Bacon (2013a), Robles and Méndez (2014a), and Øgaard (2016). Triviality for naive set theory can arise from surprising places. Slaney (1989) shows that in RWX, RW plus (N3), the naive set axioms produce triviality. This is the case even though RW rejects standard contraction principles and excluded middle is not a contraction principle for the conditional. If one rejects *all* of the contraction principles, however, then one can enjoy a non-trivial naive set theory.

Brady (1971; 1983; 1989b; 2006b; 2014) has done perhaps more than anyone in constructing models for naive set theories demonstrating their non-triviality. Brady's model constructions involve nested transfinite sequences of models that refine the interpretation of implication and membership. The outer sequence of models refines the interpretation of implication, and at each stage of that sequence, a new interpretation of membership is generated, which is used in refining the interpretation of membership. The details of Brady's model constructions are beyond the scope of this section, so we leave further investigation to the interested reader.[19] Among the non-triviality results Brady has obtained, we will highlight two: (1) One can have a negation-inconsistent, but non-trivial, naive set theory, meaning that for some formula A, both A and $\neg A$ are theorems, while there being some formulas that are not theorems and (2) for some base logics, naive set theory is simply consistent, meaning that there is no formula A for which both A and $\neg A$ are provable. Regarding (1), some of the logics Brady shows to be non-trivial have (N3) as a theorem. Bacon (2013b) has provided an alternative model construction showing that the positive fragment of a logic extending TW+(I2) supports a non-trivial naive set theory.

Although not a result in naive set theory, Slaney (1987b) proves a related result for propositional quantification. In a propositional language with propositional quantification, the naive comprehension scheme for propositional quantifiers says that, for each formula A, $\exists p(p \leftrightarrow A)$. In the comprehension scheme, it is important that p can occur free in A. Slaney shows the non-triviality of naive comprehension for propositional quantifiers added to a logic extending RW by providing a model on the real numbers.

Extensionality principles specify conditions under which sets are identical. Sets having the same members,

$$\forall z(z \in x \leftrightarrow z \in y) \rightarrow x = y,$$

turns out not to be an appropriate extensionality principle for naive set theory. Instead, the axiom

$$\forall z(x \in z \leftrightarrow y \in z) \rightarrow x = y$$

is adopted. When identity is defined using extensionality, then the latter axiom yields a stronger logic of identity. Brady and Routley (1989) and Brady (2006b) show that, for some base logics, non-triviality extends to theories of naive set theory with an extensionality principle. Identity and its substitution principles can lead to problems

[19] Field et al. (2017) explore some variations on Brady's constructions.

for naive set theory with extensionality, as noted by Weber (2010a; 2020) and Field et al. (2017).

The non-triviality arguments mentioned earlier are somewhat sensitive to the language. If fusion is not in the language, then naive set theory over DJ can be shown to be non-trivial. If, however, fusion is in the language, then a variant of Curry's paradox leads to triviality via the axiom (I2), conjunctive syllogism, which is in DJ. For the variant, define the set K as $\{x : (x \in x \circ x \in x) \rightarrow p\}$. To aid readability, we will use $\mathcal{K}$ for the formula $K \in K$. We then argue as follows.[20]

(1) $\mathcal{K} \leftrightarrow ((\mathcal{K} \circ \mathcal{K}) \rightarrow p)$, as an instance of abstraction.
(2) $\mathcal{K} \rightarrow ((\mathcal{K} \circ \mathcal{K}) \rightarrow p)$, from (1).
(3) $\mathcal{K} \rightarrow (\mathcal{K} \rightarrow (\mathcal{K} \circ \mathcal{K}))$, as a theorem of B.
(4) $\mathcal{K} \rightarrow \Big((\mathcal{K} \rightarrow (\mathcal{K} \circ \mathcal{K})) \wedge ((\mathcal{K} \circ \mathcal{K}) \rightarrow p)\Big)$, using (A3) with (2) and (3).
(5) $\mathcal{K} \rightarrow (\mathcal{K} \rightarrow p)$, using (I2) and (4).
(6) $(\mathcal{K} \circ \mathcal{K}) \rightarrow p$, using (R6) and (5).
(7) $\mathcal{K}$, by (1) and (6).
(8) $\mathcal{K} \circ \mathcal{K}$, using (7) twice with (3).
(9) p, using (6) and (8).

While (I2) does not lead to triviality in the absence of fusion, it does in the presence of fusion, as it provides a kind of contraction principle for fusion.

Since there are non-trivial naive set theories, one might wonder what they can do. We'll remark on a few of their interesting features. The first feature is that there is a universal set, namely a set of all sets, $\{x : \exists y(x \in y)\}$. If identity is in the language, then there is a second, distinct universal set, $\{x : x = x\}$. The second feature is that there are, even in the presence of the stronger extensionality axiom given earlier, co-extensional yet distinct sets, such as the two universal sets. This is a feature remarked upon by Brady (1971), Weber (2010a), and Field et al. (2017), among others. While those sets have the same members, namely all sets, having the same members does not guarantee identity of sets. The reason is that the extensionality principles are formulated with a (non-contractive) relevant biconditional, and there is a very high bar for having such biconditionals as theorems. The third feature is that naive set theories permit one to prove a general fixed point theorem: For any formula $A(x, y)$, there is a term f_A such that for all terms t, it is provable that $t \in f_A \leftrightarrow A(f_A, t)$.[21]

Weber (2010b; 2012; 2013) has developed naive set theory over a quantified logic extending TW+(I2) with the rule $A, \neg B \Rightarrow \neg(A \rightarrow B)$. He has obtained many results of classical set theory, including results on cardinal and ordinal numbers. Ordinals in this set theory have many of the prominent features they do in classical set theory, such as being well-ordered. There is an infinite set, and a form of transfinite recursion is provable. On the more non-classical side, there is a set of all ordinals, which is itself

[20] Our argument is based on that of Routley et al. (1982, 367). Mares (2021) presents alternative formulations of this paradox, using a natural deduction system and using models.
[21] See Terui (2004) for the proof.

an ordinal, and the universe of sets can be well-ordered. There is much left to explore in relevant naive set theory, as well as relevant arithmetic and other theories.[22]

13.5 Restricted Quantification

As we saw in Chapter 11, quantification can be adequately modeled, so we can turn to some important applications for quantification. The first of these is RESTRICTED QUANTIFICATION. Suppose we have variable-binding operators in the language, Π and Σ, such that $\Pi x[Ax\colon Bx]$ is supposed to be understood as saying that all As are Bs, and $\Sigma x[Ax\colon Bx]$ is supposed to be understood as saying that some A is B. We might add a third, N, such that $Nx[Ax\colon Bx]$ means that no A is B. In TV, these can be defined in terms of the material conditional and the first-order quantifiers, as $\forall x(A \to B)$, $\exists x(A \wedge B)$, and $\forall x(A \to \neg B)$, respectively. Let us consider these definitions in the context of relevant logics, using the relevant implication where TV uses the material conditional.

Beall et al. (2006) note several natural rules one might want restricted quantifiers to obey, such as

(RQ1) $\forall x A(x) \Rightarrow \Pi x[Bx\colon Ax]$, and
(RQ2) $\Pi x[Ax\colon Bx], A(a) \Rightarrow B(a)$.

However, since the universal quantifier is allowed to bind vacuously, trouble lurks. Suppose that A is derivable and doesn't contain x. Then $\forall x A$ is derivable, whence $\Pi x[Bx\colon Ax]$. If Π permits contraposition, which is to say that $\Pi x[Bx\colon Ax] \Rightarrow \Pi x[\neg Ax\colon \neg Bx]$ is accepted, then $\neg A$ will result in $\neg B$, for arbitrary B. This is a form of explosion, so we encounter an incompatibility between plausible principles, namely that if everything is A then all Bs are A and contraposition for restricted universal quantification. As one might expect, a form of weakening results in explosion with a form of contraposition.

Beall et al. consider several relations among claims involving restricted quantifiers. The classical definition of Σ is adopted, but the definitions of Π and N are are found wanting. One thing that the definition of Π seems to get correct is that Π should obey something like *modus ponens*, (RQ1). The weakening-like principle, (RQ2), seems plausible, but it will not do if the conditional of the definition is the implication of the base logic. That is simply letting in some amount of weakening for the basic implication. Instead, Beall et al. propose a new conditional to use in the definition of Π. They define the conditional, $\mapsto$ via a verification condition.

- $a \Vdash B \mapsto C$ iff for all $b, c \in K$ such that $Rabc$ and $a \leq c$, if $b \Vdash B$, then $c \Vdash C$.

Using this conditional, they define Π and N as follows.

- $\Pi x[Ax\colon Bx]$ is $\forall x(A(x) \mapsto B(x))$.
- $Nx[Ax\colon Bx]$ is $\forall x(A(x) \mapsto \neg B(x))$.

22 Research on naive set theory is ongoing enterprise. Weber (2021a) has obtained many results, albeit in a logic that has two implications, one relevant and one irrelevant.

With the Tarskian verification condition, these definitions, along with the definition of $\Sigma x[Ax\colon Bx]$ as $\exists x(A(x) \wedge B(x))$, show that several natural constraints on restricted quantification are satisfied.

A general strategy for restricted quantification is to use a new conditional, $\rightsquigarrow$, to define some restricted quantifiers. Beall (2011) proposes that to be adequate for restricted quantification, such a conditional will need to satisfy the following.

- $A, A \rightsquigarrow B \Rightarrow B$.
- $B \Rightarrow A \rightsquigarrow B$.
- $A \rightarrow B \Rightarrow A \rightsquigarrow B$.
- $\vdash (A \wedge B) \rightsquigarrow A$.

Further, the rule $A \rightsquigarrow B \Rightarrow \neg B \rightsquigarrow \neg A$ will need to fail. Beall (2011) presents an alternative to the proposal of Beall et al. (2006) that satisfies these conditions. The proposal is that

- $A \rightsquigarrow B$ is defined as $(A \rightarrow B) \vee B$.

Beall argues that this proposal delivers on the required principles while avoiding the ones that need to be avoided.

These proposals for relevant restricted quantification are promising, but there is more to be done. Restricted quantification is important in applications, such as set theory and arithmetic, and it remains to be seen how well these proposals work in those settings, or others. There has not been any work on restricted quantification in the Mares–Goldblatt framework, so it is unclear which of the constraints would be satisfied in that framework. The Mares–Goldblatt framework could, potentially, provide new options for models for restricted quantification.[23] This is an area that warrants further exploration.

13.6 Combining Modalities and Quantifiers

To bring together themes from earlier chapters, we can combine first-order quantifiers and modalities. As shown by Ferenz (2021), one can use the Mares–Goldblatt frames of Chapter 11 for the quantifiers and the binary accessibility relation for the modalities. The only addition to the framework that is required is a closure condition on Prop and PropFun to ensure that propositions to interpret necessary claims are always available. The resulting frame theory is adequate, as one would expect.

When both modalities and quantifiers are around, it is natural to wonder about the status of the Barcan formula, $\forall x \Box Fx \rightarrow \Box \forall x Fx$, and its converse, $\Box \forall x Fx \rightarrow \forall x \Box Fx$. In the frame theory sketched earlier, the converse Barcan formula is valid. The Barcan formula can be validated using a simple frame condition, as shown by Ferenz (2021).[24]

[23] A similar point holds for Fine's approach to quantification, as noted by Calasso and Logan (2023).

[24] Ferenz and Tedder (2025) study versions of the Barcan formula, and its converse, in more detail.

Tedder and Ferenz (2022) show how to give a more general frame theory for quantified modal relevant logics using neighborhoods in combination with the Mares–Goldblatt semantics. Their frames take the ternary relation to be a relation between points, sets of points, and sets of points, in the manner of Goble (2003). It would take us too far afield to go into the details of their framework, but they prove that a wide array of logics are Sound and Complete with respect to different classes of their frames.

Finally, the discussion of this brief section has so far focused on *first-order* quantification and its interaction with modalities. One might wonder about the interaction of *propositional* quantification and modality. In recent years, there has been a renewed interest in the interaction of propositional quantification and modalities. This work has largely been focused on TV-based modal logics, but it need not be so focused.[25] There has been a little work done on propositional quantification and relevant modal logic, as indicated in Chapter 8, and it seems that further studying propositional quantification and relevant modal logic would be fruitful.

13.7 Whither Relevant Logics?

To close this book, we will suggest some directions to go for future research.[26] These suggestions should not be seen as exhaustive of possible research directions.

In this chapter, we have presented an overview of work on relevant arithmetic. There is more to do on relevant arithmetic, particularly with weaker arithmetics and with weaker base logics.[27] Work has only just begun on the theory of rational numbers in relevant logics, and, as far as we know, nothing has been done on the real numbers and analysis in relevant logic.[28] Many more or less traditional topics in arithmetic, such as recursive function theory, the machinery for Gödel numbering, and provability predicates are largely unexplored. More broadly, there are questions about how to develop other mathematical theories, such as lattice theory or group theory, in the context of relevant logics.[29]

We have presented first-order and propositional quantification for relevant logics, and that naturally points the way to higher-order quantification. This is largely unexplored in the context of relevant logics. Dunn (1988) provides a notable exception, proving an impossibility result in the area, and recently Logan and Boccuni (2025) have investigated abstraction principles in relevant logics, using them to derive some basic principles of arithmetic. Logan (2025b) has provided Fine-style frames for second-order relevant logics, but adequate Mares–Goldblatt frames are yet to be investigated. In addition, frames for other higher-order resources, such as

[25] See Fritz (2024).

[26] The title of this section is a reference to the title of Avron's provocative and insightful article (1992).

[27] Tedder (2021a) has some surprising results about a weak arithmetic theory.

[28] Slaney (2025) identifies three distinct approaches to investigating rational numbers with relevant logics.

[29] Badia and Tedder (2018) explore some features of the theories of directed linear orders and real closed fields, as well as a few other mathematical theories.

λ-abstraction, are in need of investigation.[30] Before turning away from quantifiers, we should highlight a longstanding open problem, mentioned in Chapter 11: axiomatizing constant-domain quantified relevant logics that use the Tarskian verification condition for the universal quantifier.

Turning to relevant modal logics, there is certainly more to be done. Many modalities that have been studied in TV-based modal logics have not been studied much, or at all, in the context of relevant logics. In this regard, temporal logics seem like a natural direction to investigate, as well as more deontic logics, to take advantage of the paraconsistency of relevant logics. As suggested in Chapter 8, relevant logics can offer new possibilities as well, and these should be investigated. Inquisitive relevant logics have been developed for Fine-style frames but not for ternary relational frames. It would be good to know whether this is possible or whether one can prove a limitative result showing that it cannot be done.

There are many issues regarding the base logics that could do with further investigation. A particularly prominent one is the role of *rules* in relevant logics. Brady (1993; 1994) has studied rules in relevant logics, but there has not been the sort of interest in admissible rules for relevant logics that one sees for intuitionistic logic.[31] In connection with rules, further investigation of consequence relations for relevant logics seems worthwhile, as exemplified by the recent work of Øgaard.[32] Additionally, relevant logics fall within the general category of substructural logics, and it would be worthwhile to draw out further points of similarity and difference with the other substructural logics. In this vein, there are two directions to highlight. One is comparison with the logics with weakening but without contraction. The other is further investigation of and comparison with logics without the distribution axiom, such as Brady's preferred logic MC and linear logic and its relatives.[33] Some of these logics, particularly linear logic, are well studied, but many are not, despite apparently promising features.

Finally, although we have focused on more formal issues throughout this book, there are less formal, philosophical issues that are worth pursuing. To start, making connections between relevant logics and debates ongoing elsewhere in philosophy is promising, particularly in the areas of metaphysics and epistemology, as well as feminist theory.[34]

The philosophical foundations of relevant logics are undergoing a reevaluation. Apart from reassessing the extant philosophical reasons for relevant logics, alternatives are being developed. A foundational issue, worth further discussion, is whether there is, or should be, a privileged relevant logic. Anderson and Belnap favored E and R, while (certain time slices of) Sylvan favored B, and Brady favors MC. What philosophical views support adopting a particular relevant logic as the correct one?

[30] Bacon (2023) provides an excellent overview of models for higher-order logics.

[31] See, for example, Rybakov (1997), Iemhoff (2001), or Cintula and Metcalfe (2010).

[32] Badia et al. (2024) provide a promising approach to relevant consequence relations.

[33] In Chapter 9, we mentioned Lattice-R as a logic lacking distribution earlier in the book. See Brady and Meinander (2013) for defense of MC. See Restall (2000) for an introduction to linear logic.

[34] Plumwood (1993) provides a case for the connection between relevant logic and feminist theory, a connection further explored by Mangraviti (2023).

Or, is a spirit of pluralism more appropriate for the relevant logician?[35] We hope these issues will be investigated more in the future.

We will end the book by echoing the words of Mares (2004): Relevant logics are exciting, interesting, and useful, and there is plenty for philosophical logicians to do.

35 For general discussion of logical pluralism, see Beall and Restall (2006) and Russell (2014). For discussion of pluralism in connection to relevant logics, see Restall (2014) and Kouri (2016).

References

Ackermann, W. (1956). Begründung einer strengen Implikation. *Journal of Symbolic Logic*, 21(2):113–128.

Ackermann, W. (1958). Über die Beziehung zwischen strikter und strenger Implikation. Dialectica, 12(3/4):213–222.

Alchourrón, C. E., Gärdenfors, P., and Makinson, D. (1985). On the logic of theory change: Partial meet contraction and revision functions. *Journal of Symbolic Logic*, 50(2):510–530.

Anderson, A. R. (1958a). The logic of norms. *Logique et Analyse*, 1(2):84.

Anderson, A. R. (1958b). A reduction of deontic logic to alethic modal logic. *Mind*, 67(265):100–103.

Anderson, A. R. (1960). Completeness theorems for the systems E of entailment and EQ of entailment with quantification. *Mathematical Logic Quarterly*, 6(7–14):201–216.

Anderson, A. R. (1963). Some open problems concerning the system E of entailment. *Acta Philosophica Fennica*, fasc. 16:7–18.

Anderson, A. R. (1967a). The formal analysis of normative systems. In Rescher, N., editor, *The Logic of Decision and Action*, pages 147–213. University of Pittsburgh Press. Originally published as Technical Report No.2, Contract No. Sar/Nonr-609(16), New Haven, 1956.

Anderson, A. R. (1967b). Some nasty problems in the formal logic of ethics. *Noûs*, 1(4): 345–360.

Anderson, A. R. (1968). A new square of opposition: Eubouliatic logic. In *Akten des XIV. Internationalen Kongresses für Philosophie*, volume 2, pages 271–284.

Anderson, A. R. (1972). An intensional interpretation of truth-values. *Mind*, 81(323):348–371.

Anderson, A. R. and Belnap, N. D. (1962). The pure calculus of entailment. *Journal of Symbolic Logic*, 27(1):19–52.

Anderson, A. R. and Belnap, N. D. (1975). *Entailment: The Logic of Relevance and Necessity, Vol. I*. Princeton University Press.

Anderson, A. R., Belnap, N. D., and Dunn, J. M. (1992). *Entailment: The Logic of Relevance and Necessity, Vol. II*. Princeton University Press.

Artemov, S. and Fitting, M. (2016). Justification logic. In Zalta, E. N., editor, *The Stanford Encyclopedia of Philosophy*. Metaphysics Research Lab, Stanford University, winter 2016 edition.

Artemov, S. and Fitting, M. (2019). *Justification Logic*. Cambridge University Press.

Avron, A. (1987). A constructive analysis of **RM**. *Journal of Symbolic Logic*, 52(4):939–951.

Avron, A. (1990a). Relevance and paraconsistency—a new approach. *Journal of Symbolic Logic*, 55(2):707–732.

Avron, A. (1990b). Relevance and paraconsistency—a new approach part II: The formal systems. *Notre Dame Journal of Formal Logic*, 31(n/a):169–202.

Avron, A. (1992). Whither relevance logic? *Journal of Philosophical Logic*, 21(3):243–281.

Bacon, A. (2013a). Curry's paradox and ω-inconsistency. *Studia Logica*, 101(1):1–9.

Bacon, A. (2013b). A new conditional for naive truth theory. *Notre Dame Journal of Formal Logic*, 54(1):87–104.

Bacon, A. (2023). *A Philosophical Introduction to Higher-Order Logics*. Routledge.

Badia, G. (2016a). A Lindström-style theorem for finitary propositional weak entailment languages with absurdity. *Logic Journal of the IGPL*, 24(2):115–137.

Badia, G. (2016b). The relevant fragment of first order logic. *Review of Symbolic Logic*, 9(1):143–166.

Badia, G. (2017a). Infinitary propositional relevant languages with absurdity. *Review of Symbolic Logic*, 10(4):663–681.

Badia, G. (2017b). Model definability in relevant logic. *IfCoLog Journal of Logics and Their Applications*, 3(4):623–646.

Badia, G. (2019). Incompactness of the A1 fragment of basic second order propositional relevant logic. *Australasian Journal of Logic*, 16(1):1–8.

Badia, G., Běhounek, L., Cintula, P., and Tedder, A. (2024). Relevant consequence relations: An invitation. *Review of Symbolic Logic*, 17(3):762–792.

Badia, G., Cintula, P., Hajek, P., and Tedder, A. (2022). How much propositional logic suffices for Rosser's essential undecidability theorem? *Review of Symbolic Logic*, 15(2):487–504.

Badia, G. and Tedder, A. (2018). On elimination of quantifiers in some non-classical mathematical theories. *Mathematical Logic Quarterly*, 64(3):140–154.

Barwise, J. (1993). Constraints, channels and the flow of information. In Aczel, P., Israel, D., Katagiri, Y., and Peters, S., editors, *Situation Theory and its Applications Vol. 3*, pages 3–27. CSLI Publications.

Barwise, J. and Perry, J. (1983). *Situations and Attitudes*. CSLI Publications.

Beall, J. C. (2011). Adding to relevant restricted quantification. *Australasian Journal of Logic*, 10:36–44.

Beall, J. C., Brady, R., Dunn, J. M., et al. (2012). On the ternary relation and conditionality. *Journal of Philosophical Logic*, 41(3):595–612.

Beall, J. C. and Restall, G. (2006). *Logical Pluralism*. Oxford University Press, Oxford.

Beall, J. C., Brady, R. T., Hazen, A. P., Priest, G., and Restall, G. (2006). Relevant restricted quantification. *Journal of Philosophical Logic*, 35(6):587–598.

Belnap, N. (1960). Entailment and relevance. *Journal of Symbolic Logic*, 25(2):144–146.

Belnap, N. (1962). Tonk, plonk and plink. *Analysis*, 22(6):130–134.

Belnap, N. (1977a). How a computer should think. In Ryle, G., editor, *Contemporary Aspects of Philosophy*, pages 30–56. Oriel Press.

Belnap, N. (1977b). A useful four-valued logic. In Dunn, J. M. and Epstein, G., editors, *Modern Uses of Multiple-Valued Logic*, pages 8–37. Reidel.

Belnap, N. (1982). Display logic. *Journal of Philosophical Logic*, 11(4):375–417.

Belnap, N., Gupta, A., and Dunn, J. M. (1980). A consecutive calculus for positive relevant implication with necessity. *The Journal of Philosophical Logic*, 9(4):343–362.

Belnap, N., Perloff, M., and Xu M. (2001). *Facing the Future: Agents and Choices in Our Indeterminist World*. Oxford University Press.

Berto, F. (2015). A modality called 'negation'. *Mind*, 124(495):761–793.

Berto, F. and Restall, G. (2019). Negation on the Australian plan. *Journal of Philosophical Logic*, 48(6):1119–1144.

Bílková, M., Majer, O., and Peliš, M. (2016). Epistemic logics for sceptical agents. *Journal of Logic and Computation*, 26(6):1815–1841.

Bílková, M., Majer, O., Peliš, M., and Restall, G. (2010). Relevant agents. In Beklemishev, L., Goranko, V., and Shehtman, V., editors, *Advances in Modal Logic*, number 8, pages 22–38. College Publications.

Bimbó, K. (2007a). $LE^{t}_{\rightarrow}$, $LR^{\circ}_{\hat{\sim}}$, LK and cutfree proofs. *Journal of Philosophical Logic*, 36(5):557–570.

Bimbó, K. (2007b). Relevance logics. In Jacquette, D., editor, *Philosophy of Logic*, volume 5 of *Handbook of the Philosophy of Science*, pages 723–789. Elsevier.

Bimbó, K. (2009). Combinatory logic. In Zalta, E. N., editor, *Stanford Encyclopedia of Philosophy*. The Metaphysics Research Lab, Center for the Study of Language and Information, Stanford University.

Bimbó, K. (2011). *Combinatory Logic: Pure, Applied and Typed*. Taylor & Francis.

Bimbó, K. (2014). *Proof Theory: Sequent Calculi and Related Formalisms*. CRC Press.

Bimbó, K. and Dunn, J. M. (2008). *Generalized Galois Logics: Relational Semantics of Nonclassical Logical Calculi*. Center for the Study of Language and Information.

Bimbó, K. and Dunn, J. M. (2018). Larisa Maksimova's early contributions to relevance logic. In Odintsov, S., editor, *Larisa Maksimova on Implication, Interpolation, and Definability*, volume 15 of *Outstanding Contributions to Logic*, pages 33–60. Springer International Publishing.

Bimbó, K. and Dunn, J. M. (2023). Fine's semantics for relevance logic and its relevance. In Faroldi, F. L. G. and Putte, F. V. D., editors, *Kit Fine on Truthmakers, Relevance, and Nonclassical Logic*, pages 125–149. Springer Verlag.

Bimbó, K., Dunn, J. M., and Ferenz, N. (2018). Two manuscripts, one by Routley, one by Meyer: The origins of the Routley-Meyer semantics for relevance logics. *Australasian Journal of Logic*, 15(2):171–209.

Blackburn, P., de Rijke, M., and Venema, Y. (2002). *Modal Logic*. Cambridge Tracts in Theoretical Computer Science. Cambridge University Press.

Blumson, B. (2021). Relevance and verification. *Philosophical Quarterly*, 71(3):457–480.

Boolos, G. (1993). *The Logic of Provability*. Cambridge University Press.

Božić, M. and Došen, K. (1984). Models for normal intuitionistic modal logics. *Studia Logica*, 43(3):217–245.

Brady, R., editor (2003). *Relevant Logics and Their Rivals, Volume II, A continuation of the work of Richard Sylvan, Robert Meyer, Val Plumwood and Ross Brady*. Ashgate.

Brady, R. T. (1971). The consistency of the axioms of abstraction and extensionality in a three-valued logic. *Notre Dame Journal of Formal Logic*, 12(4):447–453.

Brady, R. T. (1982). Completeness proofs for the systems RM3 and BN4. *Logique et Analyse*, 25(97):9–32.

Brady, R. T. (1983). The simple consistency of a set theory based on the logic CSQ. *Notre Dame Journal of Formal Logic*, 24(4):431–449.

Brady, R. T. (1984a). Depth relevance of some paraconsistent logics. *Studia Logica*, 43(1–2):63–73.

Brady, R. T. (1984b). Natural deduction systems for some quantified relevant logics. *Logique et Analyse*, 27(8):355–377.

Brady, R. T. (1988). A content semantics for quantified relevant logics. I. *Studia Logica*, 47(2):111–127.

Brady, R. T. (1989a). A content semantics for quantified relevant logics. II. *Studia Logica*, 48(2):243–257.

Brady, R. T. (1989b). The non-triviality of dialectical set theory. In Priest, G., Routley, R., and Norman, J., editors, *Paraconsistent Logic: Essays on the Inconsistent*, pages 437–470. Philosophia Verlag.

Brady, R. T. (1993). Rules in relevant logic – II: Formula representation. *Studia Logica*, 52(4):565–585.

Brady, R. T. (1994). Rules in relevant logic – I: Semantic classification. *Journal of Philosophical Logic*, 23(2):111–137.

Brady, R. T. (2006a). Normalized natural deduction systems for some relevant logics 1: The logic DW. *Journal of Symbolic Logic*, 71(1):35–66.

Brady, R. T. (2006b). *Universal Logic*. CSLI Publications.

Brady, R. T. (2012). The consistency of arithmetic, based on a logic of meaning containment. *Logique et Analyse*, 55(219):353–383.

Brady, R. T. (2014). The simple consistency of naive set theory using metavaluations. *Journal of Philosophical Logic*, 43(2–3):261–281.

Brady, R. T. (2017a). Metavaluations. *Bulletin of Symbolic Logic*, 23(3):296–323.

Brady, R. T. (2017b). Some concerns regarding ternary-relation semantics and truth-theoretic semantics in general. *IfCoLog Journal of Logics and Their Applications*, 4(3):755–781.

Brady, R. T. (2021). Arithmetic formulated in a logic of meaning containment. *The Australasian Journal of Logic*, 18(5):447–472.

Brady, R. T. and Meinander, A. (2013). Distribution in the logic of meaning containment and in quantum mechanics. In Berto, F., Mares, E., Tanaka, K., and Paoli, F., editors, *Paraconsistency: Logic and Applications*, pages 223–255. Springer.

Brady, R. T. and Routley, R. (1989). The non-triviality of extensional dialectical set theory. In Priest, G., Routley, R., and Norman, J., editors, *Paraconsistent Logic: Essays on the Inconsistent*, pages 415–436. Philosophia Verlag.

Brotherston, J. and Goré, R. (2011). Craig interpolation in displayable logics. In Brünnler, K. and Metcalfe, G., editors, *Automated Reasoning with Analytic Tableaux and Related Methods*, pages 88–103. Springer Berlin Heidelberg.

Burgess, J. P. (1981). Relevance: A fallacy? *Notre Dame Journal of Formal Logic*, 22(2): 97–104.

Burgess, J. P. (1984). Read on relevance: A rejoinder. *Notre Dame Journal of Formal Logic*, 25(3):217–223.

Burgess, J. P. (2005). No requirement of relevance. In Shapiro, S., editor, *Oxford Handbook of Philosophy of Mathematics and Logic*, pages 727–750. Oxford University Press.

Caicedo, X., Metcalfe, G., Rodríguez, R., and Tuyt, O. (2019). The one-variable fragment of Corsi logic. In Iemhoff, R., Moortgat, M., and de Queiroz, R., editors, *Logic, Language, Information, and Computation*, pages 70–83, Springer.

Caicedo, X., Metcalfe, G., Rodríguez, R., and Tuyt, O. (2022). One-variable fragments of intermediate logics over linear frames. *Information and Computation*, 287:104755. Special Issue: Selected Papers from WoLLIC 2019, the 26th Workshop on Logic, Language, Information and Computation.

Calasso, M. and Logan, S. A. (2023). Stratified restricted universals. *Asian Journal of Philosophy*, 2(2):44.

Charlwood, G. (1978). *Representations of Semilattice Relevance Logics*. PhD thesis, University of Toronto.

Charlwood, G. (1981). An axiomatic version of positive semilattice relevance logic. *Journal of Symbolic Logic*, 46(2):233–239.

Cheng, J. (1998). Temporal relevant logic as the logic basis for reasoning about dynamics of concurrent systems. In *IEEE International Conference on Systems, Man, and Cybernetics (SMC'98)*, volume 1, pages 794–799. IEEE.

Cheng, J. (2004). Temporal relevant logic as the logic basis of anticipatory reasoning-reacting systems. In *AIP Conference Proceedings*, volume 718, pages 362–375. AIP.

Cheng, J. (2009). A temporal relevant logic approach to modeling and reasoning about epistemic processes. In *2009 Fifth International Conference on Semantics, Knowledge and Grid*, pages 18–25. IEEE.

Ciardelli, I., Groenendijk, J., and Roelofsen, F. (2019). *Inquisitive Semantics*. Oxford University Press.

Cintula, P. and Metcalfe, G. (2010). Admissible rules in the implication-negation fragment of intuitionistic logic. *Annals of Pure and Applied Logic*, 162(2):162–171.

Cintula, P. and Noguera, C. (2021). *Logic and Implication: An Introduction to the General Algebraic Study of Non-Classical Logics*. Springer Verlag.

Copeland, B. J. (1979). On when a semantics is not a semantics: Some reasons for disliking the Routley-Meyer semantics for relevance logic. *Journal of Philosophical Logic*, 8(1):399–413.

Copeland, B. J. (1983). Pure semantics and applied semantics. *Topoi*, 2(2):197–204.

Crossley, J. N. and Humberstone, L. (1977). The logic of "actually". *Reports on Mathematical Logic*, 8:11–29.

De, M. and Omori, H. (2018). There is more to negation than modality. *Journal of Philosophical Logic*, 47(2):281–299.

Došen, K. (1986). Negation as a modal operator. *Reports on Mathematical Logic*, 20:15–87.

Došen, K. (1992). The first axiomatization of relevant logic. *Journal of Philosophical Logic*, 21(4):339–356.

Došen, K. (1999). Negation in the light of modal logic. In Gabbay, D. M. and Wansing, H., editors, *What is Negation?*, pages 77–86. Springer.

Dunn, J. M. (1966). *The Algebra of Intensional Logics*. PhD thesis, University of Pittsburgh.

Dunn, J. M. (1973). A 'Gentzen system' for positive relevant implication (abstract). *Journal of Symbolic Logic*, 38:356–357.

Dunn, J. M. (1976). Intuitive semantics for first-degree entailments and 'coupled trees'. *Philosophical Studies*, 29(3):149–168.

Dunn, J. M. (1979a). R-mingle and beneath. Extensions of the Routley-Meyer semantics for R. *Notre Dame Journal of Formal Logic*, 20(n/a):369.

Dunn, J. M. (1979b). Relevant Robinson's arithmetic. *Studia Logica*, 38(4):407–418.

Dunn, J. M. (1987). Relevant predication 1: The formal theory. *Journal of Philosophical Logic*, 16(4):347–381.

Dunn, J. M. (1988). The impossibility of certain higher-order non-classical logics with extensionality. In Austin, D. F., editor, *Philosophical Analysis*, pages 261–279. Kluwer Academic Publishers.

Dunn, J. M. (1990a). Relevant predication 2: Intrinsic properties and internal relations. *Philosophical Studies*, 60(3):177–206.

Dunn, J. M. (1990b). Relevant predication 3: Essential properties. In Dunn, J. and Gupta, A., editors, *Truth or Consequences*, pages 77–95. Kluwer Academic Publishers.

Dunn, J. M. (1993). Star and perp: Two treatments of negation. *Philosophical Perspectives*, 7:331–357.

Dunn, J. M. (1995). Positive modal logic. *Studia Logica*, 55(2):301–317.

Dunn, J. M. (2022). Kripke's argument for γ. In Bimbó, K., editor, *Relevance Logics and other Tools for Reasoning: Essays in Honor of J. Michael Dunn*, pages 178–181. College Publications.

Dunn, J. M. and Hardegree, G. M. (2001). *Algebraic Methods in Philosophical Logic*. Oxford University Press.

Dunn, J. M. and Restall, G. (2002). Relevance logic. In Gabbay, D. M. and Guenthner, F., editors, *Handbook of Philosophical Logic*, volume 6, pages 1–136. Kluwer, 2nd edition.

Dziobiak, W. (1983). There are $2^{\aleph_0}$ logics with the relevance principle between R and RM. *Studia Logica*, 42(1):49–61.

Ewald, W. B. (1986). Intuitionistic tense and modal logic. *Journal of Symbolic Logic*, 51(1):166–179.

Ferenz, N. (2018). Richard Routley, "Semantic analysis of entailment and relevant implication: I". *Australasian Journal of Logic*, 15(2):210–279.

Ferenz, N. (2019). *Quantified Modal Relevant Logics*. PhD thesis, University of Alberta.

Ferenz, N. (2021). Quantified modal relevant logics. *The Review of Symbolic Logic*, 16(1):210–240.

Ferenz, N. (2024). One variable relevant logics are S5ish. *Journal of Philosophical Logic*, 53:909–931.

Ferenz, N. and Tedder, A. (2022). Neighbourhood semantics for modal relevant logics. *Journal of Philosophical Logic*, 52:145–181.

Ferenz, N. and Tedder, A. (2025). Quantified modal relevant logics II: Welcome to the neighbourhood. In Sedlár, I., Standefer, S., and Tedder, A., editors, *New Directions in Relevant Logic*, pages 153–182. Springer.

Ferguson, T. M. (2017). *Meaning and Proscription in Formal Logic: Variations on the Propositional Logic of William T. Parry*. Springer Verlag.

Ferguson, T. M. (2025). A topic-theoretic perspective on variable-sharing (from the black sheep of the family). In Sedlár, I., Standefer, S., and Tedder, A., editors, *New Directions in Relevant Logic*, pages 43–59. Springer.

Ferguson, T. M. and Ramirez-Camara, E. (2021). Episodes in model-theoretic xenology: Rationals as positive integers in R#. *The Australasian Journal of Logic*, 18(5):428–446.

Field, H. (2008). *Saving Truth From Paradox*. Oxford University Press.

Field, H., Lederman, H., and Øgaard, T. F. (2017). Prospects for a naive theory of classes. *Notre Dame Journal of Formal Logic*, 58(4):461–506.

Fine, K. (1974). Models for entailment. *Journal of Philosophical Logic*, 3(4):347–372.

Fine, K. (1976). Completeness for the semilattice semantics with disjunction and conjunction (abstract). *Journal of Symbolic Logic*, 41:560.

Fine, K. (1988). Semantics for quantified relevance logic. *Journal of Philosophical Logic*, 17:27–59.

Fine, K. (1989). Incompleteness for quantified relevance logics. In Norman, J. and Sylvan, R., editors, *Directions in Relevant Logic*, pages 205–225. Kluwer. Reprinted in Anderson et al. (1992, §52).

Font, J. M. (2016). *Abstract Algebraic Logic: An Introductory Textbook*. College Publications.

Francez, N. (2019). Relevant connexive logic. *Logic and Logical Philosophy*, 28(3):409–425.

Friedman, H. and Meyer, R. K. (1992). Whither relevant arithmetic? *Journal of Symbolic Logic*, 57(3):824–831.

Fritz, P. (2024). *Propositional Quantifiers*. Cambridge University Press.

Fuhrmann, A. (1988). *Relevant Logics, Modal Logics and Theory Change*. PhD thesis, Australian National University.

Fuhrmann, A. (1990). Models for relevant modal logics. *Studia Logica*, 49(4):501–514.

Gabbay, D. M. (1976). *Investigations in Modal and Tense Logics with Applications to Problems in Philosophy and Linguistics*. Reidel.

Galatos, N., Jipsen, P., Kowalski, T., and Ono, H. (2007). *Residuated Lattices: An Algebraic Glimpse at Substructural Logics*. Elsevier.

Galatos, N. and Ono, H. (2006). Algebraization, parametrized local deduction theorem and interpolation for substructural logics over FL. *Studia Logica*, 83(1–3):279–308.

Giambrone, S. (1985). TW+ and RW+ are decidable. *Journal of Philosophical Logic*, 14(3):235–254.

Giambrone, S. (1992). Real reduced models for relevant logics without WI. *Notre Dame Journal of Formal Logic*, 33(3):442–449.

Giambrone, S. and Meyer, R. K. (1989). Completeness and conservative extension results for some boolean relevant logics. *Studia Logica*, 48(1):1–14.

Giambrone, S., Meyer, R. K., and Urquhart, A. (1987). A contractionless semilattice semantics. *Journal of Symbolic Logic*, 52(2):526–529.

Giambrone, S. and Urquhart, A. (1987). Proof theories for semilattice logics. *Mathematical Logic Quarterly*, 33(5):433–439.

Girard, P. and Weber, Z. (2015). Bad worlds. *Thought: A Journal of Philosophy*, 4(2):93–101.

Goble, L. (1999). Deontic logic with relevance. In McNamara, P. and Prakken, H., editors, *Norms, Logics and Information Systems: New Studies on Deontic Logic and Computer Science*, pages 331–345. IOS Press.

Goble, L. (2000). An incomplete relevant modal logic. *Journal of Philosophical Logic*, 29(1):103–119.

Goble, L. (2001). The Andersonian reduction and relevant deontic logic. In Brown, B. and Woods, J., editors, *New Studies in Exact Philosophy: Logic, Mathematics and Science. Proceedings of the 1999 Conference of the Society of Exact Philosophy*, pages 213–246, Paris. Hermes Science Publications.

Goble, L. (2003). Neighborhoods for entailment. *Journal of Philosophical Logic*, 32(5): 483–529.

Goble, L. (2006). Paraconsistent modal logic. *Logique et Analyse*, 193:3–29.

Golan, R. (2023). A unified interpretation of the semantics of relevance logic. *Mind*, 132(528):1105–1125.

Goldblatt, R. (2009). Conservativity of Heyting implication over relevant quantification. *Review of Symbolic Logic*, 2(2):310–341.

Goldblatt, R. (2011). *Quantifiers, Propositions, and Identity: Admissible Semantics for Quantified Modal and Substructural Logics*. Cambridge University Press.

Goldblatt, R. and Kane, M. (2009). An admissible semantics for propositionally quantified relevant logics. *Journal of Philosophical Logic*, 39(1):73–100.

Goldblatt, R. I. (1974). Semantic analysis of orthologic. *Journal of Philosophical Logic*, 3(1/2):19–35.

Gupta, A. and Belnap, N. (1993). *The Revision Theory of Truth*. MIT Press.

Hansson, S. O. (2017). Logic of Belief Revision. In Zalta, E. N., editor, *The Stanford Encyclopedia of Philosophy*. Metaphysics Research Lab, Stanford University, winter 2017 edition.

Hendricks, V. and Symons, J. (2015). Epistemic logic. In Zalta, E. N., editor, *The Stanford Encyclopedia of Philosophy*. Metaphysics Research Lab, Stanford University, fall 2015 edition.

Hornischer, L. and Berto, F. (2025). The logic of dynamical systems is relevant. *Mind*, 134(535):670–706.

Horty, J. F. (2001). *Agency and Deontic Logic*. Oxford University Press.

Hughes, G. E. and Cresswell, M. J. (1996). *A New Introduction to Modal Logic*. Routledge.

Humberstone, L. (1982). Necessary conclusions. *Philosophical Studies*, 41(3):321–335.

Humberstone, L. (1988). Operational semantics for positive R. *Notre Dame Journal of Formal Logic*, 29:61–80.

Humberstone, L. (1990). Expressive power and semantic completeness: Boolean connectives in modal logic. *Studia Logica*, 49(2):197–214.

Humberstone, L. (2010). Smiley's distinction between rules of inference and rules of proof. In Smiley, T. J., Lear, J., and Oliver, A., editors, *The Force of Argument: Essays in Honor of Timothy Smiley*, pages 107–126. Routledge.

Humberstone, L. (2011). *The Connectives*. MIT Press.

Humberstone, L. (2013). Aggregation and idempotence. *Review of Symbolic Logic*, 6(4): 680–708.

Humberstone, L. (2016). *Philosophical Applications of Modal Logic*. College Publications.

Humberstone, L. (2019). Supervenience, dependence, disjunction. *Logic and Logical Philosophy*, 28(1):3–135.

Humberstone, L. (2021). Propositional variables occurring exactly once in candidate modal axioms. *Filosofiska Notiser*, pages 27–73.

Iemhoff, R. (2001). On the admissible rules of intuitionistic propositional logic. *Journal of Symbolic Logic*, 66(1):281–294.

Jago, M. (2013). Recent work in relevant logic. *Analysis*, 73(3):526–541.

Jago, M. (2020). Truthmaker semantics for relevant logic. *Journal of Philosophical Logic*, 49(4):681–702.

Jago, M. (2023). Disjunctive parts. In Faroldi, F. L. G. and Van De Putte, F., editors, *Kit Fine on Truthmakers, Relevance, and Non-classical Logic*, pages 167–188. Springer.

Kamide, N. (2018). Decidable temporal and sequential relevant logics. *Journal of Logic and Computation*, 28(2):403–432.

Kamide, N. and Wansing, H. (2015). *Proof theory of N4-related paraconsistent logics*. College Publications.

Kashima, R. (2003). On semilattice relevant logics. *Mathematical Logic Quarterly*, 49(4): 401–404.

Kerr, A. D. (2019). A plea for KR. *Synthese*, 198(4):3047–3071.

Kielkopf, C. F. (1977). *Formal Sentential Entailment*. University Press of America.

Knudstorp, S. B. (2024). Relevant S is undecidable. In *LICS '24: Proceedings of the 39th Annual ACM/IEEE Symposium on Logic in Computer Science*, pages 1–8.

Kouri, T. (2016). Restall's proof-theoretic pluralism and relevance logic. *Erkenntnis*, 81(6):1243–1252.

Kremer, P. (1989). Relevant predication: Grammatical characterisations. *Journal of Philosophical Logic*, 18(4):349–382.

Kremer, P. (1993). Quantifying over propositions in relevance logic: Nonaxiomatisability of primary interpretations of $\forall p$ and $\exists p$. *Journal of Symbolic Logic*, 58(1):334–349.

Kremer, P. (1997). Dunn's relevant predication, real properties and identity. *Erkenntnis*, 47(1):37–65.

Kremer, P. (1999). Relevant identity. *Journal of Philosophical Logic*, 28(2):199–222.

Kripke, S. (2022). A proof of gamma. In Bimbó, K., editor, *Relevance Logics and other Tools for Reasoning: Essays in Honor of J. Michael Dunn*, pages 261–265. College Publications.

Kuznets, R. and Studer, T. (2019). *Logics of Proofs and Justifications*. College Publications.

Lavers, P. (1985). *Generating Intensional Logics*. MA thesis, University of Adelaide.

Lavers, P. (1987). Relevance and disjunctive syllogism. *Notre Dame Journal of Formal Logic*, 29(1):34–44.

Logan, S. A. (2019). Notes on stratified semantics. *Journal of Philosophical Logic*, 48(4): 749–786.

Logan, S. A. (2021). Strong depth relevance. *The Australasian Journal of Logic*, 18(6): 645–656.

Logan, S. A. (2022). Depth relevance and hyperformalism. *Journal of Philosophical Logic*, 51(4):721–737.

Logan, S. A. (2024). *Relevance Logic*. Elements in Philosophy and Logic. Cambridge University Press.

Logan, S. A. (2025a). Nondeterministic and nonconcurrent computational semantics for BB+ and related logics. *Journal of Logic and Computation*, 35(1):36–56.

Logan, S. A. (2025b). Semantics for second order relevant logics. In Sedlár, I., Standefer, S., and Tedder, A., editors, *New Directions in Relevant Logic*, pages 183–201. Springer.

Logan, S. A. and Boccuni, F. (2025). Frege meets Belnap: Basic law V in a relevant logic. In Sedlár, I., Standefer, S., and Tedder, A., editors, *New Directions in Relevant Logic*, pages 381–404. Springer.

Logan, S. A. and Leach-Krouse, G. (2021). On not saying what we shouldn't have to say. *The Australasian Journal of Logic*, 18(5):524–568.

Lokhorst, G. C. (2006). Andersonian deontic logic, propositional quantification, and Mally. *Notre Dame Journal of Formal Logic*, 47(3):385–395.

López-Escobar, E. G. K. (1985). On intuitionistic sentential connectives. I. *Revista Colombiana de Matemáticas*, 19:117–130.

Majer, O., Punčochář, V., and Sedlár, I. (2023). Truth-maker semantics for some substructural logics. In Faroldi, F. and Van De Putte, F., editors, *Kit Fine on Truthmakers, Relevance, and Non-classical Logic*, Outstanding Contributions to Logic, pages 207–222. Springer.

Maksimova, L. (1973). A semantics for the calculus E of entailment. *Bulletin of the Section Logic*, 2(1):18–23.

Maksimova, L. L. (1969). Interpretatsiya sistem so strogoĭ implikatsieĭ. In *10th All-Union Algebraic Colloquium (Abstracts)*, page 113. (An interpretation of systems with rigorous implication).

Mancosu, P., Galvan, S., and Zach, R. (2021). *An Introduction to Proof Theory: Normalization, Cut-Elimination, and Consistency Proofs*. Oxford University Press.

Mangraviti, F. (2023). The liberation argument for inconsistent mathematics. *Australasian Journal of Logic*, 29(2):278–315.

Mangraviti, F. and Tedder, A. (2023). Consistent theories in inconsistent logics. *Journal of Philosophical Logic*, 52(04):1133–1148.

Mares, E. D. (1992a). Andersonian deontic logic. *Theoria*, 58(1):3–20.

Mares, E. D. (1992b). The semantic completeness of RK. *Reports on Mathematical Logic*, pages 3–10.

Mares, E. D. (1992c). Semantics for relevance logic with identity. *Studia Logica*, 51(1):1–20.

Mares, E. D. (1993). Classically complete modal relevant logics. *Mathematical Logic Quarterly*, 39(1):165–177.

Mares, E. D. (1994). Mostly Meyer modal models. *Logique et Analyse*, 37(146):119–128.

Mares, E. D. (1996). Relevant logic and the theory of information. *Synthese*, 109(3):345–360.

Mares, E. D. (2000a). CE is not a conservative extension of E. *Journal of Philosophical Logic*, 29(3):263–275.

Mares, E. D. (2000b). The incompleteness of RGL. *Studia Logica*, 65(3):315–322.

Mares, E. D. (2002). A paraconsistent theory of belief revision. *Erkenntnis*, 56(2):229–246.

Mares, E. D. (2003). Halldén-completeness and modal relevant logic. *Logique et Analyse*, 46(181):59–76.

Mares, E. D. (2004). *Relevant Logic: A Philosophical Interpretation*. Cambridge University Press.

Mares, E. D. (2009). General information in relevant logic. *Synthese*, 167(2):343–362.

Mares, E. D. (2012). Relevant logic and the philosophy of mathematics. *Philosophy Compass*, 7(7):481–494.

Mares, E. D. (2019). From iff to is: Some new thoughts on identity in relevant logics. In *Graham Priest on Dialetheism and Paraconsistency*, pages 343–363. Springer International Publishing.

Mares, E. D. (2020). Relevance Logic. In Zalta, E. N., editor, *The Stanford Encyclopedia of Philosophy*. Metaphysics Research Lab, Stanford University, winter 2020 edition.

Mares, E. D. (2021). An informational interpretation of weak relevant logic and relevant property theory. *Synthese*, 199:547–569.

Mares, E. D. (2024). *The Logic of Entailment and its History*. Cambridge University Press.

Mares, E. D. and Goldblatt, R. (2006). An alternative semantics for quantified relevant logic. *The Journal of Symbolic Logic*, 71(1):163–187.

Mares, E. D. and Meyer, R. K. (1992). The admissibility of γ in *R*4. *Notre Dame Journal of Formal Logic*, 33(2):197–206.

Mares, E. D. and Meyer, R. K. (1993). The semantics of R4. *Journal of Philosophical Logic*, 22(1):95–110.

Mares, E. D., Seligman, J., and Restall, G. (2010). Situations, constrains and channels. In van Benthem, J. and ter Meulen, A., editors, *Handbook of Logic and Language*, pages 329–344. Elsevier.

Mares, E. D. and Standefer, S. (2017). The relevant logic E and some close neighbours: A reinterpretation. *IfCoLog Journal of Logics and Their Applications*, 4(3):695–730.

Martin, E. P. and Meyer, R. K. (1982). Solution to the P – W problem. *Journal of Symbolic Logic*, 47(4):869–887.

McArthur, R. P. (1981). Anderson's deontic logic and relevant implication. *Notre Dame Journal of Formal Logic*, 22(2):145–154.

McNamara, P. and Van De Putte, F. (2021). Deontic logic. In Zalta, E. N., editor, *The Stanford Encyclopedia of Philosophy*. Metaphysics Research Lab, Stanford University, spring 2021 edition.

Méndez, J. M., Robles, G., and Salto, F. (2012). Ticket entailment plus the mingle axiom has the variable-sharing property. *Logic Journal of the IGPL*, 20(1):355–364.

Meyer, R. K. (1966). *Topics in Modal and Many-valued Logic*. PhD thesis, University of Pittsburgh.

Meyer, R. K. (1970). R_I – the bounds of finitude. *Mathematical Logic Quarterly*, 16(7): 385–387.

Meyer, R. K. (1971a). Entailment. *Journal of Philosophy*, 68(21):808–818.

Meyer, R. K. (1971b). On coherence in modal logics. *Logique et Analyse*, 14:658–668.

Meyer, R. K. (1973). Intuitionism, entailment, negation. In Leblanc, H., editor, *Truth, Syntax and Modality*, pages 168–198. North-Holland.

Meyer, R. K. (1976a). Ackermann, Takeuti, and Schnitt: For higher-order relevant logic. *Bulletin of the Section of Logic*, 5(4):138–142.

Meyer, R. K. (1976b). Metacompleteness. *Notre Dame Journal of Formal Logic*, 17(4): 501–516.

Meyer, R. K. (1976c). Relevant arithmetic. *Bulletin of the Section of Logic*, 5(4):133–135. Reprinted as Meyer (2021c).

Meyer, R. K. (1985). A farewell to entailment. In Dorn, G. and Weingartner, P., editors, *Foundations of Logic and Linguistics, Problems and Their Solutions*, pages 577– 636. Plenum Press.

Meyer, R. K. (1990). Peirced clean through. *Bulletin of the Section of Logic*, 19(3):100–101.

Meyer, R. K. (1998). $\supset$E is admissible in "true" relevant arithmetic. *Journal of Philosophical Logic*, 27(4):327–351.

Meyer, R. K. (2008). Fallacies of division. *Proceedings of the XXII World Congress of Philosophy*, 13:71–80.

Meyer, R. K. (2021a). Arithmetic formulated relevantly. *The Australasian Journal of Logic*, 18(5):154–288.

Meyer, R. K. (2021b). The consistency of arithmetic. *The Australasian Journal of Logic*, 18(5):289–379.

Meyer, R. K. (2021c). Relevant arithmetic. *The Australasian Journal of Logic*, 18(5):150–153. This is a republication of the abstract, "Relevant Arithmetic", by Meyer, originally published in the *Bulletin of the Section of Logic* 5 (1976): 133–135.

Meyer, R. K. and Dunn, J. M. (1969). E, R, and γ. *Journal of Symbolic Logic*, 34(3):460–474.

Meyer, R. K., Giambrone, S., and Brady, R. T. (1984). Where gamma fails. *Studia Logica*, 43(3):247–256.

Meyer, R. K. and Mares, E. D. (1993). Semantics of entailment 0. In Schroeder-Heister, P. and Došen, K., editors, *Substructural Logics*, pages 239–258. Oxford Science Publications.

Meyer, R. K. and Mortensen, C. (1984). Inconsistent models for relevant arithmetics. *Journal of Symbolic Logic*, 49(3):917–929.

Meyer, R. K. and Mortensen, C. (2021). Alien intruders in relevant arithmetic. *The Australasian Journal of Logic*, 18(5):401–425.

Meyer, R. K. and Restall, G. (1999). "Strenge" arithmetics. *Logique et Analyse*, 42(167/168):205–220.

Meyer, R. K. and Routley, R. (1972). Algebraic analysis of entailment I. *Logique et Analyse*, 15:407–428.

Meyer, R. K. and Routley, R. (1973). Classical relevant logics I. *Studia Logica*, 32:51–66.

Meyer, R. K. and Routley, R. (1974). Classical relevant logics II. *Studia Logica*, 33(2):183–194.

Meyer, R. K., Routley, R., and Dunn, J. M. (1979). Curry's paradox. *Analysis*, 39(3):124–128.

Meyer, R. K. and Urbas, I. (1986). Conservative extension in relevant arithmetic. *Mathematical Logic Quarterly*, 32(1-5):45–50.

Mints, G. (1992). *A Short Introduction to Modal Logic*. Center for the Study of Language and Information.

Mortensen, C. (1983). The validity of disjunctive syllogism is not so easily proved. *Notre Dame Journal of Formal Logic*, 24(1):35–40.

Mortensen, C. (1986). Reply to Burgess and to Read. *Notre Dame Journal of Formal Logic*, 27(2):195–200.

Moschovakis, Y. N. (1974). *Elementary Induction on Abstract Structures*. Dover Publications.

Norman, J. and Sylvan, R., editors (1989). *Directions in Relevant Logic*. Kluwer Academic Publishers.

Øgaard, T. F. (2016). Paths to triviality. *Journal of Philosophical Logic*, 45(3):237–276.

Øgaard, T. F. (2019). Non-Boolean classical relevant logics I. *Synthese*, (8):1–32.

Øgaard, T. F. (2020a). Farewell to suppression-freedom. *Logica Universalis*, 14(3):297–330.

Øgaard, T. F. (2020b). Substitution in relevant logics. *Review of Symbolic Logic*, 13(3): 655–680.

Øgaard, T. F. (2021a). Boolean negation and non-conservativity I: Relevant modal logics. *Logic Journal of the IGPL*, 29(3):340–362.

Øgaard, T. F. (2021b). Boolean negation and non-conservativity II: The variable-sharing property. *Logic Journal of the IGPL*, 29(3):363–369.

Øgaard, T. F. (2021c). Boolean negation and non-conservativity III: The Ackermann constant. *Logic Journal of the IGPL*, 29(3):370–384.

Øgaard, T. F. (2021d). From Hilbert proofs to consecutions and back. *Australasian Journal of Logic*, 18(2):51–72.

Øgaard, T. F. (2021e). Non-Boolean classical relevant logics II: Classicality through truth-constants. *Synthese*, (3-4):1–33.

Øgaard, T. F. (2022). Confused entailment. *Topoi*, 41(1):207–219.

Øgaard, T. F. (2023). The weak variable sharing property. *Bulletin of the Section Logic*.

Øgaard, T. F. and Standefer, S. (2025). Failures of γ. *Studia Logica*. Forthcoming.

Ohnishi, M. and Matsumoto, K. (1957). Gentzen method in modal calculi. *Osaka Mathematical Journal*, 9(2):113–130.

Omori, H. and De, M. (2022). Shrieking, shrugging, and the Australian plan. *Notre Dame Journal of Formal Logic*, 63(2):137–141.

Omori, H. and Wansing, H. (2017). 40 years of FDE: An introductory overview. *Studia Logica*, 105(6):1021–1049.

Omori, H. and Wansing, H. (2019). Connexive logics. An overview and current trends. *Logic and Logical Philosophy*, 28(3):371–387.

Onishi, T. (2015). Substructural negations. *Australasian Journal of Logic*, 12(4):177–203.

Onishi, T. (2016). Understanding negation implicationally in the relevant logic R. *Studia Logica*, 104(6):1267–1285.

Onishi, T. (2024). Multiset-multiset frames. *Journal of Philosophical Logic*, 53(5):1241–1264.

Ono, H. (1977). On some intuitionistic modal logics. *Publications of RIMS Kyoto University*, 13:687–722.

Ono, H. (2019). *Proof Theory and Algebra in Logic*. Springer.

Orlov, I. (2024). The calculus of the compatibility of propositions. *Australasian Journal of Logic*, 21(1):20–46. Translated by Werner Stelzner.

Pacuit, E. (2017). *Neighborhood Semantics for Modal Logic*. Springer.

Parks, Z. and Byrd, M. (1989). Relevant implication and Leibnizian necessity. In Norman, J. and Sylvan, R., editors, *Directions in Relevant Logic*, pages 179–184. Kluwer Academic Publishers.

Plumwood, V. (1993). The politics of reason: Towards a feminist logic. *Australasian Journal of Philosophy*, 71(4):436–462.

Plumwood, V. (2023). Some false laws of logic. *Australasian Journal of Logic*, 20(2):97–137.

Prawitz, D. (1965). *Natural Deduction: A Proof-Theoretical Study*. Almqvist and Wicksell.

Priest, G. (2001). Paraconsistent belief revision. *Theoria*, 67(3):214–228.

Priest, G. (2008). *An Introduction to Non-Classical Logic: From If to Is*. Cambridge University Press.

Priest, G. (2016). *Towards Non-Being: The Logic and Metaphysics of Intentionality*. Oxford University Press, 2nd edition.

Punčochář, V. (2017). Algebras of information states. *Journal of Logic and Computation*, 27:1643–1675.

Punčochář, V. (2019). Substructural inquisitive logics. *Review of Symbolic Logic*, 12(2): 296–330.

Punčochář, V. (2020). A relevant logic of questions. *Journal of Philosophical Logic*, 49(5): 905–939.

Punčochář, V. and Sedlár, I. (2017). Substructural logics for pooling information. In Baltag, A., Seligman, J., and Yamada, T., editors, *Logic, Rationality, and Interaction*, pages 407–421, Springer.

Punčochář, V., Sedlár, I., and Tedder, A. (2023). Relevant epistemic logic with public announcements and common knowledge. *Journal of Logic and Computation*, 33(2): 436–461.

Read, S. (1980). What is wrong with disjunctive syllogism? *Analysis*, 41(2):66–70.

Read, S. (1983). Burgess on relevance: A fallacy indeed. *Notre Dame Journal of Formal Logic*, 24(4):473–481.

Read, S. (1988). *Relevant Logic: A Philosophical Examination of Inference*. Blackwell.

Restall, G. (1993). How to be really contraction free. *Studia Logica*, 52(3):381–391.

Restall, G. (1994). *On Logics Without Contraction*. PhD thesis, University of Queensland.

Restall, G. (1995). Display logic and gaggle theory. *Reports on Mathematical Logic*, 29: 133–146.

Restall, G. (1999). Negation in relevant logics (how I stopped worrying and learned to love the Routley star). In Gabbay, D. M. and Wansing, H., editors, *What is Negation?*, pages 53–76. Kluwer Academic Publishers.

Restall, G. (2000). *An Introduction to Substructural Logics*. Routledge.

Restall, G. (2005). Logics, situations and channels. *Journal of Cognitive Science*, 6:125–150.

Restall, G. (2011). Models for substructural arithmetics. *Australasian Journal of Logic*, 8:82–99.

Restall, G. (2014). Pluralism and proofs. *Erkenntnis*, 79(S2):279–291.

Restall, G. and Slaney, J. (1995). Realistic belief revision. In *Proceedings of the First World Congress in the Fundamentals of Artificial Intelligence* (WOCFAI) 1995, 3–7 July 1995, Paris, pages 367–378.

Restall, G. and Standefer, S. (2023a). Collection frames for distributive substructural logics. *Review of Symbolic Logic*, 16(4):1120–1157.

Restall, G. and Standefer, S. (2023b). *Logical Methods*. MIT Press.

Ripley, E. (2018). Blurring: An approach to conflation. *Notre Dame Journal of Formal Logic*, 59(2):171–188.

Robles, G. (2013). Admissibility of Ackermann's rule δ in relevant logics. *Logic and Logical Philosophy*, 22(4):411–427.

Robles, G. and Méndez, J. M. (2010). A Routley-Meyer type semantics for relevant logics including B_r plus the disjunctive syllogism. *Journal of Philosophical Logic*, 39(2):139–158.

Robles, G. and Méndez, J. M. (2011a). A class of simpler logical matrices for the variable-sharing property. *Logic and Logical Philosophy*, 20(3):241–249.

Robles, G. and Méndez, J. M. (2011b). A Routley-Meyer semantics for relevant logics including TWR plus the disjunctive syllogism. *Logic Journal of the IGPL*, 19(1):18–32.

Robles, G. and Méndez, J. M. (2012). A general characterization of the variable-sharing property by means of logical matrices. *Notre Dame Journal of Formal Logic*, 53(2): 223–244.

Robles, G. and Méndez, J. M. (2014a). Blocking the routes to triviality with depth relevance. *Journal of Logic, Language and Information*, 23(4):493–526.

Robles, G. and Méndez, J. M. (2014b). Curry's paradox, generalized modus ponens axiom and depth relevance. *Studia Logica*, 102(1):185–217.

Robles, G. and Méndez, J. M. (2014c). Generalizing the depth relevance condition: Deep relevant logics not included in R-mingle. *Notre Dame Journal of Formal Logic*, 55(1): 107–127.

Robles, G. and Méndez, J. M. (2014d). The non-relevant De Morgan minimal logic in Routley-Meyer semantics with no designated points. *Journal of Applied Non-Classical Logics*, 24(4):321–332.

Robles, G. and Méndez, J. M. (2016). A companion to Brady's 4-valued relevant logic BN4: The 4-valued logic of entailment E4. *Logic Journal of the IGPL*, 24(5):838–858.

Robles, G. and Méndez, J. M. (2018). *Routley-Meyer Ternary Relational Semantics for Intuitionistic-Type Negations*. Elsevier, Academic Press.

Robles, G., Méndez, J. M., and Salto, F. (2010). A modal restriction of R-mingle with the variable-sharing property. *Logic and Logical Philosophy*, 19(4):341–351.

Robles, G., Salto, F., and Méndez, J. M. (2011). A weak logic with the axiom mingle lacking the variable-sharing property. *Bulletin of the Section of Logic*, 40(3/4):195–202.

Rogerson, S. (2007). Natural deduction and Curry's paradox. *Journal of Philosophical Logic*, 36(2):155–179.

Rogerson, S. and Butchart, S. (2002). Naïve comprehension and contracting implications. *Studia Logica*, 71(1):119–132.

Rogerson, S. and Restall, G. (2004). Routes to triviality. *Journal of Philosophical Logic*, 33(4):421–436.

Routley, R. (1984). The American plan completed: Alternative classical-style semantics, without stars, for relevant and paraconsistent logics. *Studia Logica*, 43(1-2):131–158.

Routley, R. (1989). Philosophical and linguistic inroads: Multiply intensional relevant logics. In Norman, J. and Sylvan, R., editors, *Directions in Relevant Logic*, pages 269–304. Kluwer.

Routley, R. and Meyer, R. K. (1972a). The semantics of entailment—II. *Journal of Philosophical Logic*, 1(1):53–73.

Routley, R. and Meyer, R. K. (1972b). The semantics of entailment—III. *Journal of Philosophical Logic*, 1(2):192–208.

Routley, R. and Meyer, R. K. (1973). The semantics of entailment. In Leblanc, H., editor, *Truth, Syntax, and Modality: Proceedings Of The Temple University Conference On Alternative Semantlcs*, pages 199–243. North-Holland Publishing Company.

Routley, R. and Meyer, R. K. (1975). Towards a general semantical theory of implication and conditionals. I. Systems with normal conjunctions and disjunctions and aberrant and normal negations. *Reports on Mathematical Logic*, 4:67–89.

Routley, R. and Meyer, R. K. (1976). Towards a general semantical theory of implication and conditionals. II. Improved negation theory and propositional identity. *Reports on Mathematical Logic*, 9:47–62.

Routley, R. and Plumwood, V. (1989). Moral dilemmas and the logic of deontic notions. In Priest, G., Routley, R., and Norman, J., editors, *Paraconsistent Logic: Essays on the Inconsistent*, pages 653–702. Philosophia Verlag.

Routley, R., Plumwood, V., Meyer, R. K., and Brady, R. T. (1982). *Relevant Logics and Their Rivals*, volume 1. Ridgeview.

Routley, R. and Routley, V. (1969). A fallacy of modality. *Noûs*, 3(2):129–153.

Routley, R. and Routley, V. (1972). The semantics of first degree entailment. *Noûs*, 6(4): 335–359.

Russell, G. (2014). Logical pluralism. In Zalta, E. N., editor, *The Stanford Encyclopedia of Philosophy*. The Metaphysics Research Lab.

Rybakov, V. V. (1997). *Admissibility of Logical Inference Rules*. Elsevier.

Salto, F., Robles, G., and Méndez, J. M. (2018). Curry's paradox, generalized contraction rule and depth relevance. In Boudouris, K., editor, *Proceedings of the XXIII World Congress of Philosophy*, pages 35–39. Philosophy Documentation Center.

Savić, N. and Studer, T. (2019). Relevant justification logic. *Journal of Applied Logics*, 6(2):395–410.

Schwarz, W. (2013). Contingent identity. *Philosophy Compass*, 8(5):486–495.

Sedlár, I. (2015). Substructural epistemic logics. *Journal of Applied Non-Classical Logics*, 25(3):256–285.

Sedlár, I. (2016). Epistemic extensions of modal distributive substructural logics. *Journal of Logic and Computation*, 26(6):1787–1813.

Seki, T. (2003a). General frames for relevant modal logics. *Notre Dame Journal of Formal Logic*, 44(2):93–109.

Seki, T. (2003b). A Sahlqvist theorem for relevant modal logics. *Studia Logica*, 73(3):383–411.

Seki, T. (2009). Completeness of relevant modal logics with disjunctive rules. *Reports on Mathematical Logic*, 44:3–18.

Seki, T. (2011a). The γ–admissibility of relevant modal logics II — the method using metavaluations. *Studia Logica*, 97(3):351–383.

Seki, T. (2011b). The γ-admissibility of relevant modal logics I — the method of normal models. *Studia Logica*, 97(2):199–231.

Seki, T. (2012a). An algebraic proof of the admissibility of γ in relevant modal logics. *Studia Logica*, 100(6):1149–1174.

Seki, T. (2012b). Metacompleteness of substructural logics. *Studia Logica*, 100(6):1175–1199.

Seki, T. (2013). Some metacomplete relevant modal logics. *Studia Logica*, 101(5):1115–1141.

Seki, T. (2015). Halldén completeness for relevant modal logics. *Notre Dame Journal of Formal Logic*, 56(2):333–350.

Shapiro, L. and Beall, J. (2021). Curry's Paradox. In Zalta, E. N., editor, *The Stanford Encyclopedia of Philosophy*. Metaphysics Research Lab, Stanford University, Winter 2021 edition.

Slaney, J. (1984). A metacompleteness theorem for contraction-free relevant logics. *Studia Logica*, 43(1-2):159–168.

Slaney, J. (1987a). Reduced models for relevant logics without WI. *Notre Dame Journal of Formal Logic*, 28(3):395–407.

Slaney, J. (1987b). A simple proof of consistency for a naive axiom of propositional comprehension. *Journal of Non-classical Logic*, 4:23–31.

Slaney, J. (1989). RWX is not Curry Paraconsistent. In Priest, G., Sylvan, R., and Norman, J., editors, *Paraconsistent Logic: Essays on the Inconsistent*, pages 472–480. Philosophia Verlag.

Slaney, J. (1990). A general logic. *Australasian Journal of Philosophy*, 68(1):74–88.

Slaney, J. (1992). Finite models for some substructural logics. *Logique et Analyse*, 35(139/140):313–333.

Slaney, J. (2022). Models of relevant arithmetic. *The Australasian Journal of Logic*, 19(1): 32–50.

Slaney, J. (2025). Relevant rational arithmetic. In Sedlár, I., Standefer, S., and Tedder, A., editors, *New Directions in Relevant Logic*, pages 453–468. Springer.

Slaney, J., Meyer, R. K., and Restall, G. (1996). Linear arithmetic desecsed. *Logique et Analyse*, 39:379–388.

Standefer, S. (2018). Trees for E. *Logic Journal of the IGPL*, 26(3):300–315.

Standefer, S. (2019). Tracking reasons with extensions of relevant logics. *Logic Journal of the IGPL*, 27(4):543–569.

Standefer, S. (2020). Actual issues for relevant logics. *Ergo*, 7(8):241–276.

Standefer, S. (2021a). Identity in Mares-Goldblatt models for quantified relevant logic. *Journal of Philosophical Logic*, 50(6):1389–1415.

Standefer, S. (2021b). An incompleteness theorem for modal relevant logics. *Notre Dame Journal of Formal Logic*, 62(4):669–681.

Standefer, S. (2021c). Translations between linear and tree natural deduction systems for relevant logics. *Review of Symbolic Logic*, 14(2):285–306.

Standefer, S. (2022a). Completeness via metacompleteness. In Bimbó, K., editor, *Relevance Logics and other Tools for Reasoning: Essays in Honor of J. Michael Dunn*, pages 394–409. College Publications.

Standefer, S. (2022b). Revisiting semilattice semantics. In Duntsch, I. and Mares, E., editors, *Alasdair Urquhart on Non-classical and Algebraic Logic and Complexity of Proofs*, pages 241–258. Springer.

Standefer, S. (2022c). What is a relevant connective? *Journal of Philosophical Logic*, 51(4):919–950.

Standefer, S. (2023a). A substructural approach to explicit modal logic. *Journal of Logic, Language and Information*, 32:333–362.

Standefer, S. (2023b). Varieties of relevant S5. *Logic and Logical Philosophy*, 32(1):53–80.

Standefer, S. (2023c). Weak relevant justification logics. *Journal of Logic and Computation*, 33(7):1665–1683.

Standefer, S. (2024). Routes to relevance: Philosophies of relevant logics. *Philosophy Compass*, 19(2):e12965.

Standefer, S. (2025a). Ignorance and the possibility of error in relevant epistemic logic. In Sano, K., Ono, H., and Hatano, R., editors, *Exploring Negation, Modality and Proof.* Springer. Forthcoming.

Standefer, S. (2025b). Variable-sharing as relevance. In Sedlár, I., Standefer, S., and Tedder, A., editors, *New Directions in Relevant Logic*, pages 97–117. Springer.

Standefer, S. and Brady, R. T. (2019). Natural deduction systems for E. *Logique et Analyse*, 61(242):163–182.

Standefer, S. and Mares, E. (2025). Symmetry and completeness in relevant epistemic logic. *Journal of Philosophical Logic*, 54(2):429–450.

Swirydowicz, K. (1999). There exist exactly two maximal strictly relevant extensions of the relevant logic R. *Journal of Symbolic Logic*, 64(3):1125–1154.

Sylvan, R. (1988). Radical pluralism: An alternative to realism, antirealism and relativism. In Nola, R., editor, *Relativism and Realism in Science*, pages 253–291. Kluwer.

Sylvan, R. (1992). Process and action: Relevant theory and logics. *Studia Logica*, 51(3-4): 379–437.

Tabakci, S. K. (2022). Subminimal negation on the Australian plan. *Journal of Philosophical Logic*, 51:1119–1139.

Tedder, A. (2015). Axioms for finite collapse models of arithmetic. *Review of Symbolic Logic*, 8(3):529–539.

Tedder, A. (2021a). Consistency and decidability in some paraconsistent arithmetics. *Australasian Journal of Logic*, 18(5):473–502.

Tedder, A. (2021b). Information flow in logics in the vicinity of BB. *Australasian Journal of Logic*, 18(1):1–24.

Tedder, A. (2023). Situations, propositions, and information states. In Bimbó, K., editor, *Relevance Logics and other Tools for Reasoning: Essays in Honor of J. Michael Dunn*, pages 410–426. College Publications.

Tedder, A. (2025). Topics in relevant logic: A semantic perspective. *Erkenntnis*, pages 1–29. Forthcoming.

Tedder, A. and Bilková, M. (2022). Relevant propositional dynamic logic. *Synthese*, 200(3): 1–42.

Tedder, A. and Ferenz, N. (2022). Neighbourhood semantics for quantified relevant logics. *Journal of Philosophical Logic*, 51(3):457–484.

Tennant, N. (2017). *Core Logic*. Oxford University Press.

Terui, K. (2004). Light affine set theory: A naive set theory of polynomial time. *Studia Logica*, 77(1):9–40.

Thistlewaite, P. B., McRobbie, M. A., and Meyer, R. K. (1988). *Automated Theorem-Proving in Non-Classical Logics*. Pitman Publishing.

Tokarz, M. (1980). *Essays in Matrix Semantics of Relevant Logics*. Institute of Philosophy and Sociology of the Polish Academy of Sciences.

Urbas, I. and Sylvan, R. (1989). Prospects for decent relevant factorisation logics. *Journal of Non-classical Logic*, 6(1):63–79.

Urquhart, A. (1972a). Semantics for relevant logics. *Journal of Symbolic Logic*, 37(1):159–169.

Urquhart, A. (1972b). *The Semantics of Entailment*. PhD thesis, University of Pittsburgh.

Urquhart, A. (1983). Relevant implication and projective geometry. *Logique et Analyse*, 26(3):345–357.

Urquhart, A. (1984). The undecidability of entailment and relevant implication. *Journal of Symbolic Logic*, 49(4):1059–1073.

Urquhart, A. (1993). Failure of interpolation in relevant logics. *Journal of Philosophical Logic*, 22(5):449–479.

Urquhart, A. (1999). Beth's definability theorem in relevant logics. In Orlowska, E., editor, *Logic at Work*, pages 229–234. Heidelberg.

Urquhart, A. (2003). *Relevant Logics and Their Rivals, Volume II, A Continuation of the Work of Richard Sylvan, Robert Meyer, Val Plumwood and Ross Brady*, chapter 11.6, pages 222–230. Ashgate.

Urquhart, A. (2016a). Relevance logic: Problems open and closed. *Australasian Journal of Logic*, 13(1):11–20.

Urquhart, A. (2016b). The story of γ. In Bimbó, K., editor, *J. Michael Dunn on Information Based Logics*, pages 93–105. Springer.

Urquhart, A. (2017). The geometry of relevant implication. *IFCoLog Journal of Logics and their Applications*, 4(3):591–604.

Urquhart, A. (2019). Relevant implication and ordered geometry. *Australasian Journal of Logic*, 16(8):342–354.

Urquhart, A. (2023). The geometry of relevant implication II. *Australasian Journal of Logic*, 20(1):88–93.

van Benthem, J. and Humberstone, L. (1983). Halldén-completeness by gluing of Kripke frames. *Notre Dame Journal of Formal Logic*, 24(4):426–430.

van Benthem, J. (2010). *Modal Logic for Open Minds*. Center for the Study of Language and Information.

van Dalen, D. (2002). *Intuitionistic Logic*, pages 1–114. Springer.

Wansing, H. (1998). *Displaying Modal Logic*. Kluwer.

Wansing, H. (2002). Diamonds are a philosopher's best friends. *Journal of Philosophical Logic*, 31(6):591–612.

Weber, Z. (2010a). Extensionality and restriction in naive set theory. *Studia Logica*, 94(1):87–104.

Weber, Z. (2010b). Transfinite numbers in paraconsistent set theory. *Review of Symbolic Logic*, 3(1):71–92.

Weber, Z. (2012). Transfinite cardinals in paraconsistent set theory. *Review of Symbolic Logic*, 5(2):269–293.

Weber, Z. (2013). Notes on inconsistent set theory. In Berto, F., Mares, E., Tanaka, K., and Paoli, F., editors, *Paraconsistency: Logic and Applications*, pages 315–328. Springer.

Weber, Z. (2020). Property identity and relevant conditionals. *Australasian Philosophical Review*, 4(2):147–155.

Weber, Z. (2021a). *Paradoxes and Inconsistent Mathematics*. Cambridge University Press.

Weber, Z. (2021b). Relevant arithmetic and mathematical pluralism. *The Australasian Journal of Logic*, 18(5):569–596.

Weber, Z., Badia, G., and Girard, P. (2016). What is an inconsistent truth table? *Australasian Journal of Philosophy*, 94(3):533–548.

Weiss, Y. (2019). A note on the relevance of semilattice relevance logic. *Australasian Journal of Logic*, 16(6):177–185.

Weiss, Y. (2020a). A characteristic frame for positive intuitionistic and relevance logic. *Studia Logica*, 109(4):687–699.

Weiss, Y. (2020b). Cut and gamma I: Propositional and constant domain R. *Review of Symbolic Logic*, 13(4):887–909.

Weiss, Y. (2022a). Revisiting constructive mingle: Algebraic and operational semantics. In Bimbo, K., editor, *Essays in Honor of J. Michael Dunn*, pages 435–455. College Publications.

Weiss, Y. (2022b). Semantics for pure theories of connexive implication. *Review of Symbolic Logic*, 15(3):591–606.

Weiss, Y. (2025). A relevant framework for barriers to entailment. *Journal of Applied Logics - IfCoLog Journal*, 12(5):1319–1347.

Williamson, T. (2006). Indicative versus subjunctive conditionals, congruential versus non-hyperintensional contexts. *Philosophical Issues*, 16(1):310–333.

Index

Printed in the United Kingdom by TJ Clays Ltd.